# Lecture Notes in Management and Industrial Engineering

**Series editor**

Adolfo López-Paredes, Valladolid, Spain

This bookseries provides a means for the dissemination of current theoretical and applied research in the areas of Industrial Engineering & Engineering Management. The latest methodological and computational advances that both researchers and practitioners can widely apply to solve new and classical problems in industries and organizations constitute a growing source of publications written for and by our readership.

The aim of this bookseries is to facilitate the dissemination of current research in the following topics:

- Strategy and Enterpreneurship
- Operations Research, Modelling and Simulation
- Logistics, Production and Information Systems
- Quality Management
- Product Management
- Sustainability and Ecoefficiency
- Industrial Marketing and Consumer Behavior
- Knowledge and Project Management
- Risk Management
- Service Systems
- Healthcare Management
- Human Factors and Ergonomics
- Emergencies and Disaster Management
- Education

More information about this series at http://www.springer.com/series/11786

Nikhil Gurjar

# A Forward Looking Approach to Project Management

## Tools, Trends, and the Impact of Disruptive Technologies

 Springer

Nikhil Gurjar
Consulting Connoisseurs
Navi Mumbai, Maharashtra
India

ISSN 2198-0772             ISSN 2198-0780   (electronic)
Lecture Notes in Management and Industrial Engineering
ISBN 978-981-10-9250-3        ISBN 978-981-10-0782-8   (eBook)
DOI 10.1007/978-981-10-0782-8

This Springer imprint is published by Springer Nature
The registered company is Springer Science+Business Media Singapore Pte Ltd.

*Dedicated to my mother, my late father
and my daughters: Riya and Arya*

# Preface

The current work provides a basis for advanced treatment of concepts in project management. In this world, where most questions are answered through Google or the internet, the knowledge element in *project management* has come under the influence of disruptive technologies. In other words, the 'number of points' given to a project manager for knowing something that is easily available/obtainable on the internet has come down. This is having far-reaching consequences. The reader needs to orient toward newer benchmarks of what is required for success in the business cases. Thus, we deal with a few advanced concepts in this book.

As the name implies, it is not an elementary book to read and get to know the first level of the topics. Rather, it is an advanced-level treatment of the subject, to be initiated after the preliminary study has already been completed.

The book is designed for practicing project managers, engineering, MBA, as well as Ph.D. students who need to understand the various dynamics that are typically encountered in a project environment. Undergraduates can use this as a manual if they are interested in pursuing a career in a projectized environment. For researchers, this book provides ample ideas on potential areas of research in the domain of applied project management.

Most projects today do not follow the conventional route ending up in parallel activities. Staffing and capability issues can be dominant in most such projects. Further, not all projects need 'equal' emphasis in all areas… some areas, therefore, are more 'mucked' than others! Therefore, all said, *practice differs from theory*! This book is all about bridging that gap and giving the project manager a strong platform to work on.

The content in the book is taken from several books and training programs. I have also tried to use discussion threads from various fora on project management. Many of the tools have been developed on the basis of modeling and simulation methods. They are specially designed by me and are copyrighted under the IP of Consulting Connoisseurs. And these were tested at several projects across the globe.

Most of the exercises in the book are actually meant for you, the reader, to perform them as you go. Therefore, this book does not have a 'read-all' and 'come back later' kind of design. On the contrary, the approach is to use the 'learning' by 'doing' paradigm, whereby the reader is expected to do the exercises at that time before reading on.

Finally, as a fellow member of the project management community, I would ask you to enjoy your work and try to gain as much as possible from this book. So, happy reading and happy learning.

Nikhil Gurjar

# Acknowledgement

The dreams of my late father ...whose blessings gave me the courage to take up this long ordeal—that's how I would put the journey that took me through the highs and lows and helped me complete this book. Naturally, such a journey was also supported strongly by my mother. And her ill-health during the early days of writing this book, really kept me going. As I waited outside the ICUs for long days without good network connectivity, the situation enabled me to insulate myself from the hassles of daily business and focus / concentrate on the book. Conviction and support from her side played a major role in ensuring that I was moving towards completing this mammoth task I undertook.

After the completion of a 100 pages, the fatigue started creeping in! It was then that my children, Riya and Arya, would come to me and motivate me to take it ahead. They seemed to make the lows seem like trivial challenges. And they were constantly proud while discussing about daddy's book, without knowing the end result and the end product. I can't thank them enough.

My wife, Pallavi, has probably been the pillar of this work. She has been the support structure that has helped me take on this kind of a project. Having written quite a few research papers, I always saw a book to be a simple extension. However, I quickly realized that the project was going to be a much larger one than anticipated. Pallavi's help, in managing the household and seeing that this project was successful, was immense and the immeasurable amount of encouragement and support, indeed, helped me move ahead.

My teacher and co-author in several research papers, Prof. S.D. Jog, was a steering/guiding light throughout this project. I would certainly find no words powerful enough to thank him for it. He has been a constant source of encouragement and has helped develop a lot of new perspectives in life.

I would also like to thank my ex-colleague, Dr. Bertram Ehrhardt, whose penchant for research is something I missed. His frank methods of management analysis have provided a lot of insight that is vividly captured in certain sections of this work. And his encouragement was immense. He always believed that I needed to do something extraordinary and I believe this book is the first installment to his belief. I would also like to thank other friends and former colleagues Markus Forsch, late Klaus Timmerbeil, Horst Janssen, and David Comstock. Dave (David) and my team members Rick Choron and Nick Aryaamir were among those who constantly pushed my creative instincts when it came to project analyses.

I would also like to thank Dr. Sanjay Pattiwar, Shirish Aradwad, Vivek Achalkar, and Dr. Pradeep Mahajan, for the opportunities in government consulting, especially the PPP models and IT infrastructure projects. Many of our frameworks were tested across verticals and domains and ratified during the process.

Well no study in project management is possible without project managers. I got to see all good and bad managers in various projects, thanks to my wide variety of experiences across various sectors. The gaps in the knowledge levels were also glaring in some cases. Interestingly, I did learn a lot from the competition, that is, from other management consultants and partners. Formal modeling and simulation tools, coupled with some creative faculties and a good number of data points and first-hand experiences helped me move forward in this compilation. When I mentioned that I had written a book to my former colleague, Kai Waldmann, he commented 'I expected this to happen earlier!' Nothing took me by surprise more than these words of his!

I would also like to thank the encouragement from my IIT alumni: Anil Risbud, Milind Deshpande, Ajay Phatak. My client, Shoeb Kuruwadwala, also a fellow alumnus, did play a major role in providing me with the platform to help transform his project-based company. I would also like to thank Shantanu Bhadkamkar, the President of the Maharashtra Chamber of Commerce, Industry and Agriculture for his encouragement.

Workshops at CII, PPMAI, NITIE, and other places, on project management helped me refine the framework better. Needless to add, these experiences helped me express the learnings and ensure that the reader understands the same in a better way.

Thanks also to Springer for taking this book to the global audience.

Nikhil Gurjar

# The Rationale Behind This Book

Most books on project management enrich the subject by providing a smooth treatment to what is commonly accepted as the 'norm', the 'global practice', or the 'convention'. While this is an enormous exercise, it often deviates from the actual content/subject matter expertise required of the project manager. This usually translates into certain forms of *gaps that project managers face sometime during their careers (viz. the frustration in mid-career/post-mid-career individuals!)*. The need, therefore, is less on understanding the literature that is there, but rather, having a firm handle on the driving reasons behind a particular practice in a given industry. This book addresses this need. It questions the readers on the logic they use and helps them gain a perspective as they move along.

During my several years of experience in various types of projects, the one aspect I identified was *the lack of alignment between the team members of a particular project*. Though different schools argue about this phenomenon in different ways, my perspective has been that this is attributed to the gaps in the knowledge levels of the team members. For instance, in practice, the interface between the Chief Risk Officer and the Project Manager is not understood or defined in most organizations. Such huge gaps need to be covered by providing the readers with a more comprehensive framework. Most books available today either do not have the depth of coverage required to describe the dynamics of the logic involved or are too specialized to have a cross-faculty approach. I have tried to balance the needs by going a level deeper into most areas in this book.

Since this book was primarily written to look at *Modeling and Simulation approaches in Project Management*, most of the frameworks are also different from the conventional literature. In doing so, I have tried to maintain an 'advanced' outlook whereby I have consciously tried to *avoid describing the Basics where possible*. We expect the readers to do a preliminary reading (as part of their under/graduate curricula or their certifications) and then, come over to understand the perspectives involved in typical practical situations.

We have also *avoided typical topics* that pit us against a plethora of competing literature! In doing so, I have tried to look at those issues that are pertinent to the project manager. Rather than sticking to the paradigm of a 'nice-to-know' approach, I have tried to focus entirely on the topics that are often left-out in conventional treatment of the subject. And in doing so, I have also tried to look for references that are readily available on the internet, in addition to books. By this strategy, I have tried to make it simple for the reader to refer to topics elsewhere, while also trying to focus on the core content in this book.

The book is *extremely demonstrative*; in the sense, that it provides (a) *simple exercises*, (b) *plain / straightforward questions* and (c) *evaluative perspectives*. This is essential to enable the reader to reflect on what is being said and, at the same time, allows a wider audience to read the book. In doing so, I have also tried to make it easy for the international reader to follow the details of the book. Having worked in three different geographies myself, viz. Europe, US, and India, I believe the language is easy-to-understand for the common reader.

The book is structured to start looking at *the lacunae in practice that are immediately visible*. The reader, therefore, needs to be quick in understanding the lacunae; and also, my own consulting experience has proved that this is a great starting point for the learner/learning process. Naturally, finding faults is, at times, a painful exercises; and, the reader needs to be ready to go through that pain (be it painfully simple or painfully complex!) to ensure that he has the right perspective and is able to condition his thinking toward substantial gains at the end of the read. In other words, it is not a hobby or a light read book by definition. So, anyone who just wants to spend time reading bits and pieces as a quick read would not gain from the content here. A large part of the content has also been validated in training programs and consulting engagements where senior managers and executives have finally realized the dynamics of businesses and their own organization's symptoms in the project management domain.

A major part of the book contains *unique and easy- to -use frameworks and models*. These Quick Frameworks, developed at Consulting Connoisseurs, help the reader quickly assess the situation, since most of them are based on the portfolio approach in consulting. The models and simulations further help the reader understand the interrelationships and the rigor in the decision-making processes, something that would be required to consider the major aspects influencing a given situation.

One striking aspect of the book is that it is *Forward Looking* throughout, thereby preventing discussions pertaining to blame games. In other words, the situation at hand is typically analyzed in each section and the history is discounted as far as possible! That way, the consultative approach to problems is leveraged.

Unlike conventional literature, the emphasis in this book is on four fundamental questions: *What is necessary to trigger conceptual thought-processes, How to analyze a given situation, How to apply new concepts in an eco-system, and When to apply?* Traditional books like PMBoK, etc., stress on the thought process of 'What is Project Management Theory' and 'Where is it normally applied'. Both of them are like 'God given' realities that the Project Manager is expected to imbibe. Our approach differs drastically from this line of thinking.

In most project situations, due to the information age, there has been a rise in alignment issues at the workplace. While information is readily available via the internet, it also is not always optimized to give the project manager the right perspective. At the same time, information within the organization is often not shared freely due to various reasons. This causes a difference in viewpoints. These mis-alignments that occur at the tactical, operational, and sometimes, strategic levels, cause huge losses of energies within the organization. In this book, our endeavor is to minimize *the Loss of Energy between the Project Managers and the Business Owners*. I have tried to assist the Project Managers to understand what they know and a possible reasoning on what they need to know in their given situations.

The book combines faculties across *the domains of management, technology, and social sciences*. In doing so, I have tried to stick to focusing exclusively on the project issues that are pertinent. The reader is given the freedom to read additional topics by looking up the source or additional literature, where required.

Since there is a multitude of combinations possible while combining the three domains mentioned, *the book takes on an iterative flow rather than a linear one*. In other words, the same concept is dealt with iteratively. This allows the user to understand the concepts better. It also enabled me to delve deeper into each concept with each iteration and, at the same time, provided allowance for showing multiple perspectives on each topic. In doing so, I could give simpler concepts for readers looking for restricted take-aways and more advanced concepts for readers who are looking for more complex issues, requirements, and needs. Thereby, I have also made it easier for the reader to read and use, without compromising on the needs of the wider spectrum of readers.

The book is very hands-on in its approach and is, therefore, *very applied*. It is change oriented and is *fearless in criticizing some of the populist views*.

As with any advanced treatment, the book is *open ended*. Any reader who reads this book might want to reread and rework his own answers that need to be up to his own satisfaction. In that process, the reader would be able to reflect more on the concepts and would be in a better position to absorb various dimensions mentioned in the book, and, also add value to them by including a few of his own.

# Contents

# List of Figures

# List of Tables

# Part I
# A Reality Check

# Chapter 1
# Introduction

**Abstract** In this chapter, we systematically try to understand the genesis of the debate on the understanding of project management as an *Art* or as a *Science*. In doing so, we go back to the evolution of project management as a faculty and the current practices today. We also question the perspectives from the limitations that we see in the project experiences across the world. To keep the focus, we stick to *business perspectives first, and methods later.* We try to evaluate the approach from an action perspective rather than a reporting perspective. Two important introductory components that are touched upon here are Consulting Connoisseurs models on (a) *the types of project environments and (b) the classification of projects (types of projects).* Limitations of the conventional literature only emphasize the need to have a better system to structure and leverage from the definitions of project environments and the projects. We question the usability of the classification system from the perspective of the project manager who is in the field / on the job. In order to make such exercises more meaningful, we define alternative frameworks for both the project environments as well as the project classifications. In the process, we have also delved into the distinction between the two approaches: project environment versus project itself. The *Ownership Model for Classifying Project Environments,* introduced in this chapter, touches on the way project environments drive the success of a given manager in a given project/environment. The *Business Abstraction Based Model for Projects* defined in this chapter typifies projects by helping the project manager understand the broader controls associated with the project contexts. The chapter opens a vast range of potential perspectives that the project manager could develop for his specific situation. We also touch upon the use and misuse of stereotyping in such applications.

**Keywords** PMBoK® · Art versus science debate · Layered approach of PMBoK® based approaches · Types of project environments · Ownership model for classifying project environments · Types of projects · Business abstraction-based model for project types

© Springer Science+Business Media Singapore 2017
N. Gurjar, *A Forward Looking Approach to Project Management,*
Lecture Notes in Management and Industrial Engineering,
DOI 10.1007/978-981-10-0782-8_1

## 1.1 Introduction

Project management is one of the most popular branches in the management discipline. Most people argue that project management, at some point, borders between an art and a science! While the 'art' component is arguably true with any management discipline, the science cannot be undermined by this argument. Having said that, after working in different project environments globally, I saw that the fundamental understanding of project management is often not up to the desired mark. This phenomenon (of lack of knowledge) makes most managers overrate the 'art' component. We differ from this view in this book.

## 1.2 How Project Management Evolved

Project management is highly application oriented like any other management science. Any practicing manager can, therefore, appreciate that there needs to be little gap between 'Theory' and 'Practice'. One of the first attempts to formalize Project Management theory from a practice perspective, was the Project Management Institute's PMBoK®. This was a wonderful compilation of the 'best practices' across various types of projects. It helped lay the foundation for a 'practice standard'. It also triggered the 'Project Professional' revolution. Everywhere one went, one found people referring to the 'standards' mentioned in PMBoK®. This has had two fallouts:

1. People have been reading the project management literature and
2. They believed in implementing the same where possible.

From an organizational perspective, this was helpful too. The organizations were then able to improve their practices with relative ease, as there was a formal 'text' of 'standards' available. Thus, PMBoK® was indeed a welcome innovation.

### 1.2.1 PMBoK® as a Guidance Versus PMBoK® as a Standard

With the practice going that way, innovation in practical environments was relatively limited. The standards/compliance view was the larger focus. Every professional started viewing any project situation as something that was either being compliant with the PMBoK® standard or not being so. Unfortunately, despite this, most managers had still been seeing cost and time overruns. So PMBoK® as a standard had serious limitations. Moreover, due to its popularity, the understanding of project management education has been transformed to be restricted to management of operational processes.

This is like the situation in a rugby match, where players start pouncing on the ball. The end result...there are too many layers to deal with! Like the 'pile of players', one has to really sift through the 'stack' to find what one really needs. With the shift

**Fig. 1.1**  State of PM
education today

**An Example of Layers and the Real Thing!**

in the focus to primarily processes, most times the core management principles were left 'deep down under' and the statistics of cost and time overruns only reinforce this finding.

Another example is described in Fig. 1.1. Here, we have a word, and one could start with MA and keep adding alphabets. Each time there is an addition of alphabets, the word changes in its meaning. The deeper one goes, one sees the actual word and the meaning. In other words, with each layer, there is a value addition, but it could still be far from the objective. One needs to actually penetrate through these layers and take it forward from there. Just like the word shown in the figure, it would be clear that to get the right gist, one needs to drill deeper! Our attempt in this book, therefore, is to try and work things from the core issues of Project Management (the complete word as shown in figure), layer by layer, as far as possible. We are going to isolate the 'skin' or 'outer' process layers and ensure that we have a better understanding of the project environment and how it connects with the other faculties of management. This also means that we would be approaching a few topics *iteratively* rather than *linearly* like in most other books.

### 1.2.2  Navigating Through the Layers

In the present work, we have tried to help managers understand their project contexts and look at various other tricky and dicey issues that are not often within the subject matter of conventional education. Therefore, in this endeavor, we are looking at aspects where modeling and simulation methods help managers in redefining their style of working. So in addition to the 'layering' approach, we are going to look at an 'integrating' approach, across a wider array of concepts, in this book.

The second important aspect is the 'ease of implementing'. While advanced topics are often difficult to comprehend and implement, our tools and techniques are optimized to be relatively easy to work with. But to do that effectively, we often need to 'push' you …the reader …out of your comfort zone. Our 'Quick Tests' and 'Short Questions' help the reader understand the context and navigate through the complex set of issues involved. Again, as a reader, it would be useful to work out the questions and the tests and to revisit them once one has finished reading the chapter. That would help build the appropriate perspective giving two benchmarks (an initial one and a post reading one).

## 1.3   How Are These Methods Different?

### 1.3.1   The Micro-Drill Effect

While existing methods are 'operational/functional' in their orientation, they tend to falsely impress upon the readers and students that their methods are unique and specialized. Often times, this leads to a very dangerous situation. The project manager tends to believe that he is doing something specialized and goes out of sync with the management/other functional areas in his organization. Further, the lack of integration of the concepts leads to suboptimal decision making. This leads to huge gaps in the expertise and the delivery. This is what we refer to as the *micro-drill effect*, where there seems to be significant depth without any breadth/coverage/active interlinkage between the various elements.

The methods discussed in this book are far more involved because they try to consciously overcome the shortcomings of the micro-drill effect. That is, they look at the interrelationships between the various functions. We also focus on tools and techniques that enable this integration and the decision-making criteria that go along with the use of these tools.

### 1.3.2   Characterizing the Change

At any point, the advanced techniques described here are meant to:

1. Be factual rather than fictitious extensions of opinions.
2. Be tailored to decision making rather than interpreting situational information.
3. Be useful in deriving results across a wide range of solutions rather than having a limited applicability.
4. And, most importantly, combine different faculties of management, thereby, making them extremely powerful.

While this is true with most types of advanced techniques, the ones in the existing literature are notorious for quickly moving into the highly differentiated territories of their specialties. In other words, if one were to look at a specific topic like HR in projects, after one reads a conventional book, it would be very difficult to relate the knowledge, to say, Corporate Project Risk Management. Hence, though these are designed to be advanced treatment, the linkages with existing faculties are often left to the *extensions of the reader's opinions* rather than the factual mode. The vast majority of literature is tailored to interpreting. Very few books actually have explicit decision variables (for instance, what should be the notice period in the contracts with the employees). Further, the applicability requires a generic treatment. For instance, Kenneth Blanchard's *One Minute Manager* provides a quickie for typical situations. Most project situations have *atypical handles* forcing the reader to ensure the balances between the differentiated pockets of knowledge and the mundane causal behavior (integrated knowledge). This is where the conventional literature fails to provide a suitable platform for the manager.

And when one advocates such methods, one needs to look at business perspectives first, methods later. This is what we have attempted in this book. For instance, estimating the risk in projects is an elaborate exercise in itself. However, what are the management decisions involved? If you look at the manager's perspective, the decisions are fairly simple:

1. Is the project within the acceptable 'risk appetite' of the organization?
2. What needs to be done to control the risks (other than holding meetings!)?
3. How to define control points to ensure that the risks are being managed?
4. How much of the risks need to be controlled? How and why?
5. Is there a refinement of the business case imminent?

Now, in such a case, the traditional approach fails…mainly due to the fact that most 'traditional' approaches stop at 'reporting' and they believe in passing the 'strategic' information to some other stakeholder. This is far from what is required from any manager! The end result is that this orientation, developed using conventional literature, quickly turns political; and the managers, in practice, tend to 'control' the extent and type of information that goes to the stakeholder thereby trying to 'cover' themselves up, rather than exhibiting superior project management expertise.

And this posture actually trivializes the whole concept by 'painting' pictures that are different from the reality. Using Modeling and Simulation *actually* helps in such a situation. We will demonstrate this in subsequent chapters. Modeling and Simulation help integrate complex 'traditionally differentiated' knowledge areas, thereby helping the project manager gain meaningful insight into the use of such information.

## 1.4 Types of Project Environments

There are multiple perspectives used to define the types of project environments. The most popular ones categorize projects by:

1. Sector
2. Function

3. Organizational size
4. Value
5. Duration, etc.

In fact, these perspectives are so popular that most recruiters and recruitment advertisements seem to overly stress on the need to know them. Again, this is the 'process'-centric approach described earlier. While there are significant merits in knowing the sector, the function, the value, etc., they actually do not 'facilitate' the management process of the project.

### 1.4.1  What Is the Business Issue?

The cardinal question to be asked by the manager is the *purpose* of such a categorization or classification. What project managers need to really ask is the following:

**What happens if I don't know the categorization or the classification of the project environment?**
**Can I not manage the project then?**

### 1.4.2  Limitations of Existing Attempts

The business issue above needs to be clearly understood by the reader. This actually gives us a few critical aspects to ponder upon! These aspects have been debated at length in various project management interest groups and discussion fora. However, such interest groups and fora do not give us the 'entire' perspective. And that is essentially due to two reasons:

1. It is 'driven' by project managers, who are only looking at their own 'side' of the whole situation. Therefore, whatever the results or findings of these groups, they do not have an 'inclusive' view involving the other 'stakeholders' in the decision.
2. Due to the overemphasis on the 'process-centric' view, there has been a divide in the project management community. The first group within the community 'differentiates' between the 'management' skill and the 'technical' skill. And the second group treats the 'technical' and the 'management' skill as a continuum. Hence, the process view actually tends to vary from one organization to the other as not just the skill sets, but the scope of the job description of the *designated project manager* also vary; thereby, causing a varying perception on the role of the process-centric orientation needed by the manager.

So, in effect, the search for the answers to the business issue that we have identified/are speaking of are reinforcing the traditional classification and categorization perspectives of sectors, functions, etc., due to the above two reasons. However, *we do not subscribe to this view in this book.*

### *1.4.3   Refining the Business Issue*

Simplistically put, a project manager can (and should be able to) manage any project! His level of competence is actually governed by his ability to understand specifics of his business role and the understanding of his project environment. Hence, it is essential to understand the characterization of his project environment from the perspective of his 'potential competence'.

Hence, the cardinal question now shifts to:

**Will I succeed in a given project environment?**

Hence, the focus quickly moves to the outcome/the performance. Ultimately, both the questions are similar, yet different. The first one is more focused on the manager and his own competence and internal decision framework. The redefined business issue is more direct and more relevant for both the project manager and the organization to understand. Apparently, it goes way beyond the competence parameter as we will soon discover.

### *1.4.4   An Alternative Framework*

There are two parameters that help define the high-level classification of a particular project environment. We now introduce the *Ownership Model for Classification of Project Environments*. They are (a) The ownership at the operational level and (b) The ownership at the strategic level. We must note that it is not responsibility that is mentioned, rather, the ownership of the project. While responsibility is an assignment parameter (that is defined in a job description), ownership is actually an amalgamation of:

1. The formal assignment
2. The manager's willingness to 'own' the tasks, perform them independently,
3. His competence
4. And the management support given to him.

Therefore, ownership has both the manager's as well as the stakeholder's component as shown in Fig. 1.2. These include elements that progressively increase the ownership levels from:

1. Job Description
2. Job Requirements
3. Competence
4. Willingness, and
5. Management Support.

The ownership, therefore, could be broadly checked for two dimensions, i.e., strategic level and the operational level.

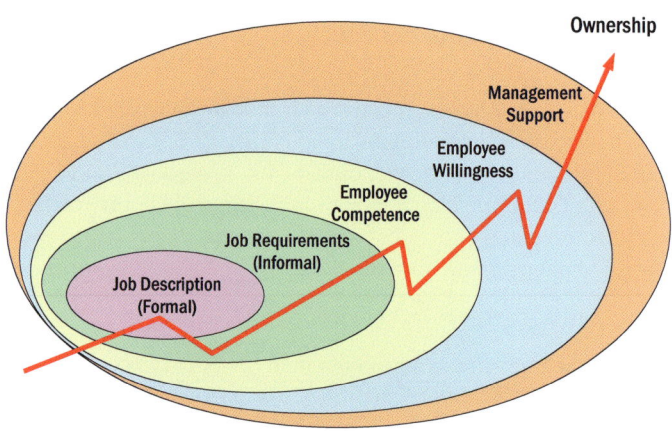

**Fig. 1.2**  Elements of ownership

**Fig. 1.3**  Types of project environments: Ownership Model Portfolio

These can be practically visualized using the grid (portfolio) shown in Fig. 1.3. For a project manager, it is of utmost importance to understand his own competence parameters. In an environment that has low requirements of 'ownership', both at the strategic as well as the operational level, the designated project manager actually acts more like a 'resource' rather than a 'management' expert. This typically occurs in 'large'-sized organizations where many people work on a particular project at the same time. The project manager, in such cases, is typically a hierarchical place holder. The hierarchy and the sheer numbers of the 'large' organization make the decision-making process rather diffused. The end-effect of such a scenario is that the administration of decision-making skills is random. In other words, driven by *politics* rather than a defined criteria. When the projects in such environments fail, it is usually due to the 'political' dynamics within the team rather than any other parameter and the general environment is often that of a political minefield.

The second type of environment is where a project manager has a high level of ownership at the operational level, but a low level of ownership at the strategic level. Most recruitment advertisements focus on these kinds of requirements. In such environments, the project manager has a lot of 'expectations' to match (in terms of his work fulfillment.) However, he is either not trained or not aware or not involved with the strategic picture. In other words, he is required to focus on the 'tasks' in any given project. His primary role is to ensure that the tasks are correctly defined and executed to fulfillment. Such managers will have to rely on the senior management for direction, each time there is a nonoperational decision at work.

Such project environments often fail because the manager channelizes his energies to the tasks rather than the strategic objective. Therefore, he is, many times, found 'foolish' at the time of any 'crisis' because he is driving himself against a set of impossible targets, or is perceived as someone 'unable to understand' his strategic role very well (for reasons beyond him). However, he is respected for his knowledge when it comes to operational issues and clarifications. An overburdened project manager typically also ends up as one in the second environment. Such environments, needless to say, prefer project managers with experience: someone who understands when to go slow and when not to!

The third type of environment is one in which a project manager has a high level of ownership at the strategic level, but a low level of ownership at the operational level. Such managers typically have good skills in general management and finance. He is often the 'dreamer' who does not like the 'details'. Such managers usually have a distaste for 'technical' issues and are very 'management' focused. They typically have, up their sleeves, a few 'recipes for disaster' and a few more 'recipes for covering up', at all times. The environment typically provides for technical resources who manage the nitty-gritty issues of each case.

Such project environments usually fail when the manager is 'continuously' manipulating or changing the strategy. Or when the manager is just not adequately aware of the implications of his decisions. And oftentimes, such managers start isolating themselves from the problem, thereby becoming a part of the problem rather than being a part of the solution.

The fourth and last type of environment is one which has high levels of ownership in both parameters. Here, it is not just the processes, but also the strategy that has to be balanced. Such an environment is an ideal one for a strong project manager because he understands both the strategic as well as the operational parameters well. However, in order to succeed, the manager needs to understand how he has to integrate the operational inputs with strategic inputs. This kind of an environment requires advanced tools and techniques that help the manager succeed. This book focuses on those advanced concepts that are required in such a situation.

While we have defined the environments, we have also identified a few skills of the project manager that are required in each of these environments. Our attempt here is to try and clarify the various kinds of environments from the perspective of our cardinal questions. More the ownership component, the easier it is for the manager to maneuver through the demands: operational or strategic. And once that is clarified, it does not matter as to what the sector, geography, size, value, or any other parameter of the project is. Our idea is that the span of control of the manager is very well characterized by these parameters and makes it an easy framework for decision making for the reader. And with *Prof. Google* helping every manager today, the knowledge component boils down to reading articles on the internet rather than having to face the music and gathering experience everytime! In other words, the knowledge and experience component are many times understood and misunderstood to be, by and large, substitutable with each other, with relative ease (in terms of time and effort); especially if the project manager makes a sincere effort to complement the same and has the skill to learn and adapt.

### A Quick Test

If you are a practicing manager, before you go any further, you need to ask yourself a few questions. These questions are designed to help you understand the quadrant in which you are operating.

There are essentially different levels of 'ownership' that we would like to present here. A low level of ownership is one in which the project manager does not decide these by himself. Or even if he decides, he just cannot implement them without approval. In other words, he provides 'someone else' the input for decision making. In the case of a high level of ownership, a project manager is able to decide these *on his own and take necessary steps to implement the same on his own*. In other words, he takes an approval where needed: usually a token approval that is rather quickly granted, or even one on a 'back-dated' basis (an overdriven extension), or does not even need to take any approval from anyone. Critical issues here are those of scope adjustment, costs, manpower, and project infrastructure. In other words, while one looks at implementing decisions, the key is to identify where one lies on the ownership spectrum as shown in Fig. 1.4.

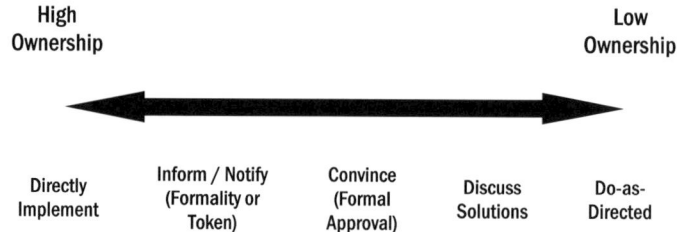

**Fig. 1.4** Positions on the ownership spectrum

Understanding these positions, the reader must try to look at the following questions to ascertain the characterization of their current role within their organization:

**As a manager, can you change the sequences of activities and tasks in your project? If yes, are you required to take any approvals?**

**Can you increase the manpower required for your project independently? If yes, are you required to take any approvals?**

**You might be given a budget for your project. Do you know the detailed calculations of how they were arrived at? If not, can you get to know them and benefit from this knowledge?**

**Do you know how this project affects the short-term or the long-term profitability of your company? If not, can you get to know them?**

**Do you know the investment-spread of other candidate projects in the company? If not, can you get to know them?**

**Do you have signatory powers of your company for large transactions and contractual requirements? If not, are you likely to have them for your next project?**

**Can you invest in any infrastructure required for this project (software or building or any other)? If yes, are you required to take any approvals?**

**Have any of the decisions, taken by you and mentioned above, not been approved by your management? If yes, how often does it happen? And do you understand the reasons for the same?**

These are some of the questions that could help you characterize your project environment. The questions are more focused toward understanding the 'management' perspective rather than your own. Often times, good managers tend to ignore these management issues and are too focused on their work and their aspirations. This could lead to tricky situations.

## 1.5 Types of Projects

We now move on to project types. The classification of project environments and projects is not a new practice in literature as discussed in the earlier section.

### 1.5.1 Cardinal Questions

Many different perspectives exist for project types as well. However, one needs to know *where and how to use the classification*. Several authors have tried to provide pointers to this issue. However, it is important to know the relative merits and demerits to any classification that *applies to your specific context*. At Ground Zero, I believe that there is little additional input that one would get by defining project types.

To draw a parallel, one could refer to the concept of stereotypes. The application of stereotypes is not always fruitful. In fact, while using stereotypes, one has to many times be careful about the potential problems that one could face.

## 1.5.2 Conventional Thinking on Types

The vast majority of project professionals do find a level of comfort in project classification. We refer to two different research papers to explain the vastly different views of researchers in this area. But before we do that, we will quickly provide an overview of the existing ways in which projects are classified. This is beautifully summarized by Archibald [1] in his publication. The publication covers this in reasonable depth.

At the high level, Archibald states:

Many papers and books, and much research, deal with project management in a general sense, but only a few to date examine the projects themselves: the common denominators for the discipline of project management. How are these various types of projects the same, and how are they different? Which aspects of projects can be standardized for all projects, versus those aspects that can be standardized only for specific project categories?

He goes on to elaborate on the need to classify as:

Crawford et al. (2004), in their recent PMI funded research, concluded that all organizations that have large numbers of projects must and do categorize them, although the categories are not always immediately visible. This pervasive *de facto* categorization is often taken for granted: 'That's the way we always do it.'

The logic that there is an existing practice that needs to be given direction is then given in his work as:

The basic question here is not whether or not projects should be categorized, but *How can they best be categorized for practical purposes?* Two closely related questions are:
*What are the purposes of project categorization?*
*What criteria or project attributes are best used to categorize projects?*

Crawford et al. (2004) state that it is dysfunctional to try to categorize projects without knowing what purpose will be served by the categorization.

My own submission here is that a categorization should help the manager in some practical way. If that does not happen, it cannot be useful to an organization, except in reporting and building stereotypes. The purpose and uses as given in the work are listed as follows:

1. Definition of strategic project portfolios and their alignment with growth strategies
2. Selection and development of the best project life cycle (or life span) models
3. Identification and application of best practices for:

   (a) Project selection and prioritization
   (b) Planning, executing, and controlling methods and templates
   (c) Risk management methods
   (d) Governance policies and procedures
   (e) Development of specialized software applications.

4. Building of specialized bodies of knowledge
5. Selection and training of project managers and project management specialists
6. Focusing and improving PM education and training
7. More effective individual PM certification and career planning
8. More focused research efforts
9. Organizing paper presentation tracks at professional meetings
10. Plus additional benefits not yet identified.

The typical attributes used in the categorization take cues from the evolution of Special Interest Groups (SIGs) within PMI. The publication indicates the following as the dominant ones:

1. Application area or product
2. Stage of life cycle
3. Grouped or single
4. Strategic importance
5. Strategic driver
6. Geography
7. Scope
8. Timing
9. Uncertainty
10. Risk
11. Complexity
12. Customer
13. Ownership
14. Contractual.

**A Quick Test**

In order to *beat the convention*, let us do a little of brain-storming to understand how this works. Such an exercise is meant to give us an 'open' set of views. While developing these views, one also has to revisit the robustness of the viewpoint. For instance, I am taking a general example that is relatively easy for everyone to follow. If one were to say that a Wok would be ideal to cook chicken, the robustness should look at two counter questions: (a) What is the specific reason that makes the Wok ideal for the chicken (or where do I have to necessarily use a Wok) and (b) What is the next best alternative to the Wok and why. When one looks at both these questions, one gets an insight into the 'robustness' of the concepts used to develop the initial understanding viz., *A Wok would be ideal to cook Chicken.*

I am also using both classification and categorization in this section. While they are two different things, both involve the systematic grouping of projects. One is an exclusive group (classification) while the other is more diffuse (categorization). In other words, a project could be in multiple categories, but is usually restricted to a particular classification. Nevertheless, when I mention classification or categorization, I am referring to both of them in this section.

**What according to you could be a new and a good basis to classify/categorize projects (other than the one presented here)? Explain the same in detail. What are the relative merits and demerits of the proposed basis over the existing method?**

**How should one choose the basis to classify/categorize projects? Explain the same in detail.**

**As a project manager, why should you know the type of the project according to your own schema? Where do you see a potential failure if you do not know the type of the project? How does it help you in your work, if you know the classification/categorization?**

**What methods are used in your current organization to classify projects? Explain why they might be following the current practice. Do you think you could redefine their current practice and bring in improvements? Explain how and why.**

These are a set of open-ended questions that touch upon the concept of project classification/categorization.

### 1.5.3   Why Is Classification Not Useable?

While the concept of project classification/categorization may sound great on paper, I have hardly seen it being used that way in the practice of the project manager. There are two fundamental reasons:

1. The fundamental problem in most projects is that of employee turnover. To clarify, the employees most times leave projects for better opportunities or when things turn 'sour'. In such a situation, the correlation of the employee success (perceived or actual) is a greater determinant than any of the other factors. But when the agent

linking and using classification to practice changes, it just snaps off! :-). Hence, my view still holds that the classification of project environments is an important one and more fundamental. However, anything manager centric, in my opinion, would be additional paperwork at this stage!

2. The other important reason is that the wide variety of projects in any organization may not be recognized by the program office at a given point in time. For instance, if a company is expanding by constructing facilities, the various construction projects would be recognized as individual projects. However, the same program office would shy away from listing out projects like some IT upgradation occurring at the purchase function at a particular location, although these too are technically, projects! In other words, not all activity under the project definition is normally taken into the compilation of the project lists. Going by the literature, for a portfolio management system to be effectively used, the first step is to compile/list out ALL the current projects in the organization. Hence, the concept appears to be overdriven when one wants to account for all these different projects and take necessary steps to 'define' and 'optimize' the portfolio. In practice, the program office would not bother to define 'projects' that they think are unimportant. In other words, some projects get prejudged as 'low priority'/'out-of-the-radar' projects even before one meaningfully applies the tool of portfolio management. Hence, the very purpose and use of classification becomes questionable.

Due to these two reasons, it is often clear that the intended application of the concept of project classification, most times, does not serve its end objective.

### 1.5.4  An Alternate View

To describe the precise linkage between classification and the application, I have defined two fundamental parameters that are of interest to the reader. I call this the *Business Abstraction-based Model.* If one looks at any project context, the following two parameters are important to understand:

1. The impact of the project on the business
2. The time when this impact is going to be felt by the business.

Naturally, a high-impact project will be an important one for the management. In other words, in such projects, the success of the project manager is going to be more important or critically viewed. Again, in such projects, the ones that are going to impact the business in the near future would lay an even greater emphasis on the success of the project manager. Thus, the ranking would be expected to be higher in that quadrant. So, one could define 4 Ranks or classes of projects.

1. Low impact, Distant Future
2. Low Impact, Immediate Future
3. High Impact, Distant Future, and
4. High Impact, Immediate Future.

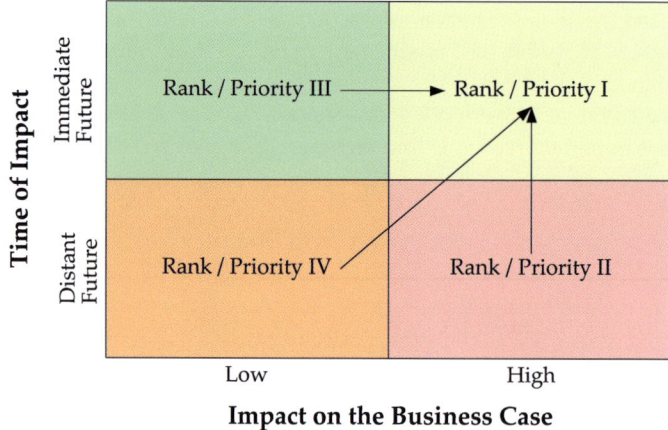

**Fig. 1.5** Ranking/priority of project success as given to projects

However, as the saying goes, *'nothing succeeds like success'*, the project manager's success is always given top priority! So regardless of where one would start or where one is positioned, all the quadrants seemingly converge to that containing the rank/priority 1 as shown in Fig. 1.5. It, therefore, becomes a political decision rather than a management one, that potentially jeopardizes the effort to classify and use the portfolio effectively.

This view is reinforced in the second research publication by Dvir et al. [2]. The purpose of the study by Dvir was to combine the theory of project success factors with the search for a natural project classification. The authors have tried to cover a very wide range of variables like:

1. Organizational Environment: This factor covers variables like (a) Existence of unit spirit, (b) Managers as role models, (c) Social activities out of working hours, (d) Room for professional growth, and (e) Possibilities for consulting with experienced professionals.
2. Manager's Style: This Factor covers Variables like (a) Exact specification of tasks, (b) Personal supervision of performance, (c) Involvement with workers, (d) Acts to increase workers' motivation, and (e) Involving workers in decision making
3. Communication Style: This Factor covers Variables like (a) Open communication, (b) Frequent updating of status, and (c)Involvement of managers in day-to-day problem solving.
4. Flexibility in Management: This Factor covers Variables like (a) Encouraging new ideas and (b) Willingness to consider changes and new approaches.
5. Delegation and Authority: This Factor covers Variables like (a) Setting general policy and goals and (b) Technical issues managed by the professionals.
6. Organizational Learning: This Factor covers Variables like (a) Participation in professional seminars, (b) Constant follow-up of technological developments, and (c) Application of lessons learned during project execution.

7. Team Characteristics: The penultimate Factor covering Variables like (a) Key personnel in the project for its entire duration, (b) High technical level, (c) Key Personnel with strong managerial qualifications, and (d) Some team members with operational experience.
8. Manager's Qualifications: The final Factor covering Variables like (a) Professionally experienced, (b) Technical Leader, (c) Extensive managerial experience, and (d) Canonical correlation between success measures.

In the conclusion of the work, the authors give interesting insights that are based on the characterization of the project environments. However, the level of detailing is too high and general statistical correlations are good for theorizing, but not strong enough for quick and practical decision making. Therefore, one has to orient oneself to a more rigorous causal model rather than the empirical one.

Our framework tries to simplify the findings by providing the reader/manager a quick evaluation model. Although the entire work of Dvir argues against universality, our framework tries to leverage the 'intelligence' of the project manager as he analyzes and evaluates an abstraction of these specific factors. Therefore, our factors are reasonably universal and yet cover the required details to help the project manager understand the decision variables in a given situation.

## 1.6  The Distinction Between the Two Approaches

We briefly mentioned stereotyping while discussing the concepts of classification and categorization. The reader needs to validate his own experiences with potential positives from both the approaches viz. project environments versus projects themselves. This could be fairly easy.

**A Quick Hands-On**

**In order to make a past failed project successful, what approach would give you a better assessment of the alternative plans of action available? Explain the various parameters. Attribute the factors to the model and compare them.**

Once the reader understands these aspects, it is relatively easy to understand which stereotype would provide the manager with better information to guide him through a more robust decision-making process.

To summarize, therefore, the context of stereotyping is important to understand. Slinger [3] has given a useful overview to this question. We use the analogous interpretation of the advantages and the disadvantages on each of the perspectives we have been using viz. the classification of environments and the classification of projects per se.

## 1.6.1  Advantages of Stereotyping

To assist your answering the questions, we list the potential advantages here. While doing so, the reader is suggested to verify the degree of applicability of the advantage or disadvantage on both the perspectives. In case you find additional advantages, feel free to add to the lists. They are only meant to help you understand the various parameters to evaluate the perspectives.

1. Stereotypes can be useful if you are in a new situation and need to make a quick judgment and fast decisions. For example, if you have never been around a member of royalty before, you might stereotype them as being formal and reserved, which will help you to respond to them in a respectful manner to follow their behavior.
2. Stereotyping can be seen as simplifying our surroundings so they are easier to understand. Stereotypes enable you to categorize people into groups, which allows you to form expectations about people and situations making life more predictable and easier to understand.
3. Even though the media may stereotype celebrities in a bad way, they may also stereotype them in a good way. They will make them look themselves and use them as role models to others. This may make the celebrity more popular and liked by the public which will increase their self-esteem.
4. Some people may like to be stereotyped as they want their looks and how they dress to come across clearly to other people, therefore they instantly want them to judge them, how to act toward them and what people's reactions are toward them, e.g., 'chavs' may want to show their dominance and want to gain their self-respect by showing this. This may be influenced by how they dress or even how they walk, which is what they like.

If one observes closely, these advantages are provided for by the understanding of the project environments. One might barely be able to actually link the project categorization with meaningful stereotypes on the decision-making criteria. Most of the books, therefore, adopt a prescriptive view to the stereotypes. For instance, even in the work by Archibald, a lot of the findings were prescriptive, without much of the formal understanding of how 'exactly' would the lack of use of the categorization impact the project.

## 1.6.2  Disadvantages of Stereotyping

Again, as in the above subsection, the reader is advised to verify for himself as to how these could be interpreted in the case of the two perspectives presented.

1. The downside of using a stereotype to make a judgment about someone is that it might be completely misguided and incorrect, causing you to act differently toward the person which can offend them.

2. The media may have changed the way the person actually is on a normal daily life to how they are on TV. This may be because it will draw more attention from the public eye as it is making the person, normally a celebrity, to look stupid. This will therefore increase the amount of viewers that a program gets.

3. For example, Paris Hilton is involved with a show called Paris Hilton's Best Friend. Some people may perceive this as a silly program and a waste of time because why make a TV program just to make a 'best friend'. This makes Paris look slightly silly and blonde because she is the one that has agreed to go on the show. The only reason she may have done this is because she wants to get seen by the public but she is too blonde to realize that people may be 'laughing' at her. It can be argued that are we as 'silly' as Paris Hilton? This is because we are the ones stupid enough to watch the programme, therefore we are the reason how she makes money. People may stereotype her as being false, spoilt, vain, and an attention seeker but is this what the producers want us to see? My point is that the media may have changed her to look how they want her to look instead of how she actually is.

4. Some people may assume that the same group of people share the same ideals and personality traits just because they have something in common like their dress sense, culture, or taste in music. Just because they may look the same, it does not mean they feel the same or think the same.

5. This is a form of prejudice and can be distressing for the person who is being pre-judged, who might feel that they are misunderstood, particularly when a stereotype is racially motivated. The saying *Don't judge a book by its cover* demonstrates that appearance is not enough to understand who a person really is. Stereotyping can create problems in many social situations like the workplace, at school, or in the local community. For example, in the workplace if a female boss were to manage a group of men, she might feel that she has to prove that she is capable to do the job because of the negative stereotyping that suggests that women are less capable, and the men might in turn assume that she is bossy and incompetent, this then creates negativity for her in the company.

The project categorization would actually tend to pre-empt decisions without giving the project an individual flavor. We have touched this issue especially in the context of R&D projects in later chapters. As far as the project environment is considered, the framework allows adequate subjectivity to maneuver through the model and keep the reader away from the potential pit-falls or disadvantages mentioned here.

## 1.7  Key Takeaways

Hmm, so now, most of you would appreciate that project management is less about processes, although there is a lot of emphasis on that aspect from both the literature as well as the current trends in recruitment. While we are not trying to revolutionize the 'profession' or the 'industry' here, we are definitely making a point that there is a need to be cautious while understanding and applying any methodology in the project environment. Our submission is that the science needs to be better understood.

The understanding of a project environment is changing today. With the information revolution, the project environment has moved from a 'congregation' of many subject matter experts (SMEs) and 'business perspectives' to the more significant aspect called *ownership*. Identifying the project environment is probably the most important starting point for any manager. This is highlighted in the Ownership Model that we have presented here.

We also have moved away from the conventional approach to project manager where the classification/categorization of projects is considered a sacred and useful exercise. Rather, we have dissected this to understand what exact information it provides to the project manager. The cardinal objective is to discard any layer of 'knowledge' that does not directly affect the way one works. Like we have pointed out the psychological basis to our argument, the reader is free to exercise his wisdom in optimizing the availability and application of the knowledge faculties to his specific situation. This is highlighted in the Business Abstraction-based Model that has been described in the chapter. The reader is encouraged to understand the differences when one applies both, or either selectively or none-at-all and gain a strong understanding of his own situation. Pitfalls of stereotyping also need to be factored in, while one re-reads the chapter.

# Chapter 2
# Basic Misconceptions in Projects

**Abstract** The chapter begins by questioning the *fundamental distinctions between projects and operations*. While there are significant differences between the two faculties, the reader is also made aware of the marked similarities between them. This helps the reader align better in using the literature and management developments in other areas that are in the other knowledge faculties. The chapter then covers the basics of the interface between the project and the business viz. the *project feasibility study (PFS)*. We then try to understand *the common problems of the PFS* from the perspective of the project manager (who is subsequently responsible for the success of the project). In doing so, we identify the areas of improvement of PFS analyses in practice. An overview of the *financial perspectives of the PFS* is then presented. This is followed by a cursory treatment of *financial appraisal methods*. As a commonly observed conceptual gap in project management, the chapter then goes on to explain the *basic accounting principles in deriving cash flows*. The chapter concludes with an integrated model to evaluate projects using a multiperspective criteria.

**Keywords** Projects versus operations · Misconceptions in projects · Project feasibility study · Common problems in project feasibility study · Financial perspectives in PFS · Financial appraisal methods in PFS · Time value of money · Discounted cash flows · Net present value · Benefit–cost ratio · Internal rate of return · Basic accounting principles for cash flows · Cash flow principle · Incremental principle · Long-term fund principle · Multiperspective criteria for project evaluation · Profit · Spend · Time to first revenue · Exposure · Stakeholder orientation · Corporate positioning · Environmental considerations

## 2.1 Hello World!

Most books on project management start with the definition of projects. Well, we too would like to begin with the same. To put matters in perspective, the definition is taken from the PMI PMBoK®. The definition is simple, lucid, and well written.

A project is a temporary endeavor undertaken to create a unique product or service.

© Springer Science+Business Media Singapore 2017

N. Gurjar, *A Forward Looking Approach to Project Management*,
Lecture Notes in Management and Industrial Engineering,
DOI 10.1007/978-981-10-0782-8_2

Again, literature also elaborates the fundamental differences between projects and operations. This helps the readers understand that projects are a specialized faculty as they are distinctly differentiable from the operations and other faculties. In other words, it is a case of an evolving specialization as a result of differentiation. While this is a must know for any manager, one must also look at the similarities between projects and operations. In most books, this seems to be an ignored area. We try to look at both the sides of the coin as we introduce projects.

### 2.1.1  Drawing the Line Between Projects and Operations

Let us touch upon these aspects in a little more detail. The first point, therefore, of our discussion, is that the projects are *not very different* from operations unlike what most books seem to say. Every activity does come to an end, even in continuous process plants! Even if one tries to argue otherwise, this is definitely true especially for maintenance work. There are objectives that are specific to a period like the production volume for the period, maintenance downtime for the period, etc. Hence, both projects and operations have a specific objective. In other words, everything is indeed temporary.

Second, it is often said that the project team does not outlive the project …Well, as shown in a lighter note, that is not entirely true (Fig. 2.1). From a crew perspective, there isn't any difference. A contractor crew is often 'static' in every sense and does the same kind of work over and over again! The only aspect is that they do it at different sites (possibly). So, an electrical crew does wiring at every site in more or less the same way, using the same skill. However, the wires used might be different and the specifications could be different. An offshore IT team, for instance, would be

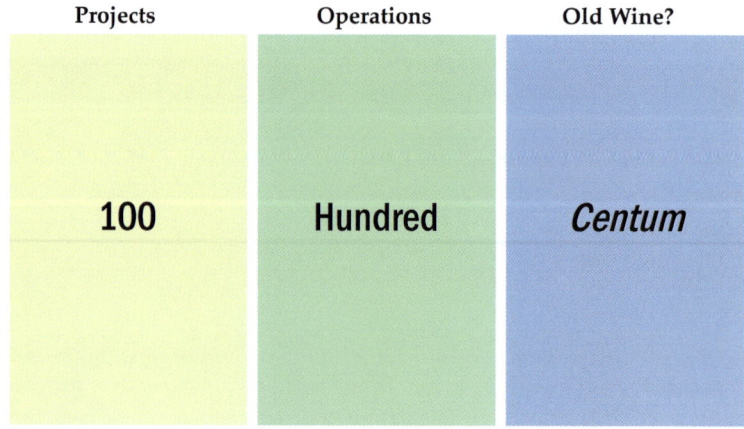

Spot the Differences!

**Fig. 2.1**  Are projects really different?

working on different projects, but the same way, using the same skill, and at the same location! So, strictly speaking, the project team often outlives the project. Hence, this too is similar to the general understanding of operations.

Third, despite what the literature says, projects are unique, so are operations. One does not have to argue about this. The product mix at a manufacturing company can always be different from its previous month. Hence, it is, in every sense, unique! Like no 2 days in any person's life are the same, so are no two periods. Each has different objectives, different set of challenges, and different ways of overcoming or being successful. So, any particular time-bucket is actually a unique one in itself.

And so, one can argue a whole lot about the similarities between projects and operations. This is, indeed, a dilemma. One could argue both ways and see different outcomes each time. Yet, if one asks a project manager, he does understand that the two are different. And I agree that they are different. However, what I find in the literature is a kind of an overdrive, whereby, authors keep harping about the differences a little too much. This is not a very positive sign. We will just see why.

### 2.1.2 Similarities Between Projects and Operations

What is essentially important to understand is that:

**A project is usually on the 'spend' part of the equation with a potential for a future 'revenue or benefit' of some kind. The gap, therefore, between the spend and the revenue is longer than the typical operations cycle in the company.**

Unlike that, operations, are an integral part of the value chain of a company. Hence, they are closely related to the revenues of the current accounting cycle or period.

**Why Is This Important?**

Barring the fundamental difference given above, there is hardly any difference between the project environment and the operations environment. Although, one would read a lot on the same in literature, let us wisen up to the fact that they are very similar. This has far-reaching consequences.

The first consequence is a simple one. If one truly understands the fact that they are similar, one could stop pushing oneself from going in the wrong direction. Afterall, the 'overdrive' is only going to create a kind of 'alter ego' in the project manager! It is time to work around the same effectively.

Second, submitting to the fact that there are similarities actually provides the project manager to leverage from research conducted in allied management areas. This eases a lot of pressure, because the project manager can now use existing repositories of knowledge from allied areas to help improve his own decision making. In other words, although different, one does not need to reinvent the entire wheel. I see a lot of project managers argue that projects are more 'difficult' than operations.

There is some degree of agreement there. However, let us look at another example to put things in perspective. Suppose one were to give a task to cook chicken and eat it! This would be pretty challenging for many as they need to learn the recipe, get the chicken, prepare for the cooking and, finally, if everything goes well, eat it (with pleasure, of course!). In other words it is difficult, like a project. Now, imagine that you have to cook and eat the chicken the same way everyday! Which of the two situations would be more difficult for you??? While the latter is arguably an operations scenario, I do believe it would be very hard to motivate oneself to cook and eat the chicken the same way everyday. One could try this and ask oneself when one reaches the third day! It does sound challenging and more difficult than the first situation! So, without delving into such sensitive topics, I just want the reader to understand that the similarities could be leveraged.

Third, the similarities do enable us to leverage the fundamental concept of operations viz. the concept of the process. Simplistically stated, a process is defined as a

> sequence of interdependent and linked procedures which, at every stage, consume one or more resources (employee time, energy, machines, money) to convert inputs (data, material, parts, etc.) into outputs.

This concept helps us align our thinking to the classical questions of 'How to' and 'Why'. Obviously, the 'Whats' are the foundation stones of the projects.

Hence, we do have an existing platform to springboard ourselves into the world of projects :-). I know that some of you might be in a state of 'shock' by now, but if project 'experience' really counts in the industry, then, the project world is *simplistically put...more similar to the operations world than we think!*. In short, the concept of project experience contradicts the premise of the claims of uniqueness. What good is an experience if it were so unique??? So, let us stop scaring people with technical jargons to prove that we are 'different' and start appreciating that our type of work is similar, but our conditions of work are different and that is what differentiates our world from the rest. In short, let us not be under **the misconception that projects are very different from operations**. I hope the above discussion has given it a meaningful closure.

In this book, our approach is precisely going to see how we could leverage existing repositories and add value to the project manager. While doing this, I have tried to make the transition go down with 'effortless ease' for the reader. As an advanced level book, it would many times work as an advantage for the reader to go back and do some reading on topics mentioned here, especially from books that are from allied branches. For instance, while talking about risk management, the reader could try and understand risk management from a financial perspective, from a statistical perspective and from an operations perspective *before* delving into the project perspective. That would assist the reader understand the commonalities and the specific aspects regarding the application in a project environment.

## 2.2 Projects in Business

The difference we mentioned earlier is certainly not a small one. So, it is required to delve into the same a little deeper now.

The first and foremost thing is to understand what a business is! Most people find it difficult to get back to fundamentals. However, we need to make this a practice in order to remind ourselves of the fundamentals that we often encounter. A business is summarized well as:

1. Something that keeps us busy
2. Something that helps us meet our needs
3. Something that adds value to the society

In other words, it is an economically relevant activity that requires time, money and resources. Therefore, most projects are a part of some business activity. In other words,

**It is a business need that translates into project activity. And most times, the project has a longer lasting effect than the operations, which implies that the business need creates an asset, in financial accounting terms, at the end of the project.**

The understanding of a project in a business environment, therefore, requires an understanding of both the business as well as the project environment. Such a study is called the **Project Feasibility Study (PFS)**.

In other words, a project feasibility study is an important document that bridges the project with the business environment. It looks at the need of the project and links it with the fulfillment of the Requirement / need. The vehicle of fulfillment is nothing else but the actual project. Since a feasibility study is carried out in the beginning, it is like the *foundation stone* of the project.

For most managers, the feasibility study is a document that loses its utility once the project gets approved! **This, again, is a very common misconception.** One might debate that this isn't so.

**A Quick Hands-On**

However, ask any project manager, who has been working on a project, the following:

**What exact inputs has he given for the preparation of the PFS?**
**Does he agree with the estimates of the time and cost in the feasibility study?**
**How often is the PFS referred to during the course of his work?**
**Does the PFS give him inputs on how one should decide the issue at hand?**

And lo! Most often than not, you are bound to find that the PFS is probably an irrelevant document for a project manager in his normal course of work!

Most practicing managers, therefore, believe that the PFS is a document that is normally required for an approval. Typical approvals are from senior management or the stakeholders or some bank or some regulatory body. In literature, the PFS is designated as one of the 'exit gates' of the project. In practice, it is indeed an 'exit gate' that tells the project manager to move on with his work. But given the fact that it is an exit gate, it is a very important activity. And the importance lies in the fact that:

> **From a project management perspective, the Project Feasibility Report is the document that provides** *the baseline of the business case that is fulfilled by the project*.

What a PFS should, therefore, intrinsically contain are the following:

1. Is the business need identified correctly?
2. Does the project satisfy the business need fully?
3. What is the business risk attached to the project?
4. Is the project return going to be worth it?
5. Are the conditions right to undertake the project?
6. How should the project be undertaken?
7. What are the 'hard' and 'soft' constraints of the project?
8. A possible description and guidance note on decisions that need to be made at a later date in the project.

---

**A Quick Test**

Project Feasibility Studies typically should have the following components:

1. Market and Demand Analysis
2. Technical Analysis
3. Financial Analysis
4. Project Cash Flows
5. Project Model
6. Project Appraisal

Now take six different PFS reports and study them in detail. Next focus on the key questions mentioned above and rate the quality of the reports. Mention all the details you feel are missing in these reports. Develop a detailed format for the PFS report and keep it for your reference.

---

You will find that the industry norm deviates significantly from the expectation and is more in sync with 'filling up a format'. This will also explain why the PFS is not taken seriously by most managers! And therefore, the genesis of the second big misconception that is being fuelled by this phenomenon in reality. At this point, therefore, what needs to be understood is that, although a PFS may not be of the quality it should be, it *still has a few details that are critical for the manager*. Therefore, one should not undermine the value of the PFS and must ensure to use it with the right perspective and objective. A good quality of the PFS is definitely a good start for the project.

The PFS must implicitly contain a detailed understanding of the project, a priority framework for the objectives, an understanding of the 'boundary conditions' and the 'initial conditions' in the proposed business case and, a meaningful way of fulfilling the needs of the business. A good feasibility report will have a proper guidance note on the execution and would draw a framework for decision-making during the execution.

## 2.3 Project Approvals are Complex Financial Decisions

Having understood the genesis of projects, and the basic document that interfaces it with the business world, we now delve into the specifics of the Financial Decision. This is important because projects are usually high-value items. And the comfort level in parting with USD 1 is certainly going to be different from the comfort level in parting with USD 1 billion! The more the value, the more the effort in 'ensuring' success (again a concept we have discussed earlier).

One of the challenges for project managers, therefore, is to ensure that they understand the three different perspectives involved in any typical project situation viz. the engineering/technology perspective, the financial perspective, and the accounting perspective as shown in Fig. 2.2. They are like three different languages that the project manager must know. They are the three foundation stones for project managers to understand the perspectives involved in financial decisions. While the

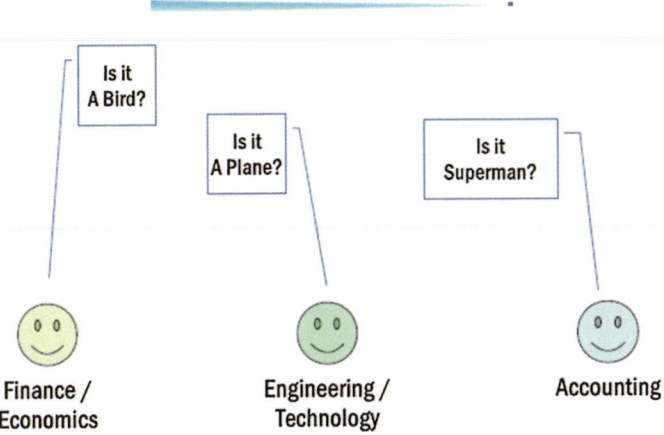

**Fig. 2.2** Developing project perspectives

engineering/technology perspective helps the manager to interface well with suppliers, clients, contractors, and customers, the financial perspective is normally more interesting to the stake holders, the markets, the lending agencies, and the layman (who often do not understand the technicalities). The accounting perspective is meant to support the financial perspective in compliances and the technological perspective with the laws of the land. It, therefore, feeds 'clean data' to the financial perspective to enable good-quality financial decision-making. This leads us to **the third misconception** among most managers. They believe:

> **One doesn't always need to understand ALL the three perspectives to be a good manager. Nothing could be farther from the truth! To be a good manager, one needs to be able to understand and balance the perspectives involved.**

But this thinking is fuelled by the way the project managers climb the corporate ladder. Some of them come from the execution side, having a lot of field experience. They are at the receiving ends where all the delays and inappropriate decisions are directly loaded on them! The planners typically interface with the technocrats and the execution team. The project finance group interfaces more with the funding agencies. Each of these experiences focuses on a specific kind of perspective. The perspective, therefore, continues to be their dominant one. While this expertise is the basis for initial promotions, the manager is soon required to develop the other perspectives adequately. The impact of this misconception surfaces at such times.

We will give an overview at this stage. This overview is intended for those project managers who are *not* from a finance background. Taking the financial perspective forward, most project situations work similar to capital budgeting principles of financial management. Therefore, it is important to understand the logic of the management principles as applied to capital budgeting. While a detailed treatment of this is given in most financial management textbooks, we just touch upon one of the tools viz. the cost–benefit analysis as an example. Needless to add, the reader is advised to go through more detailed textbooks in financial management to allow a better understanding of the concepts and the subject.

### 2.3.1  Cost–Benefit Analysis

While our objective here is not to give a detailed financial perspective, we are going to give a 'conservative' perspective to the project manager. The cost–benefit analysis is a financial management procedure to *translate* costs and benefits to *develop* cash flow values. It is, therefore, the first step in any investment modeling process. The more robust the methods are in deriving the cash flows, the more robust the subsequent analysis would be.

> **Any cost or benefit needs to be shown as a cash flow. That is, it needs to be shown as 'money' coming in or going out of the organization with a definitive amount, at an identifiable point in time.**

An elementary treatment of the cost–benefit analysis from the financial perspective would reveal the use of certain accounting principles like: Cash Flow Principle, Incremental Principle, Long-term Funds Principle, Interest Exclusion Principle, and the Post-Tax Principle. As a project manager, one needs to know what happens if we choose not to use any one or more of these principles. Moreover, one needs to try a small *test case* to understand how the cash benefit analysis works in practice.

Most importantly, one needs to also stress on understanding the limitations of the cost–benefit analysis. We leave that to advanced reading in financial management rather than discussing it over here. At the same time, one would find it interesting as to how the cash flows are derived in one's own organization. This usually spans multiple functions and the data are collated by the accountant/finance team at the organization. Moreover, every organization has a different method of calculating and assessing their financial situation. Hence, due to the multiple variants possible, we are not going to 'glorify' any single method, thereby misguiding the reader!

Though a project manager is many times not expected to have the skill or develop a cost–benefit analysis model, he needs to know the ways to do it. Oftentimes, the project manager needs to develop models along similar lines. These are covered in subsequent sections in the book. Therefore, in the interest of his own normal 'course of work', it is advisable to learn these techniques or at least understand how these are applied in real life. It would also help the project manager interface better with other functional teams within the organization.

### 2.3.2 *Appraise the Investment*

While project appraisal is a crucial activity that is included in conventional literature, it is **a common misconception that appraisal is a project management activity**. In fact, as you would see shortly, appraisal is actually the final activity in the business strategy chain. It incorporates details from not just the project phase, but also from the operations phase. It is, therefore, an activity involving the integration of concepts across strategy, marketing, technology, projects, finance, operations and, lastly, accounting. We will delve over this shortly.

**Fig. 2.3** Appraisal methods

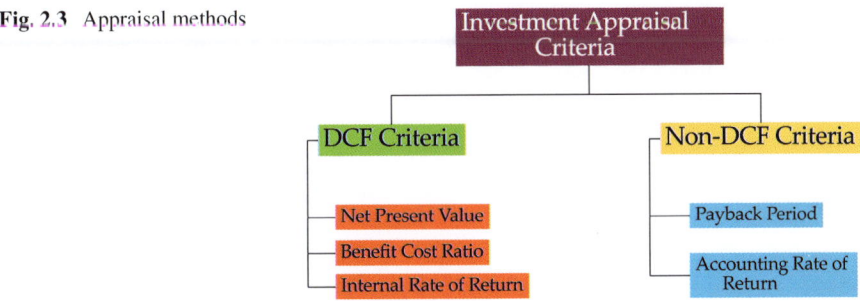

Coming back to our discussion, the next step in the process is an appraisal of the investment. There are essentially two types of appraisal criteria that are widely used viz. **Discounted Cash Flow Methods and Non-Discounted Cash Flow Methods**. The fundamental difference between the two is straightforward: One uses the time value of money to appraise investments while the other does not. The time value of money is covered in detail in the subsequent sections.

A quick review of some of the methods along with the key characteristics is given below. Let it be reiterated at this point that this is not an exhaustive treatment as these techniques are covered in-depth in books on Financial Management.

The generic classes of methods are shown in Fig. 2.3. A brief is provided below. We have restricted our treatment to Discounted Cash Flow (DF) only.

1. *Net Present Value (NPV)*

   The time value of money states that an amount X received after a year has a different 'value' Y today (present day). The amount Y is called the present value of X as it 'displaces' X over time and looks at its value today (present day). The Net Present Value is defined as the sum of the present values (PVs) of incoming and outgoing cash flows over a period of time. Incoming and outgoing cash flows can also be described as benefit and cost cash flows, respectively.

   The NPV is based on the assumption that the intermediate cash inflows of the project are reinvested at a rate of return equal to the firm's cost of capital. Cost of capital refers to the opportunity cost of making a specific investment. It is the rate of return that could have been earned by putting the same money into a different investment with equal risk. Thus, the cost of capital is the rate of return required to persuade the investor to make a given investment. For example, for an individual, to invest in a business is riskier than having fixed deposits in the bank. Hence, the business has to give a little more return than the fixed deposit. This return expected by the individual from the business is nothing but the cost of capital of that business. The NPV of a simple project monotonically decreases as the discount rate increases. The decrease in the NPV, however, is at a decreasing rate. The main merit of NPV is that it takes into account the time value of money. It considers the cash flow stream in its entirety. It squares neatly with the financial objective of maximization of wealth of the stockholders (also the key economic objective). NPVs of different projects can be added (additive property).

2. *Benefit–Cost Ratio (BCR)*

   BCR is the ratio of the benefits of a project or proposal, expressed in monetary terms, relative to its costs, also expressed in monetary terms. All benefits and costs should be expressed in discounted present values. While the NPV gives a USD figure as the net present value, the BCR is a ratio (it is just a number) having no units.

Since this criterion measures NPV per unit of outlay, it can discriminate better between large and small investments. Under unconstrained conditions, the BCR criteria will accept or reject the same way as the NPV. With budgetary limitations, BCR gives a better mechanism to rank projects.

It, however, cannot aggregate projects. So, it is not additive like the NPV. And when cash outflows occur beyond the current period, BCR is unsuitable.

3. *Internal Rate of Return (IRR)*

This is another tool that leverages from the time value of money and considers cash flow streams in their entirety. The IRR is defined as the discount rate often used in capital budgeting that makes the net present value of all cash flows from a particular project equal to zero. Generally speaking, the higher a project's internal rate of return, the more desirable it is to undertake the project.

It makes sense to businessmen who prefer thinking in terms of rate of return and find an absolute quantity like the NPV somewhat difficult to work with.

The IRR, however, may not be uniquely defined. If a project has streams with more than one change in sign, there is a possibility of having multiple IRRs. Moreover, the IRR calculation cannot distinguish between lending and borrowing and hence a high IRR need not necessarily be a desirable feature.

The IRR can be misleading when choosing between mutually exclusive projects that have substantially different outlays. In such cases, it is advisable to go for marginal IRRs. More details are available in financial management books for the reader to go back and check.

### 2.3.3 Recap of Basic Financial Management Concepts

**The Time Value of Money**

A commodity worth USD 100 today would cost different after a year. This cost difference is typically called inflation. At the same time, if one is to invest USD 100 in a bank, one gets some interest at the end of a year. Usually, there is a complex economic relation between the inflation and the interest. If one is to borrow USD 100 from a bank, and if one has to pay back the loan after a year, this amount would be a little more than the interest obtained by investing in the bank. The additional amount that one has to pay in order to get USD 100 from the bank is also referred to as the Cost of Capital of the USD 100. It is usually expressed in terms of a percentage rate, very similar to the interest rates.

Very simplistically put, it means one is borrowing cash for one's business. So, like a compound interest formula, one has

$$\text{Principal} = \frac{\text{Amount}}{(1+r)^n} \qquad (2.1)$$

**Table 2.1** Future value of the principal $P$ for a cost of capital $r$

| Year | Amount in the beginning | Interest for the year | Amount at the end of the year |
|---|---|---|---|
| 1 | $P$ | $Pr$ | $P(1+r)$ |
| 2 | $P(1+r)$ | $Pr(1+r)$ | $P(1+r)^2$ |
| 3 | $P(1+r)^2$ | $P(1+r)^2r$ | $P(1+r)^3$ |
| $n$ | $P(1+r)^{(n-1)}$ | $P(1+r)^{(n-1)}r$ | $P(1+r)^n$ |

**Table 2.2** Cases of cost outlay for the USD 100 Million project

| Case | Spend in year 1 | Spend in year 2 | Spend in year 3 |
|---|---|---|---|
| 1 | 30 | 30 | 40 |
| 2 | 20 | 40 | 40 |
| 3 | 50 | 30 | 20 |
| 4 | 40 | 20 | 40 |
| 5 | 20 | 30 | 50 |

where the Principal is the amount taken from the lending agency and the Amount is the Principal and the Interest that is returned to the agency; $r$ is the rate of the cost of capital.

If one divides the time periods and understands the value of the amount, it will show us a trend as shown in Table 2.1. This appears trivial when put in the equation; however, in reality, one needs to understand the impact of different values of $r$ on the amounts that need to be paid. This is a more complex issue for a project manager. In a project, the greater the initial spend, the greater is the amount to be repaid. And longer the project, greater the amount to be repaid.

In order to get a good handle of the issue, one needs to pick values and establish these relationships. Needless to say, a graphical approach works best here. Let us assume a project that runs for 3 years. Let us assume it to cost USD 100 million. Let us assume that there are three scenarios for the cost of capital viz. 6, 12, and 24 %. Now, depending on the outlay of the project, several scenarios are possible. The potential spread of cost outlays is shown in Table 2.2. While this is purely an academic exercise, most projects will have a distribution that can be easily derived and treated as a baseline for any 'analysis'.

For the purpose of our illustration, let us now assume that the spend is done in the beginning of the year. Therefore, using the relationships of the Time Value of Money, we get a value spectrum for the various rates of the cost of capital as shown in Table 2.3. One needs to note the trend: the earlier one spends, the more the value and the larger cost of capital, the larger the value. The variance in the case represented here is around 40 % with a change in the cost of capital. If one is doing an international project, one would understand the implications easily. Some

**Table 2.3** Value of the USD 100 Million project at the end of 3 years

| Case | Calculated with cost of capital 6% | Calculated with cost of capital 12% | Calculated with cost of capital 24% |
|------|------|------|------|
| 1 | 111.84 | 124.58 | 152.93 |
| 2 | 111.16 | 123.07 | 149.24 |
| 3 | 114.46 | 130.28 | 166.26 |
| 4 | 112.51 | 126.09 | 156.62 |
| 5 | 110.53 | 121.73 | 146.26 |

countries have a low rate while others have a much higher rate. Even the changes in the outlay could bring a change of anywhere worth of 3% (and upto 20%).

### Basic Accounting Principles

In a simplistic sense, assume one spent USD 100,000 when one invested in a fast-food business. Now let us assume that the total expenditure to use that asset was USD 20,000 per year and the amount of revenue/sales was USD 32,000 per year. In short, one had a return of USD 12,000 per year. The financial analysis is simple.

In the corporate world, however, there are certain real flows that are obtained after the application of notional elements. For instance, one could have to pay tax for the business mentioned. However, the tax rate allows for a notional cost factor called the depreciation to be treated as an expense. This *reduces* the tax liability. The project world is a similar one. While doing the analysis of the overall project and operations, the end result of the project is treated (most times) as an asset that can be depreciated or amortized over a longer period in time, reducing the tax. Thus, cash flow analysis for the cost–benefit ratio needs to take this into consideration. This is shown in Fig. 2.4. The derivation of real cost elements, the cash flow, involves data that have to come from the project manager's estimates, the business case information (from the marketing and operations people), and the accountant's desk (to affix the notional values and estimates). Thus, accounting information needs to be processed by taking into consideration the wide range of notional as well as real flows.

While a project manager is technically sound, he also needs to understand the right terminology to ensure that the right message is being communicated with the financial controllers and other stakeholders in his project. This is often a point of contention because most project managers tend to use terms rather 'loosely', thereby making the jargon-centric CFOs rather suspicious of their capabilities.

Fortunately, this is an extremely well-researched area and there are many books that deal exclusively with this subject. One of the best resources for understanding the various perspectives involved while using these principles is given in the resource of the Stern School of Business authored by Damodaran [5]. The treatment is fairly elaborate. In this book, we are only touching the overview and not demonstrating the detail.

The Basic Accounting Principles involved in the Financial Decision Making of Projects are the following:

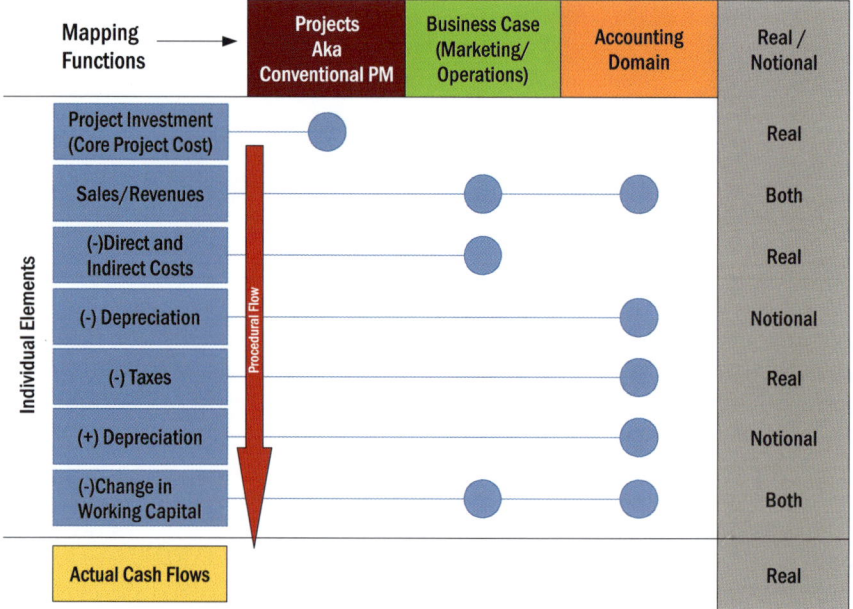

**Fig. 2.4** Process and function flow in deriving cash flow

1. *Cash Flow Principle*

   The cash flow principle states that one has to take the *actual costs and not the expenses* as per the accounting convention. This therefore, involves a serious analysis of credit sales and purchases as well as items like depreciation. At the end of the day, the project manager must think as plainly as the cashier in a bank and look at the actual cash that is paid out, rather than those that are notional in nature.

   In accrual accounting, one shows the revenues when the products are sold or the services are provided and not when they are actually paid for. Similarly, it shows expenses associated with these revenues rather than restricting oneself to the cash expenses.

2. *Incremental Principle*

   While understanding the costs associated with a project, the project manager must also understand these cost heads from an incremental perspective. That is, how much of the cost is actually due to the current project. This is particularly important because some companies tend to 'load' the project with other costs and that tends to squeeze the budget, making it hard for the project manager to justify the costs. This analysis helps to choose the best project by taking into consideration the accurate reflection of the cash flows based on an accurate reflection of the allocated funds.

3. *Long-term Funds Principle*

   In a project, there are tricky issues regarding the reference points for the funds.

In other words, while the manager thinks like the 'cashier' he needs to know the components of his kitty to start with. There are four distinct points of view and managers could use any one of them to define their cash-boxes viz. (a) equity point of view, (b) long-term funds point of view, (c) explicit cost funds point of view, and (d) total funds point of view. The equity point of view considers only the equity component in the project investment and uses that as a basis for decision making. In the long-term funds point of view, the equity and long-term debts are taken as a combined entity to evaluate the projects. Explicit cost funds point of view considers both equity and long-term debts and also incorporates short-term bank advances. These short-term bank advances are funds taken on a short-term basis to cover certain 'spikes' in the fund requirements. The total funds point of view also considers the creditors that are supplying material to the project.

While a project manager might just say he is looking at the 'entire' cash, the general convention is to restrict the perspective to the long-term funds. Long-term funds comprise of equity and long-term debt which are efficient instruments for project financing. Bank advances and creditors would be more expensive during the course of the project. This would disturb the cost of the capital assumptions of the project. Hence, while calculating the costs and the benefits, the reference point is the supplier of the long-term funds.

That being said, today, large organizations tend to leverage from the total cost perspective. For instance, contract conditions where the equipment suppliers need to provide credit to the buyer are not uncommon. In such situations, the cash outflow for the buyer is minimal, but the total fund perspective would indicate a large spontaneous liability. This trend has not percolated into common business practice, especially among mid-value projects. Hence, one should normally restrict the analysis to the long-term funds point of view.

4. The other principles are interest exclusion principle whereby the interest is not separately calculated for the LT Debt as it is already a part of the cost of capital rate and the post-tax principle whereby the cash flow must incorporate the tax flows as well to ensure that it is accurate in terms of the representation of the cost. Both these are incorporated in our analysis structure outlined in Fig. 2.4.

While interfacing with other functional teams, the project manager needs to understand these dynamics so as to be able to read the accounting and financial statements in a better way.

**A Quick Test**

To get a firm handle of the various elements involved, it is advisable to try out excel-based models for NPV, BCR, IRR, ARR, and Payback Period. This helps the manager to understand the specific formulae and the calculations. It would be interesting for any manager to take the last 10 strategic projects of the organization and look at how the parameters panned out. This would also give a feel for how the stakeholders are deciding in the company. It is very important for any manager to understand how the stakeholders decide so as to enable him to align his thinking along their priorities.

While you do so, also try to derive an excel sheet to calculate the payback period with unequal cash flows.

## 2.4  A More Balanced Perspective of Project Evaluation

It is indeed a fact that the appraisal methods mentioned above are used across a very wide spectrum of industries. However, there are a lot of 'company-specific' criteria that go beyond the realms of these conventional tools. These tools are many times used in program management as well as portfolio management. We call this framework the Balanced Project Evaluation Framework. In consulting, we often tend to advise clients on projects based on these factors as shown in Fig. 2.5.

1. *Environmental Considerations*

   This is fundamentally an 'external-focused view'. It looks at the competitive strategy of the company and the variables in the external environment. Critical issues of project selection include aspects like:

   **How is the current project going to help the organization compete?**
   **Does the success have a direct linkage to the success of the organization?**
   **How does one circumvent the challenges posed by factors like high infla-tion rates in the economy?, etc.**

   In other words, this is the proposed or potential interface with external entities.

**Fig. 2.5** Perspectives in project evaluation

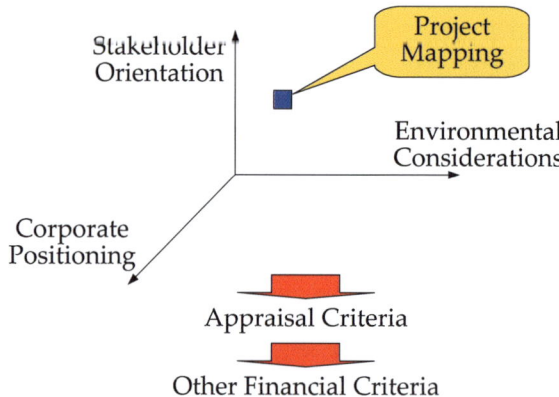

2. *Corporate Positioning*

This is an 'internally focused view'. In other words, it looks at the corporate strategy of the organization and the variables in the internal environment. In other words, issues that are under consideration here include:

**How does this project synergize with the existing operations?**
**Is the organization having the required technological preparedness?, etc.**

This is again, a detailed perspective of how a project will be interfacing with internal entities.

3. *Stakeholder Orientation*

While strategy (both competitive and corporate) is one side of the equation, the other side actually is the stakeholder orientation. This aspect has been covered in the stakeholder management faculty in most books. However, it is an important selection criterion. Stakeholders may or may not be aligned with the strategy. One often sees this phenomenon in several organizations, where the projects are well aligned with the strategic objectives, but the stakeholders simply do not want to digest them for some reason (aka *gut feeling*:-)).

Thus, stakeholder orientation is extremely important. Most literature in project management tends to ignore this issue. We are bringing this up as a separate dimension because this could 'defy' logic at times. Hollywood movies too seem to promote this idea! Essentially, stakeholders could cause a lot of 'surprises' in conventional business practices (John Grisham's Pelican Brief as an example). We deal with this issue again in subsequent chapters where we discuss stakeholder management.

4. *Normal Appraisal Criteria*

This is nothing but the criteria explained in the previous section involving the conventional financial analysis. It is always advisable for the project manager to choose *two of the criteria* to ensure a better quality of decision making.

5. *Exposure*

This is one of the fundamental factors that one checks immediately after the application of the appraisal methods, the maximum cost (or capital or reputation, etc.) that is at stake through the lifecycle of the project. So, even if a project passes through the previous 'factor-tests', it might just get rejected because of the exposure it warrants.

6. *Profit/Spend*

This is a relatively new metric where companies are looking at the spend in terms of costs, manhours, etc. Many recent acquisitions have shown that companies are intrinsically unable to justify the *spend at the projects phase* and prefer, therefore, to acquire existing facilities. In short, the project dynamics is changing in today's business environment.

7. *Time for First Revenue*

With business environments becoming more and more unpredictable and dynamic, the time when the project actually starts 'showing' its benefits is becoming more and more important among business circles. This is particularly true with investments in third-world countries like India, China, Brazil, etc. where the regulatory environment is less static and tends to pose problems to the project sponsors.

At this stage, we would also like to make it clear that, many times, project managers are well aware of these factors and they have a tendency to 'present' beautiful reports to ensure their projects go through seamlessly. It is very likely that such situations snowball into huge issues and the projects could be prematurely terminated. All said, this rampant practice is something that the reader needs to be aware of…to start with!

That being said, a whole lot of organizations do have a well-structured and a well-managed selection process. So let us not paint all the cars black! One needs to be cautious while working on project selection and appraisal, and I believe this has been reasonably conveyed at this stage.

## 2.5  Key Takeaways

The project world is actually not very different from the operations world. Accepting this opens a new world of possibilities to use and leverage from the innovations in the faculties of general operations in the project world. We will be seeing how to practically implement this in subsequent chapters.

A strong project-to-business interface is defined by the Project Feasibility Study. However, there are reasons why this report is not very 'usable' in everyday life. This again hovers around the short-cycle that treats the project feasibility study as a necessary-for-approval document rather than a necessary-for-manager document.

Understanding basic principles of Finance is an important point for most project managers as they need to relate to the perspectives used in decision making at the stakeholder level. Developing the right perspective is an essential part of project management education. However, this is not always appreciated in practice. We have tried to touch upon this misconception and have provided a few guidelines for the users to enable them take it up with the right perspective. In doing so, one needs to understand the finance world as well as the accounting world and the mechanism/manner in which they impact financial decision making.

Unlike the conventional approach where the project manager focuses on financial evaluation alone, we have presented a more realistic framework to assess the project and identify whether they are 'indeed' feasible in the true sense. This is discussed toward the end of the chapter.

# Chapter 3
# A Quick Honest Beginning!

**Abstract** This chapter is meant for the reader to come up with specific issues and cases that are observed in the projects where he is involved. In doing so, we ask the reader to come up with a rigorous evaluation of his issues. Rigorous evaluation would help the reader understand: (a) the current style of thinking in his project environment, (b) the way issues have impacted projects, (c) the application of conventional wisdom in identifying issues, (d) the potential of using conventional wisdom in solving the issues, (e) the reasons these issues were not identified earlier, (f) the probability that they would recur in future projects, (g) the probability that they would be contained in subsequent projects (at the cost of another potential problem), and (h) a potential alignment of objectives with the objectives of this book.

**Keywords** Project integration management · Project scope management · Project time management · Project cost management · Project quality management · Project quality management · Project human resource management · Project communication management · Project risk management · Project procurement management

## 3.1 All's Well!

When I train professional project managers and MBA students, the one thing that seems to bother me is the lack of the application mindset. Most people we meet show an enormous talent in the 'knowledge' mindset.

This raises several fundamental questions…

What is the purpose of learning and education?

How often do you feel you should change things?

Are you really here to apply and 'change' something?

And this is why I often ask people to do the following:

**Are you truly learning?**

Let me explain this question from a few caselets that are popular in Project Circles. The caselets are meant to give the practicing manager a starting point to look toward his project from a different perspective.

## 3.2   Caselet 1

The CEO has asked for a baby real quick…Get nine ladies pregnant, and we will have one in a month…hopefully!

Do you know what actions trigger your schedules? Please list them in detail for your current role or project. Discuss it with your colleagues and find out if you have missed any.

## 3.3   Caselet 2

The estimate for the tender was USD 30 m. The schedule was meant to be for 2 years. With good negotiations, the price was brought down to USD 25 m. However, as the award was late by 4 months, an acceleration package of USD 10 m needs to be released…

Do you know what actions trigger your costs? Please list them in detail for your current role or project. Discuss it with your colleagues and find out if you have missed any.

## 3.4   Caselet 3

'Our manager says that the technical team has not ordered the crane.'
'Our manager says that the cranes are the responsibility of the building systems team!'
'Well, now that the cranes are not ordered, let us blame the procurement team!!!'

Is your project scope well managed? List at least one element of your project that is possibly in the 'gray' zone. Discuss it with your colleagues and find out if you have missed any more….

## 3.5   Caselet 4

'Regardless of the system, I don't believe in paperwork. I prefer instructing my people directly!'

'The project plan is theoretical. I believe in being practical!'
'How many people are in your team? 30 people OR 5 engineers, 5 supervisors, 3 draftsmen and 17 workers?'

Do you have the right kind of people to make your project a success? List and rank candidates that you feel need to have superior skills, attitudes, habits, or knowledge.

---

**Another Quick Test**

List two instances in your recent project where you thought the project management was not satisfactory.

**How do you think they should have been done if they were to be satisfactory?**

**What were the consequences of not being satisfactory in your last project?**

Our attempt in this book is to try help the readers make an honest beginning in the world of project management. Most times we find that we read a lot, but are unable to start and work in the concepts of project management. Hopefully, with this exercise, things would slowly start changing and falling in place.

---

## 3.6 Real Life Cases: As Narrated by a Project Manager-1

For one of our projects, the production schedule was lagging behind and the actual production completion dates had gone ahead of the schedule. The production schedule was not proper and simply clashed with the schedules of other projects.

**Reason**: Two or more projects had delivery dates that were too close to each other and, hence, could not be completed/processed simultaneously. The main reason was the shortage of resources viz. skilled labor and faster raw material procurement.

In another project, the design had issues and therefore, the production was put on hold. These design reports were late due to delays in client approvals and comments. But the client insisted on blaming the vendor. The client had a tight schedule and hence, was unable to provide for an extension of the production completion.

Due to late completion, the material was air-lifted and this resulted in an escalation of costs and there was a wave of 'unsatisfactory' performance at the client end.

**Corrective action that should have been taken**: One needs to review the current project portfolio before acquiring new projects. Projects awarded should not have delivery dates that clash with existing schedules and cause difficulties.

## 3.7   Real Life Cases: As Narrated by a Project Manager-2

One of our projects for a leading consulting company at their Mumbai office had severe problems related to design freezing and detailing. The details in design kept changing all the time leading to cost and time overruns as well as deficiencies in the final site at the time of handover. While it was also the architect's and the client's job before contracting the work, everyone blamed the contractor for the 'bad' work. So, we should have been extra diligent with the detailing and the approvals.

The coordination with the other contractors at the site was also a major problem. The contractors for electrical and security works were directly appointed by the PMC which led to lot of schedule conflicts as well as some redoing of work (i.e., ceiling, partitions, etc.). This should have been a more open and coordinated process.

Eventually, it was the perception of my company that was damaged leading to long 'snaglists' and payment delays.

## 3.8   Real Life Cases: As Narrated by a Project Manager-3

### 3.8.1   Case 1

In an ongoing petroleum project, there are delays in various activities as the cycle times estimated were very aggressive initially to meet the project schedule. Project management should have estimated the 'time' that was realistic and developed robust plans with in-built contingencies. If this had been done far in advance, even if this did not meet the schedule, the projection would have instigated the higher management to think and take certain decisions regarding resource augmentation.

It is now too late to take decisions on resources and the project has anyway got delayed. The customer will have cost overruns much more than our own additional resource costs.

### 3.8.2   Case 2

In a recently completed project, the customer has come back with a huge amount of claim as consequential damages due to the delay. These penalties, on account of consequential damages, are much more than late delivery penalties. It is approximately 40 % of the project value.

**Corrective actions that should have been taken**: The project sales team should have studied the contract document in detail. They should have sought appropriate legal opinion before accepting the contract terms and conditions. Just to meet the sales figures, the project scope was not studied; the clauses, terms, and conditions which were not favorable were agreed upon.

# Part II
# Reviewing Projects: First Level

# Chapter 4
# The Systems View of Management-I

**Abstract** In this chapter, the systems view of project management is touched upon. We begin with the understanding of the basic *systems thinking approach as applied to a project situation*. In doing so, we attempt to clarify the similarities and the differences between the systems thinking approach from the conventional approach. The systems approach helps refine the practical business and management level evaluation for a given project. However, when one further breaks down the project, one can evaluate the *project management context*. Two important concepts are touched upon: *Project Phases and Project Life Cycles*. We cursorily touch upon the concept of project phases. However, the more crucial concept of the project life cycle needs to be elaborated in greater detail. The *types of life cycles* are then focused on. We cover three 'standard' models, viz., the linear model, the incremental model, and the rapid application development model. The uniqueness of this chapter is that it shows the reader the *application of simulation methods to determine optimal life cycles for a given project situation*. A simulation case is presented to help the reader understand variants and factor them in to determine the appropriate life cycles. While real-life cases would be complex, the simplified versions in this chapter help the reader understand the method of application and enable the reader to take this up in the real world with an improved basis. A *generic conceptual framework* is also presented in this context, to help the reader align his thinking to life cycle modeling and simulation.

**Keywords** Systems approach · Systems approach versus conventional approach · Project management · Project phase · Project life cycle · Types of project life cycle · Waterfall/linear model · Incremental model · Rapid application development · Conceptual framework for simulation of life cycles · Generic framework for simulation

## 4.1 Introduction

Well, the first and foremost aspect is to understand why we are speaking of a 'different' view called the systems view of management. For this, one needs to understand the project environment a little better. The beauty of any typical project/business environment can be summarized as follows:

© Springer Science+Business Media Singapore 2017
N. Gurjar, *A Forward Looking Approach to Project Management*,
Lecture Notes in Management and Industrial Engineering,
DOI 10.1007/978-981-10-0782-8_4

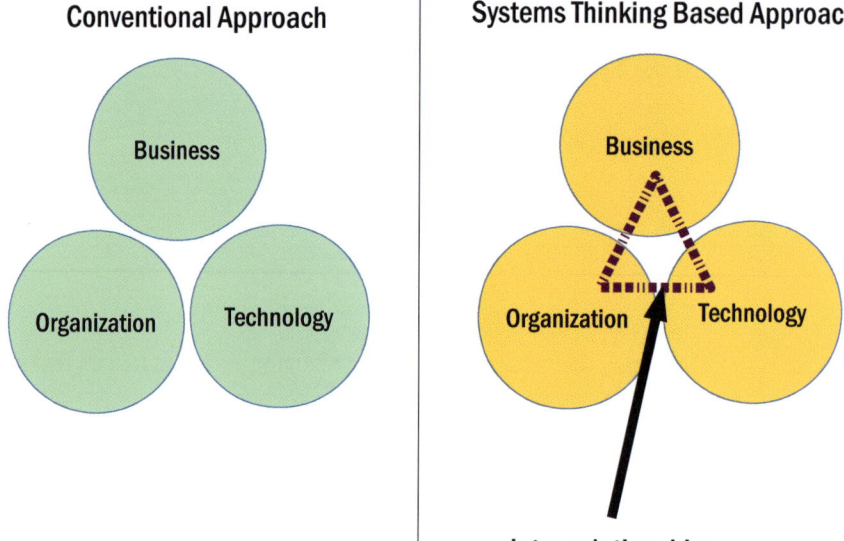

Fig. 4.1  Key differences in the approach

1. Projects cannot work in isolation. In other words, one would never find a project that is completely cut-off from the rest of the world! Therefore, projects MUST operate in a broad organizational context. In other words, there are interfaces and relationships with other entities (systems).
2. Within a project, one could have a similar organization which again means that there are interfaces within the project, i.e., possible subsystems.

Therefore, project managers need to take a holistic or systems view of a project and understand how it is situated within a larger organization.

This aspect is very essential for a project manager to understand how he needs to maneuver in a given project environment. In other words, the project world is embedded in the organization that is again embedded in the business world. The fundamental thought process is that **every entity interacts with every other entity** as shown in Fig. 4.1. Therefore, the crux of the application is on defining the entities and the interactions they have with each other. A systems approach, thus, describes a more analytical approach to management and problem solving.

The three parts of the systems view include:

1. *Systems philosophy*
   Where one has to view things as systems, interacting components working within an environment to fulfill some purpose. The system could be a person, a team, a project, an organization, an event, an action, a decision, or any other thing.

2. *Systems analysis*

   This part essentially looks at the problem-solving approach. Although it reads 'analysis', it is a combination of a 'breakdown process' and a subsequent 'build-up process.'

   Therefore, in a project interaction between a project manager and a contractor, it could be the individual personalities that are interacting. This includes the positives and negatives of two persons who are working together. A conflict between them could arise due to an 'ego' issue. However, when one looks at the project as a whole, the conflict has little relevance if it does not affect the deliverables of the project, viz., on time completion, within budget completion and in line with the agreed scope. Thus, depending on the choice of the analysis and the build-up, every phenomenon could have different ramifications at different levels. The systems analysis helps one understand such phenomenon in a better way.

3. *Systems management*

   This is the third part that actually starts addressing business, technological, and organizational issues before making changes to the situation.

   Going by the previous example, it would be wrong to ignore the conflict and deem it unimportant. Rather, one needs to understand the dynamics and ensure how it needs to be dealt with. The astute systems thinker would understand the dynamics and ensure that the conflict does not jeopardize the entire project. Using systems management, relevant issues need to be identified and the management perspective needs to be reinforced.

## 4.2 An Example Framework

Well, we take a simple technology upgradation project as an example here. Let us say that we have a system upgradation of a LAN solution. One of the simplest frameworks to evaluate the situation is the three-sphere approach: business, organization, and technology. This is a framework that is aimed at modeling the stakeholder decision. Since one does not have the relative weightages of the three spheres, they are all treated as elements of the overall decision. Now, going by the systems view, one can divide these decisions into subdecisions. This is shown in the schema below. Each decision is attempted to be broken down into business-relevant issues.

In the technology sphere, some typical detailing could look as follows:

1. What operating system is ideal for the upgraded system?
2. What application software would be ideal?
3. What kind of hardware changes are we speaking of?
4. How will the hardware affect the LAN and the internet usage?
5. Should the system be a single component or distributed/integrated components?
6. What kind of protocols are to be used?

There could be several others as one delves deeper.

In the organization sphere, some typical detailing could look as follows:

1. Is the upgrade for everyone in the organization or just for a few buildings/areas?
2. How is this project going to affect departments who already have upgraded LANs in the facility?
3. Who is going to train the people about the new system?
4. Who is going to administer and maintain the system?

Similarly in the business sphere, some typical detailing could look as follows:

1. What is the cost of the project?
2. How is the cost going to be distributed?
3. What is the support cost going to be?
4. How will the costs affect the overall functioning of the organization?
5. Will it reduce the overall costs?
6. Will it increase the market potential? etc.

What is important to realize is that the above is the *conventional breakdown* of the business issues in the project. While both conventional and systems thinking approaches appear seemingly similar, they are very different in application. Systems thinking is typically iterative in nature. In other words, loops are characteristic of the systems thinking process. When one has to apply systems management, one needs to look at the interrelationships. The above framework tries to keep the three spheres *independent*. To demonstrate the dependence, we will ask the first question of the technology sphere in conjunction with the first question of the business sphere and check the results. You could now focus on the four questions given below to demonstrate progressively the degree of dependence from a basic systems thinking perspective.

1. Which is the ideal LAN system? [independent]
2. Which is the best option given a budget (say X)? [With Dependency 1]
3. Which is the best option considering the problems of the current system? [With Dependency 2]
4. Which is the best option considering the problems of the current system and a given budget? [With Dependency 1 and Dependency 2]

While these might appear to be similar questions, the overall analysis of the questions yields 'different' results. And answering the questions would need one to revisit the assumptions in each of the independent questions that have been shown in the conventional approach. Hence, the outcome is rather different.

In real life, the framework will be far more elaborate than the one discussed here. However, the framework shown would give the project manager a simple method of analysis. It should be noted that some of the questions may not be easy to answer. Hence, there is a need to determine a more detailed and a causal model for subsequent analysis.

## 4.3 The Project Management Context

In the previous chapters, we covered the project environment in general. Thereafter, we looked at the general interfaces of a project with the business. We now move to the third layer of project management that deals with the project management context.

There are essentially three elements of interest in the project management context at this stage. They are as follows:

1. The phase
2. The life cycle
3. The organizational context

In the present chapter, we will be covering the first two elements in detail and would take up the organizational context again in a subsequent discussion.

## 4.4 Project Phase

A phase is a set of activities in a project that is done to mark the completion of a business-relevant deliverable. A deliverable is a verifiable, tangible work product, such as a feasibility study, a prototype, design documents, etc.

Phases are normally defined to:

1. Assist decision making on whether the project should go to the next phase
2. To detect and correct errors cost-effectively.

Most companies have phases defined in terms of functions. However, there are other variants possible. We will not delve into those details here because the 'phase' concept is strongly dependent on the orientation of the top management. Hence, the degree of freedom in defining any arbitrary methodology to determine the phase is virtually negligible.

The logical sequencing of project phases gives the project the characteristics of the project life cycle. The network of project phases in the project life cycle is called the project process model. Let us understand this in greater detail.

## 4.5 Project Life Cycle

Some quick recall from biology is indicated in Fig. 4.2. We take the example of the butterfly. It goes through multiple stages viz.

1. Eggs
2. Caterpillar
3. Chysalis
4. Baby Butterfly
5. Adult Butterfly.

**Fig. 4.2** Life cycle as we know!

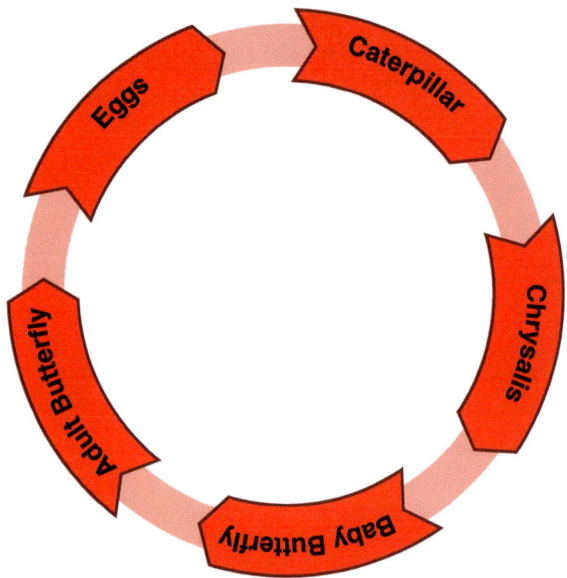

In Biology, this is exhibited as a case of alternation of generation. A project most times is like a typical 'alternation of generation' lifecycle, whereby the project phase completes and the operations begin (in most cases).

The concept of the life cycle in a project is similar to that of life. It brings in the concept of 'birth, growth, maturity, and death.' However, just like 'age' is just a number until one knows the 'use' of the number, a project manager must understand what the 'childhood' and the 'youth' of the project really mean. This is often difficult because it requires some basic wisdom on what the life cycle is.

---

**A Quick Hands-On**

So, the cardinal question for the reader is to assess the usefulness of this information through the question:

**How does the knowledge of the life cycle practically help the project manager?**

---

If you are a project manager who is forward looking, you might want to 'leverage' your past experiences and apply them to your present, to obtain better results in the future. At the same time, if you are a project manager who is pretty 'static' in your approach, you might be looking at your present to 'eat, drink, and be merry!' At this stage, you probably might be thinking that this is a philosophical question rather than a management issue. Hence, before you get swayed further (due to our examples), let us try and get to the core of the issue.

**Essentially, the management style depends on the knowledge of the life cycle**. While PMBoK® speaks of characteristics of the project, the project manager needs to understand this situation from two different perspectives:

1. Impact on Strategic Direction
2. Impact on Operational Process

The project life cycle contains crucial information on the same. The PMBoK® covers examples of life cycles in a concise manner. However, what is fundamentally missing in the treatment is the fact that a project manager must steer his project with adequate caution. What does that mean? An unclear understanding of the life cycle could lead to situations where the project is either **over-engineered or under-engineered**. Both the situations are dangerous. The former being so because there is an inappropriate utilization of resources that could be unnecessary. The latter situation is dangerous because the 'ignorance' becomes 'bliss' and the manager is soon playing the 'political minefield' to survive! We will just see how this works in a practical situation.

In other words, the detailing norms change along the life cycle and the objectives of the detailing also change. Both these aspects deserve a more detailed treatment and we will clarify this a little more as we go along.

## 4.6   Types of Project Life Cycles

The discussion in the previous section would appear pointless if we do not support it with some more information. With the advances in engineering practices, project management also has changed its context. Traditionally, projects were pretty linear in their life cycles. It was *believed* that activities followed each other as a simple set of tasks to do: One begins, completes, and then, the next begins. However, this was seldom the case.

In reality, the project could have activities run in parallel. So, the concept of 'fast tracking' is probably a 'natural' phenomenon in most projects. Not just that, in most cases, the activities could be 're-done' by design. For instance, a prototype might be made and improved upon each time. This is a planned exercise. The project manager knows that the first prototype is NOT the final one. In other words, steps are going to be repeated. What we are saying in other words is that, the concept of repeating steps or iterating is more common than what one would like to believe. This means that the steps are not always 'completed' and yet, the project moves to the next step. Rather, the steps are 'revisited or even repeated' at times.

These variances were first formally analyzed and incorporated in the knowledge industry, viz., software. The traditional construction industry still believes in the 'linear' mode of working and has not formally worked on process models (at least not comparable to the software industry). In the early 2000s, however, people started understanding the applicability of these models even for other industries. One needs to understand that process models are often important to understand and you need to choose the right one. **The choice of the life cycle model will determine the degree**

**of impacts on both the strategic direction as well as the organizational process**.
One of the best treatments on this subject was given by Pressman [6] in his work on
software engineering. He has systematically characterized and defined each of the
process models in significant details. Hence, for technical clarifications, the reader
is advised to refer to that work for more details. In this work, we are focusing on
how the reader should evaluate and improve upon the models practically.

### 4.6.1   SDLC: The Linear Model

The SDLC is often 'touted' as the traditional model. In this model, by design, the
project has a simplistic approach that is similar to a waterfall. So, a lot of time is
spent in requirement gathering so as to ensure they are complete and frozen. The
next step of engineering then takes up the requirements and completely defines all
the engineering specifications. Then, it moves to the next steps as shown in Fig. 4.3.

This model is used when:

1. Requirements are fairly well understood
2. Work flows from concept to completion in a fairly linear fashion.

Also called the classical life cycle model, this is the most popular process model you
will ever see.

There are a lot of project managers who are not actually in the 'fan club' of the
SDLC approach. However, one might have witnessed massive / large construction
projects incurring huge cost and time overruns due to geotechnical issues! Soil conditions force the entire civil engineering team to rework the strategy. In such cases,
actually, the SDLC would probably have served their business interests better.

Another interesting situation was seen when I was in the process industry. The civil
contractor was pressurized to perform to an aggressive schedule. He worked day-
and-night and was casting foundations. However, the process team realized there was
some correction required (due to a revision in the calculations). So, the foundation
drawings were changed for some of the equipment. This led to the rework of the
foundations! They had to be broken / destroyed and re-cast to newer specifications.
This is, therefore, another situation where the SDLC would probably have protected
their business interests better.

**Fig. 4.3** Most popular: the
SDLC or waterfall life cycle

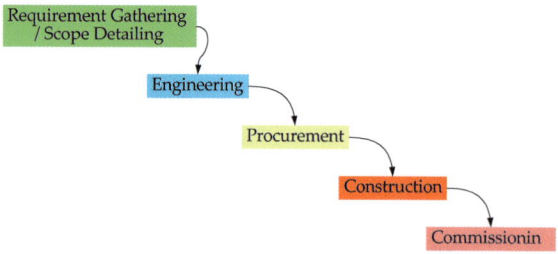

Waterfall Model: See the Cascade Effect!

Unlike other life cycle models, the SDLC leverages from its degree of detailing and the degree of completeness. The SDLC is typically more 'project operations centric' than the strategic direction. Once the project kick-starts, there is no turning back. All the strategic factors are considered **before** the project kick-starts. In other words, the initiation process *should be far more rigorous if one is going for the SDLC or the waterfall approach.* Modeling and simulation tools are often useful in such situations.

So, in the portfolio position it is in the quadrant shown in Fig. 4.4. This means the following to the manager:

1. The project philosophy involves a very 'strong' line of control as far as the budgets are concerned. Rigid budgets and, at times, shoe-string budgets are often observed in some cases. Though the budgets may not be accurate, any attempt to 'revise' the budget would be viewed counterproductive to the manager's capabilities!
2. Field experience is given a lot of weightage. Usually, innovation is at the operational level.
3. From a management perspective, project surprises are not expected and are, most times, unwelcome.
4. Due to the isolation of phases, there is a better chance to optimize the project schedule and to engineer the overall operations effectively. This is regardless of the phase and has significant advantages in terms of benefits from scaling-up activities.
5. Schedules, like the budgets, are also given due weightage. The major task of the manager is to have a good and productive resource support, most decisions being operational.
6. The manager needs to be adept at understanding the devil in the detail and should be extremely comfortable with the details of the project. This is probably the most important attribute of the manager.

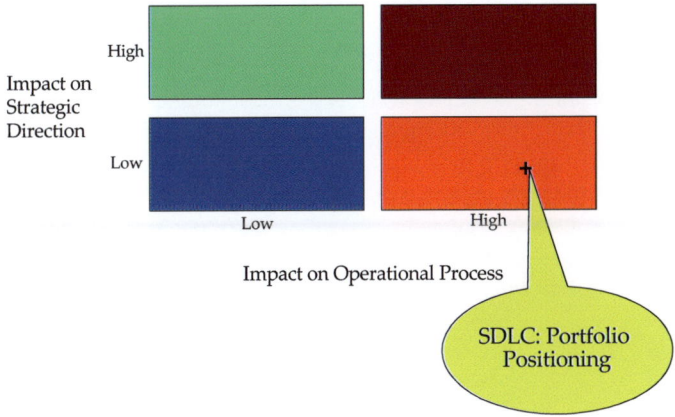

**Fig. 4.4** Portfolio positioning of SDLC/waterfall model

7. At the same time, he needs to interface effectively with the top management. Which means, translating the requirements into the 'strategic' language is often an important aspect. At the same time, the top management overseeing the project is most often required to be technically astute to ensure the project is successfully meeting the stakeholder's requirements.
8. Strong processes are required in this case to effectively manage the project.
9. When the phases are contracted, all the contractors need be to on the same performance level; in other words, the best-of-the-best in terms of output quality instead of the price. This is often very critical, and ignored due to 'budgetary constraints.'

So, in our analogy of the life cycle, these project managers are the ones that follow the motto: *you only live once!* And are often uncompromising on their standards.

It is not easy to terminate such a project. Once it starts, it is often 'strategically reassessed' only after a significant cost and time. Therefore, these projects are the most 'dangerous' from their strategic risk perspective due to their cost and time overruns.

---

**A Quick Test**

At this point, you might want to evaluate five of your recent projects and check if the waterfall model could have been a good life cycle to adopt for any of them. In case they were not, you might want to focus on the following:

1. **What conditions needed to have been different so as to have made them good candidates for the waterfall model?**
2. **Why did the management choose a different model?**
3. Additional factors you feel are relevant for the project manager who is using the waterfall model
4. The benefits of using the waterfall model in each of the five cases that were absent because they used some other alternative.

---

All said, the waterfall model is probably the most popular of all models. It is like the pro-Samuelson and the anti-Samuelson phenomenon in economics. The waterfall model is among the most widely used models due to its simplicity; and, if not for the entire project, for parts of projects or in subprojects that are the divisions of large programs. Most software systems in project management are specifically designed for the waterfall model as they are the easiest to model mathematically. Despite the love–hate relationship with practicing professionals, every project manager at the beginning of his career has used this model at sometime or the other.

### 4.6.2  The Incremental Model

It is the waterfall model used in an iterative process. This model is used in the following cases when:

1. You want to deliver products with increments on functionality
2. The client needs the capability or service early
3. The client needs to evaluate the product and provide feedback for subsequent increments
4. There is a staffing issue.

By design, each of the process 'cycles' follows the previous one. At times, if it is purely a staffing issue, the cycles can run in parallel as shown in Fig. 4.5. Or in the project management parlance, they can be 'fast-tracked.' The fundamental characterization in this model is that the project delivers in phases and in increments that are *all functional and fulfilling the business need*. Many other models like Prototyping deliver 'attempts' that are aimed at the same scope elements. In other words, they are stripped-down versions of the final product. This model, on the other hand, ensures the delivery of a unit that is fully functional.

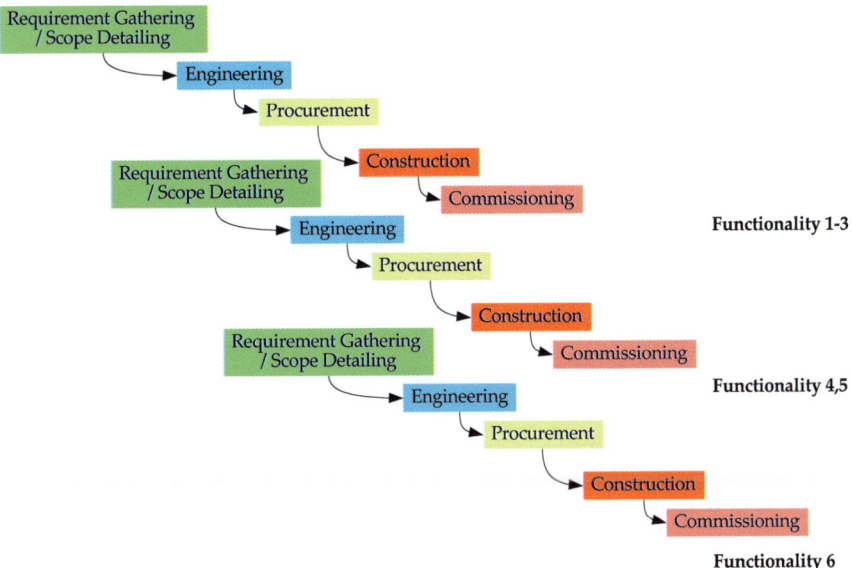

**Fig. 4.5**  Typical incremental model

**A Quick Hands-On**

In the evolutionary chain of the project life cycle, this model is the first 'inno-vation' that can be thought of from the project management perspective. What is critical from the project manager's perspective, however, is the following:

**How to understand when one has to go for the incremental model?**
**What are the failure points of this model?**
**How should one distribute the increments?**
**How to balance the 'strategic' and the 'operational' postures in this case?**

These are some of the fundamental questions required to be answered by any project manager. Let us try to provide an evaluation perspective to answer them.

While the first question has been answered in the beginning, it needs a little more clarification at this stage. From a management perspective, any activity that is repeated gives the doer an advantage. This advantage is called *the learning curve*. Simply stated, the learning curve 'eases' a task by a natural 'learning' that occurs. Typically, performing a task becomes more efficient as the cumulative number of times the task gets repeated. At the same time, repeating a task after a period also has its own disadvantages. The person, therefore, on such a project, which goes into an iterative cycle, has two aspects to consider:

1. The advantage of the 'iterative' environment due to potential learning.
2. The disadvantage of the repetition with a delay factor. The *bringing-one-upto-speed* factor! This is often discussed in meetings as a 'coordination' challenge.

So, although there is a gain due to the learning, depending on the quantum and the complexity, it could easily be offset by the 'coordination' loss. As a manager, one has to understand the impact of both these factors well.

### 4.6.3   A Quick Simulation

Let us consider an example project and the overall activity of that project. Let us say that the entire project model is under consideration. Let us assume that this activity takes 100 % of its scheduled time for 100 % of the scope. Now Table 4.1 shows us how the time elements look like for partial scope. It also tells us the coordination effort that is required since this is important for the iterative nature of the incremental model. The total effort is the sum total for that particular iteration.

Note that for scopes greater than 70 % of the functionality required, the schedule time requirement is equal to or even *greater than* that required for 100 % of the scope. This is a rather tricky situation. Of course, the effects vary on a case-to-case basis. But our aim here is to explain the importance of deriving this kind of curve if the manager is going for a partial implementation as shown in Fig. 4.6.

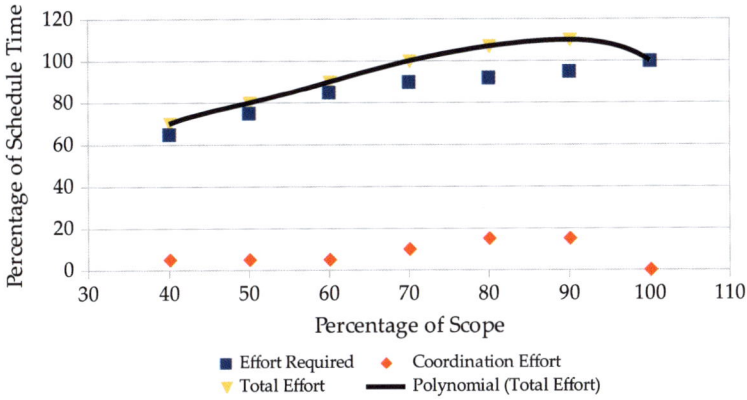

**Fig. 4.6** Graphical representation of the table

**Table 4.1** Scope and schedule correlation for the first iteration

| Percentage of scope | Percentage effort required for activity | Percentage effort required for coordination | Total effort as a percentage |
|---|---|---|---|
| 40 | 65 | 5 | 70 |
| 50 | 75 | 5 | 80 |
| 60 | 85 | 5 | 90 |
| 70 | 90 | 10 | 100 |
| 80 | 92 | 15 | 107 |
| 90 | 95 | 15 | 110 |
| 100 | 100 | 0 | 100 |

So, in our case, if the manager has decided to give 60 % the first time, he is actually spending 90 % of the scheduled time in the first pass itself.

The table shows the values behind the curves in the figure.

Let us plug some values for a still better understanding. Let us say that the total effort is 4 weeks for the *entire scope of the first iteration*. Let the project manager decide to have 60 % in the first iteration. Therefore, from a schedule perspective, his first pass of the project will now take a good 20 × 0.9 or a good 18 days (each week assumed to have 5 working days).

Now, in the second iteration, let us assume that the scope itself changes from the currently assumed overall scope. In other words, if we assume, the scope of the first iteration as our baseline, the new scope in the second iteration will have a quantum of say 120 %. And the revised schedule estimate, when compared with the current estimate could well be 150 %. In other words, with the revised scope, the project schedule would actually need 6 weeks. However, since some 60 % of the scope was already covered in the first pass, and if a good 50 % of the original scope of 100 % is usable, then our requirement is to understand how the Remainder 70 % (120 % new scope—50 % of usable first iteration achieved) can be fulfilled.

**Table 4.2** Scope and schedule correlation for the second iteration w.r.t. the first iteration

| Percentage of scope | Percentage effort required for activity | Percentage effort required for other elements | Total effort as a percentage |
|---|---|---|---|
| 60  | 100 | 25 | 125 |
| 70  | 110 | 25 | 135 |
| 80  | 120 | 25 | 145 |
| 90  | 130 | 25 | 155 |
| 100 | 140 | 25 | 165 |
| 110 | 145 | 25 | 170 |
| 120 | 150 | 0  | 150 |

So, in the second iteration, the following time elements are to be considered:

1. Time required to decipher the useful parts of the previous iteration
2. Time required to decode the documentation of the previous iteration and plan a meaningful course ahead
3. Time required to deliver the percentage of the scope that is of interest
4. Time required for the 'coordination' for subsequent iterations.

The typical combinations of the indirect items 1, 2, and 4 typically account for 25 % of the time of 4 weeks. So, in other words, it would take 5 days. So, now, the revised table could look like Table 4.2 following this discussion. Since, in our case, the balance is 70 %, if the project manager decides to go for the entire revised and balance scope, then he will be spending 135 × 4 weeks, that is 27 days. Now, we see that the total project activity itself has taken 27 + 18 or 45 days! In other words, the additional time required is 50 % more than the total time required for the second iteration. However, if we take the sum of both the iterations, the total time required was 20 + 30 days or 50 days using 100 % SDLC for both. However, due to the way we designed our project life cycle, we have saved 5 days in the process. This is an important conclusion. However, the conclusion is *strongly dependent* on the extent of scope one considers in the first iteration and the extent of change in the scope with the next iteration and the amount reusable after the change.

While this is only the project perspective, what is important is to understand the business aspects involved in this situation. A detailed treatment of how this is done is reserved for later chapters, but the project manager may make a note to revisit this section once he has read about the advanced concept.

A project manager should be able to develop such a 'living' or a 'simulated' environment and 'toy-around' with the various parameters before he goes in to decide the type of project life cycle he would use. A simple spreadsheet-based simulation or even a simple scheduling-based simulation would help evaluate the situation in a fair amount of detail. It is utmost essential for the project manager to do this evaluation with adequate rigor to understand his 'cost' elements in a better way, more importantly, to have a better handle on the appropriateness of his life cycle evaluation decisions.

Incremental models are extremely popular in R&D projects. In such projects, the review of the output at each stage is the most vital part of the entire project cycle. Hence, between the iterations, significant scope changes could be expected.

Most business situations, however, tend to avoid iterations of *complete life cycles*. The preference, therefore, is to go for shorter points that could save the time and the cost of the overall project and ensure good alignment for business success. This has given room to lots of innovations where the waterfall model could be used with 'partially static' conditions to evaluate potential changes and accelerate business outcomes.

### 4.6.4 Rapid Application Development

Most of the construction programs follow this kind of a model. Essentially, they divide the program into smaller projects or subprojects, each of which follows the incremental model. So, for instance, a refinery program might have a crude distillation unit and then, secondary units for water, power, and air. All these are an integrated group or part of the whole, but from a project perspective, they are independent units, each of which is 'incremental' in some sense, but without dependence. The RAD model helps in managing technical risks in a better way. It is a typical situation where 'break-down' or 'analysis' helps in covering the overall risks in a better way.

The key to being able to apply RAD is that the final deliverable can be broken down into components to expedite the process. In other words, *Modularization must be possible!* As soon as we start speaking of modularization, we need to understand two critical aspects:

1. Interfacing: That is ensuring that the modules have well-defined boundaries and the 'joints' can form a smooth continuum.
2. Integration: Where one needs to understand how the integration of the modules has to be optimized so that they can function as a complete business deliverable.

While this is a logical optimization of the waterfall model, one also needs to be cautious while applying it. However, the success of this model depends on several factors:

1. The requirements need to be clear at the start. If that is not the case, the RAD would often lead to greater chaos and would underperform significantly.
2. The biggest problem with large-scale projects is that they start-off with a kind of an RAD model, but then, start crumbling as they go ahead. One of the biggest issues causing this problem is that one needs to have and ensure the right number of teams to make this work. Most RAD efforts fail because there is a shortage in the number of teams, thereby causing a lot of loose ends to be taken care of.
3. One should not always 'push' the limits of technology. In an attempt to modularize, the project managers tend to overdo the concept. This often defines too many interfaces and makes the integration effort disproportionately large!

4. One also needs a proactive customer–supplier relationship. Any RAD methodology involves several 'turn-overs' to the customer. That is saying that they are continuously delivering some component of the final deliverable. This often requires a good handle at the customer's end where they have to proactively help the agency realize their business objective and manage the deliverable better. One sits on a potential risk when that does not happen! Feedback from the customers is used to adapt subsequent deliverables where required, thereby making the product an improved version of 'conforming' functionalities and deliverables.

5. Another important aspect to understand is that interfaces in any deliverable or system may compromise performance significantly. Hence, the RAD is not a suggested model for new technologies.

## 4.7   A Generic Treatment

At this time, the reader must be aware that there are several different models in the literature for project life cycles. A detailed treatment of the models would only lead to a 'jargon-centric' discussion that we would like to avoid at this stage. *In practice, process life cycle models are hybrids of those read in books. Hence, understanding a process model requires a more 'customized' skill.* By hybrids, it also means that the variants are so many that the models could be based on the 'whims and fancies' of the manager rather than being a derived and evaluated option. So, one needs to be careful about the distinction between the two.

**A Quick Hands-On**

To take this forward, we would like to equip the manager to understand the variables involved and the method of analysis required while answering the following cardinal questions:

**Which is the most appropriate model for the project?**
**What is the most cost-effective method for the project?**
**What are the critical failure points of each model?**
**How does the organization normally function? Is it really ready for the model being proposed?**
**What is your own competence level for the particular model?**

While these are basic questions, we would like to help the readers understand how to evaluate these questions in their own organizational context. We have already demonstrated a few of the insights in the treatment of the earlier models. Now, we would like to discuss a few 'generic' considerations that are required to be understood in this context.

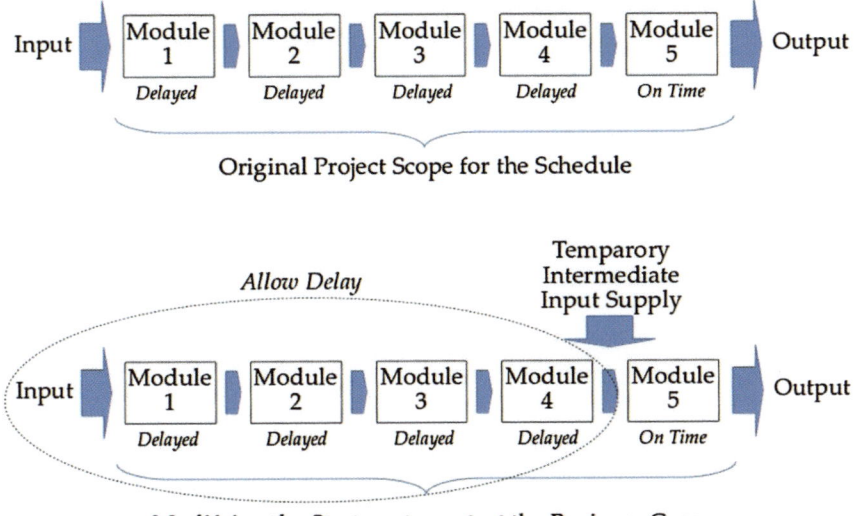

**Fig. 4.7** Case for meeting business requirements in delays

Understanding the project life cycle requires a thorough understanding of how a project life cycle is modeled. One needs to evaluate the business need and see if it is fairly clearly understood. For instance, a company has a greenfield project that is making an integrated plant. However, the project is undergoing serious delays in most of the modules of the integrated plant. Yet, the final module is still on schedule. In such a situation, the company can think of a solution by bringing in the intermediate output to feed into the final module so that they can enter the market as per their original requirement as shown in Fig. 4.7. Not every business requirement may be so. Therefore, the point to be noted is that the project manager must clearly understand the business requirement in terms of the business case of the project. We will further develop this concept in subsequent chapters. However, as a preliminary treatment, this limited understanding would suffice for this section.

In many projects, the business requirements may be 'observed' as 'changing'. In such cases, there is a possibility of seeing a lot of claims and lawsuits in the project. Such situations need to be modeled from a project life cycle perspective using the following understanding:

1. Requirements becoming clearer over time: In this case, the requirements are becoming increasingly clearer as the project progresses.
2. Requirements changing over time: In this case, the requirements are changing over time.

The first is like an efficiency factor while the second is like an effectiveness factor. If the change is marginal, it may be treated as an operational issue. However, if the change is bigger, it could have the potential to affect several other parameters. In the incremental model simulation, we demonstrated how a reasonably large change (of 20%) affects the iteration.

At this moment, it is time to give the reader a word of caution. Let me give you an analogy. I go to the hair studio and tell the hair dresser that I am looking for a specific style for my hair. The stylist works on my hair and gives me an output that is different from what I wanted. This is, in my view, a problem with the stylist! He goes by his own stereotype and skill and works on what he 'feels' is right and 'knows' to do best!! This situation is similar to the first case of efficiency. The second case is when he actually cuts as per my expectations, I find that the style isn't that attractive and request him to change it. This is the second case of effectiveness. Although most project managers are professional, they tend to lean to their first instincts and quickly end up in the 'efficiency trap'. On the other hand, if they are truly efficient, the change is driven from the customer rather than the manager.

While toying with the project life cycle model, a project manager needs to understand these elements in terms of probabilities and quanta. He needs to have a concrete understanding of values to be able to make a meaningful model. In doing so, a manager must attempt at understanding the *known unknowns!* At Consulting Connoisseurs, we have developed a three parameter framework for rqeuirement characterization. The key parameters are shown in Fig. 4.8. Let us clarify them a little more:

1. *Quantum and the Nature of Change*
   The quantum of change is actually the modification made in the overall scope of the project. If the quantum of change is marginally small, it may be a good candidate for the SDLC model. For instance, if the networking in the chairman's office is to be along the east wall and not along the northern wall, the change with respect to the whole building is negligible and can easily be incorporated. This is a case of utilizing a 'buffer' in the productivity.

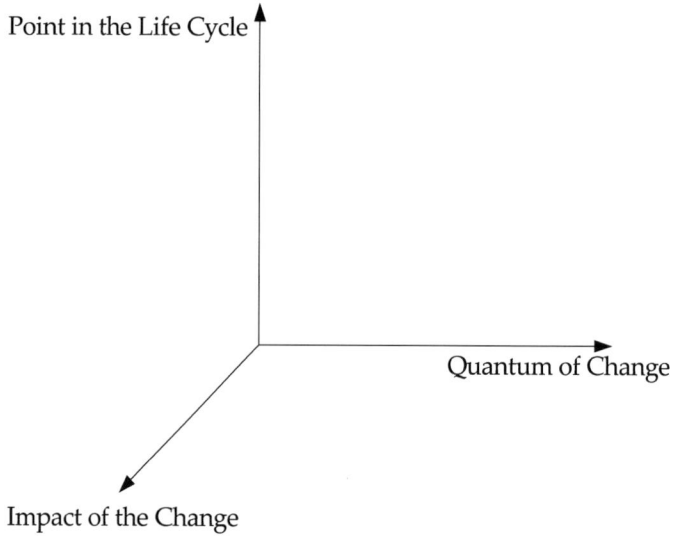

**Fig. 4.8**  Key parameters for characterizing requirements

In another project, if the crude quality is changed, the entire process parameters change. This could be termed as a high quantum.

So, the project manager must have a fairly good understanding of the *probabilities of such changes in his project*. This is a key parameter.

2. *Point in the Life Cycle*

The above parameter is important, but what is vital is to understand the point in the life cycle where these changes are introduced. If they are introduced early enough (for example: Geotechnical data in construction, or GUI formats in the case of application development), the choice of the iteration trigger point can be fixed appropriately. Again, here too, the *probabilities of the events* are crucial to evaluate the various models.

3. *Impact of Change*

While the quantum of change is a parameter, its dependency is to be considered from two perspectives. Impact of change on the past activities and impact of change on the future activities. In other words, the impact of change is a parameter that determines the extent of rework for a given quantum of change. So, it is not just dependent on the first parameter, but also on the progress of the project (or the point in the life cycle).

For instance, if a readymix concrete facility at a site has a changed capacity (installed a lower capacity than planned earlier), it will not affect the engineering, but would affect the site logistics. So, such a quantum of change would impact only the construction. Hence, one would not require to rework on past activities. It is sufficient to look at optimizing the future construction activities in this case.

Impact of change on past activities is a critical parameter to consider the option of whether one needs to go for an iterative project life cycle model. The project manager must be able to foresee these possibilities fairly well right at the beginning.

From a modeling perspective, it is sufficient to have a rough handle of the potential values of the *probabilities of the impact*. This will help the manager develop appropriate scenarios to evaluate the applicability of the appropriate life cycles.

Further, the time to make the corrections is also a critical parameter that needs to be determined by the manager. Some corrections are to be incorporated, but are not priority items. Some others are priority items. So, again here, one needs to understand how to work around the issue. The crucial aspect here is to know if there is a freedom to choose the time when one should incorporate the changes.

In general, these parameters are important to characterize the various scenarios from the requirement perspective. While the impact of change is possibly the most 'discussed' aspect in any project management literature, we have defined the other two parameters to provide the manager with the required perspective for building a scenario with adequate understanding of the dependencies. Most stakeholders typically focus on just the quantum of change. However, operational project managers are more concerned about the point of introduction and the impact of the same.

The next important aspect is to understand the delivery characteristic. This basically is a relationship between the scope coverage, effort, and duration.

**Table 4.3** Scope realization for a sample project

| Period block | Cumulative realization in pure costs (USD Million) | Cumulative cost of capital for period (USD Million) | Total cumulative cost of the project (USD Million) |
|---|---|---|---|
| 1 | 10 | 0 | 10 |
| 2 | 30 | 1 | 31 |
| 3 | 50 | 1 + 3 | 54 |
| 4 | 75 | 1 + 3 + 5 | 84 |
| 5 | 85 | 1 + 3 + 5 + 7.5 | 101.5 |
| 6 | 100 | 1 + 3 + 5 + 7.5 + 8.5 | 125 |

1. The scope coverage relates to the effort and duration as has been shown in the iterative model. The effort and duration was nominally assessed in the simulation there.
2. The effort and duration are also dependent on the type of the process model and the type of infrastructure, the know-how of the employees, etc.

Each of these need to be plugged in and the scenarios need to be evaluated. This helps the manager understand the appropriateness of the model for a given situation. In doing so, however, it is not just the economic/financial parameter that drives the decision, rather it also needs to understand the dependency of the various work elements involved.

To illustrate this, we define a project that has six activities, each taking a quarter (for simplicity). The activities have a cost realization characteristic as shown in Table 4.3. Note that the cost of capital is 10 % of the cumulative spent cost till the previous quarter. This is not very far from the truth as a mobilized site always costs money and does not come for free, regardless of the progress made at the site.

For simplicity again, let the dependency matrix be simple such that P1 precedes P2 that precedes P3 … and so on. Now our gut feeling is that this project could experience a change of around USD 10 Million. The probability of this change is only 20 %. But if this 20 % was the reality, the project could be planned using the waterfall model as shown in Table 4.4.

Let us also assume that this project change occurs between the third and the fourth period block.

So, now we need to identify the type of process model that would fit best with such a scenario. Since three periods are already complete, there is a rework in the second and the third periods. In other words, the impact of change extends to up to two periods in the past. At the same time, it also stretches beyond into the future/remainder of the project.

Now the impact or the distribution of the quantum of change is given in Table 4.5. However, let us assume there are two options (assuming) for the project manager:

1. The Project Manager *immediately* incorporates the change. This has two fallouts:
   (a) He ends up paying a premium of 20 % due to the 'revision' in his schedule. So, as Period 2 costs USD 20 Million, the impact of the change is actually going

**Table 4.4**  Scope realization of the sample if it were known in the beginning

| Period block | Cumulative realization in pure costs (USD Million) | Cumulative cost of capital for period (USD Million) | Total cumulative cost of the project (USD Million) |
|---|---|---|---|
| 1 | 10 | 0 | 10 |
| 2 | 31 | 1 | 32 |
| 3 | 53 | 1 + 3.1 | 57.1 |
| 4 | 81 | 1 + 3.1 + 5.3 | 90.4 |
| 5 | 94 | 1 + 3.1 + 5.3 + 8.1 | 111.5 |
| 6 | 110 | 1 + 3.1 + 5.3 + 8.1 + 9.4 | 136.9 |

**Table 4.5**  What the change means!

| Period block | Quantum of change on specific period (USD Million) |
|---|---|
| 1 | – |
| 2 | 1 |
| 3 | 2 |
| 4 | 3 |
| 5 | 3 |
| 6 | 1 |

to be USD $20*(20\%) + 1$ Million. That is the revision premium + quantum of change cost.

Further, each change, howsoever small, will take the entire period. That means, the change in Period 2 is just USD 1 Million worth, but would take a duration equal to that of Period 2!

2. The project manager incorporates the change by adding *a fresh partial cycle at the end of the project*. This has two fallouts: (a) He ends up paying a premium of 5% of the period cost as a revision coordination activity, and the duration of the change/correction activity in each period is reduced to half the duration. This means that the correction of activity in Period 2 will now be USD $20*(5\%) + 1$ Million in this case, and the duration is going to be half of Period 2.

Taking these possible variants into consideration, we find the following two scenarios taking shape as shown in the Tables marked as Scenario 1 Table 4.6 and Scenario 2 Table 4.7. We find that there is a difference of around USD 5 Million in the total cost, and a difference of USD 16 Million in the pure project costs in the two methodologies. It is evident that the delayed incorporation of changes is beneficial in this case.

It is the job of the project manager, therefore, to understand and evaluate the type of process model that suits his need best. The example given here is only an illustration of how this needs to be done in practice so as to ensure that you have the right *hybrid* for your own specific business situation.

Of course, real-life situations can be fairly complex. However, while considering a specific project, a gross level analysis like the one shown in the example here would

**Table 4.6** Scenario 1: instant incorporation of changes

| Period | Cumulative cost (USD Million) of the base case | Cost by period of the base case in USD Million | New cost by period of the changes in USD Million | New cumulative costs in USD Million | Cumulative cost of the capital in the new case (USD Million) | Total of new case in USD Million |
|---|---|---|---|---|---|---|
| 1 | 10 | 10 | 10 | 10 | | 10 |
| 2 | 30 | 20 | 20 | 30 | 1 | 31 |
| 3 | 50 | 20 | 20 | 50 | 4 | 54 |
| Inst correction 2 | | | 5 | 55 | 9 | 64 |
| Inst correction 3 | | | 6 | 61 | 14.5 | 75.5 |
| 4 | 75 | 25 | 33 | 94 | 20.6 | 114.6 |
| 5 | 85 | 10 | 15 | 109 | 30 | 139 |
| 6 | 100 | 15 | 21 | 130 | 40.9 | 170.9 |

Base model with instant correction (premium is 20 % of period cost, delay = period duration affected in the past)

**Table 4.7** Scenario 2: delayed incorporation of changes

| Period block | Cumulative realization in costs (USD Million) | Cost per period (USD Million) | Cumulative cost of capital for period (USD Million) | Total cumulative cost of the project (USD Million) |
|---|---|---|---|---|
| 1 | 10 | 10 | – | 10 |
| 2 | 30 | 20 | 1 | 31 |
| 3 | 50 | 20 | 4 | 54 |
| 4 | 75 | 25 | 9 | 84 |
| 5 | 85 | 10 | 16.5 | 101.5 |
| 6 | 100 | 15 | 25 | 125 |
| Delay 2 and 3 | 105 | 5 | 35 | 140 |
| Delay 4 and 5 | 112.75 | 7.75 | 45.5 | 158.5 |
| Delay 6 (1/2 quarter only) | 114.5 | 1.75 | 51.1375 | 165.6375 |

make sense. At anytime, please also check the robustness of your model. This is again treated in detail in subsequent chapters.

Note that the probability of the event occurring is only 20 %. Hence, the total expected cost can be calculated considering the probabilities of each scenario. If one takes the pure project costs, the first scenario yields something like USD 106 Million while the second is around USD 103 Million.

Such type of 'quickie' analysis using the generic approach is often useful in guiding project managers on the type of life cycles they need to choose. They also are generic enough and help one understand the impact of certain changes, especially when they occur at a later stage in the project. The reader should be informed that most of the innovations in the project life cycle models are actually based on these kinds of analyses. Some of them have been formalized to a great extent which others are 'customized' variants that are being developed in most organizations. All said, one needs to also use the rigor to evaluate the model and should beware of people who tend to use models and variants without sufficient knowledge, insight, or basis. A good project manager would normally evaluate for himself and be convinced of the way he needs to move ahead before he indulges in mindless modification or blind adoption of any project life cycle model.

## 4.8  A Popular Issue

After all this discussion, the one aspect that requires significant coverage at this stage is the dichotomy that arises in a typical project environment. A typical management view brings in the following perspectives as shown in Fig. 4.9:

1. The Awarding Agency
2. The Executing Agency
3. The Project Manager

A large number of problems in claims management originate due to these perspectives. What this perspective shows is quite plain and simple. A contracting agency, most times, could have a different understanding of their project lifecycle when compared with that of the executing agency. The project manager is the third entity that could have an even more different view of the whole situation.

This situation is a potentially dangerous one. Let us see how this works out. If the client is expecting an SDLC methodology, while the executing agency is looking for an incremental model and the project manager is expecting an RAD approach in the given situation, the resulting project would, simplistically said, be extremely chaotic. Such kind of situations are not unique and are very often encountered in

**Fig. 4.9**  Three factor view

claim management, where the 'executing agency' is perceived as an 'inexperienced' one as they are *unable to tow the lines along an SDLC approach.* At the same time, there is also a 'quick-and-dirty conclusion' that the contractor is potentially trying to inflate his bills. This is another dangerous situation and could result in a lot of 'ugly discussions,' most time ending up into an *analysis of the contractual obligations.* It is often interesting to see how people go to *any lengths to justify* their versions of the life cycle model. We will touch upon these aspects again in detail in later chapters.

But the problem gets compounded when the project manager thinks along a completely different path! Even if he is well justified and reasonable to do so, his views could quickly be misinterpreted. In other words, the one singular skill a project manager needs to have and develop is that of *sales and marketing*! While this might sound ridiculous to most veterans in project management, this ability is appreciated by those managers who have worked in multiple environments. In any case, it is time to reflect on the same while taking this forward. However, I would leave it to the 'imagination' of the reader to take it to the right conclusion whatever situation the reader might be in.

## 4.9  Key Takeaways

After a relatively long chapter (:-), it is time to review the key takeaways from this chapter. The first aspect to understand the context was to understand the systems view of management. This essentially provides us with a concrete understanding of the interfaces, the basic behavior of the 'project entity', and the nature in which the project is organized.

The systems view is often treated as parallel to that of the 'initial condition' or the 'boundary condition' concept of engineering. Therefore, a solid understanding of the systems view is required for every manager. This is provided by the project management context. As we have seen, the context itself is divided into phases, life cycles, and the organization. The project manager needs to understand each of these concepts separately as well as in combination. While the current chapter dealt with the basics on the lifecycle, the subsequent one does a quick check on the organization.

The most important takeaway is the understanding of the contextual elements from the perspective of appropriateness and decision making. The sample models are meant to provide the reader with a good insight into variants that are popular in the industry, without at the same time, being too biased with theoretical applications. In other words, it is important to understand the concept, but more importantly, derive a smooth modeling framework that is sufficiently generic, yet has the potential to incorporate specifics for the given management situation.

Possibly, the one key takeaway that comes as a corollary at this stage is that, regardless of the appropriateness of the technique, the project manager needs to be a good 'sales guy'! That apart, your own simulations would yield good results that would help evaluate, plan, and take things forward.

# Chapter 5
# The Systems View of Management-II

**Abstract** This chapter focuses on the organizational system in the project management context. It begins with the identification of *four frames of reference* in the context of the *project organization*. These four include: *(a) The structural frame, (b) The human resource frame, (c) The political frame, and (d) The symbolic frame.* In line with conventional literature, the chapter restricts itself to the structural frame and defers the *tacit treatment* of the remainder of the frames in other sections within the book. The focus then moves toward the conventional understanding of the three types of organizations viz. *functional, projectized, or the matrix* types. The chapter then delves into making an informed decision on *how one should choose the type of the organization*. This is followed by an interesting new paradigm called the *Biological Entrepreneurial Model* for the structural frame of the project organization.

**Keywords** Project organization · Structural frame · Human resource frame · Political frame · Symbolic frame · Types of project organizations · Functional organization · Projectized organization · Matrix organization · Deciding on the project organization · Biological entrepreneurial model

## 5.1 The Project Organization

This is another important element in the project management context. If one looks at the context carefully, the organization can further be classified into four key dimensional elements as shown in Fig. 5.1. We call these elements the key frames of project organization.

### 5.1.1 The Structural Frame

The first frame is called the *Structural Frame*. This frame is typically defined using the organizational chart. It has the elements of roles and responsibilities. A role is either the person's position in the business or their particular job that dictates their role

© Springer Science+Business Media Singapore 2017        71
N. Gurjar, *A Forward Looking Approach to Project Management*,
Lecture Notes in Management and Industrial Engineering,
DOI 10.1007/978-981-10-0782-8_5

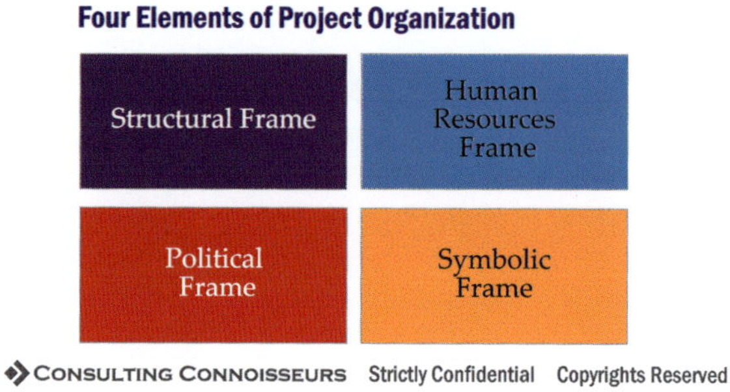

**Fig. 5.1** Elements of project
organization

**Four Elements of Project Organization**

| | |
|---|---|
| Structural Frame | Human Resources Frame |
| Political Frame | Symbolic Frame |

◆ CONSULTING CONNOISSEURS   Strictly Confidential   Copyrights Reserved

in the business. For example, a teacher's main role is the education of his students,
a doctor's role is looking after the health of his patients.

A responsibility on the other hand, is the set of key tasks that a particular worker has
to carry out. For instance, the teacher: teaching classes, marking homework, pastoral
care, covering for absent colleagues, parents evenings, etc; the doctor:- performing
surgery, diagnosing illnesses, treating illnesses, home visits, on-call, etc.

The structural frame also talks about coordination and control in the organization.

## 5.1.2   The Human Resource Frame

The HR frame focuses on the harmony between the needs of the organization and
the needs of the people.

In other words, this is an important area of alignment and is integrally involved
with:

1. The design of HR policies and
2. Developing the team concept in organizations.

## 5.1.3   The Political Frame

This frame actually assumes that organizations are *coalitions composed of varied
individuals and interest groups.*

Hence, the major thrust of this frame is on two elements:

1. Conflict and
2. Power.

Hence, this frame is involved in understanding the alignment of interest groups in
the context of the project and understanding how these interests bring in different
degrees of shifts of the efficacy in the project management context.

As a project manager, you need to understand this frame as well, for it could make or break your career:-)

### 5.1.4 The Symbolic Frame

The symbolic frame actually focuses on symbols and meanings related to events. In other words, it looks at the *cultural* aspects of the organization.

As a project manager, one needs to understand this frame in the organization. It could be a significant criterion to understand one's fitment in the overall organization.

## 5.2 Project Organization: Types of Structural Frames

Although all the four frames are important, most books focus only on the structural frames. In other words, the literature tends to swing in favor of the 'formal organization' rather than look at the dicey issues involved in the other frames.

In this book, however, we will touch upon all the four frames as we go along. While we restrict our treatment here to the structural frame, we will deal with the others in subsequent sections of the book.

Essentially, there are three types of project organizations that are defined in this context.

1. Functional Organizations
2. Matrix Organizations
3. Projectized Organizations.

We will briefly touch upon these types now. As these are already discussed at length in every elementary book, a detailed treatment at this stage on these topics is not intended.

## 5.3 Functional Organization

This is also touted as the *Classical Model* of project organization. In this model, each employee has one superior. The organization is grouped by specialty as shown in Fig. 5.2, and the projects are 'bound' by the limits of the function, from the employees' perspective! A beautiful illustration is given in PMBoK® on this model.

Most companies 'define' small projects for their KRAs and Appraisals... These are supposed to be only a 'part' of the employees commitment. Such projects usually have a functional organization. They are the most common projects in maintenance and small scale improvements faculty.

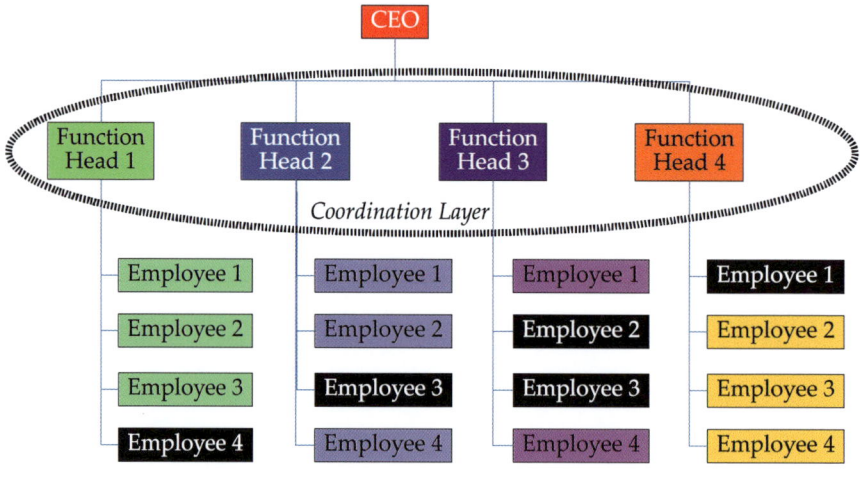

**Fig. 5.2**  Schematic of functional organization

Certain 'upgrade' projects like knowledge management, development of specialized IT platforms, implementation of ISO systems, etc. are managed using the functional approach. Coordination is done by the superiors in their own coordination meetings.

## 5.4  Projectized Organizations

This model too is characterized by each employee having only one superior. However, in this case, the project is managed by crews or teams. So, each team is typically a multispeciality one. They come together and complete the entire project. Note that certain functions may be centralized like marketing, procurement, human resources, etc. even in projectized organizations as shown in Fig. 5.3.

Unlike the functional organization, where the project is often an 'additional activity', the projectized team has complete commitment to the project. Project managers have more independence and authority by the basic frame in this case.

From a project manager's perspective, this appears to be a better organization as he directly manages his team. The project manager assumes responsibility for the appraisals and career development of the members with complete line authority. Most times the resources are exclusive and not shared in this case. The structure typically is disbanded at the end of the project.

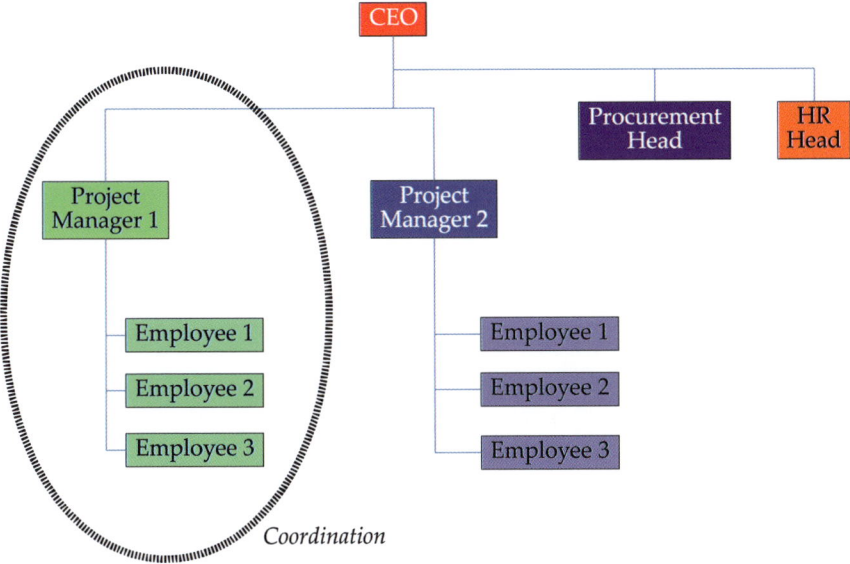

**Fig. 5.3**  Schematic of the projectized organization

A common criticism is that the project 'locks' down the personnel over the 'entire' life cycle of the project. This is not entirely true though. In most cases, after the phase is complete, the crew is cut-down to keep just the bare minimum on board.

Another criticism is that the 'knowledge and learning' especially for short projects, does not seem to be leveraged by this model as the members complete their work before even the technology is transferred. This is, however, true only in the case of *large* resource pool companies. Most mid-sized companies will have select resources and hence, the knowledge pool still develops among the existing groups.

## 5.5   Matrix Organizations

Just as we spoke of 'hybrid' models of project life cycle, here too a similar hybrid model started becoming more and more popular. Matrix organizations are, therefore, a blend of the functional as well as the projectized model.

Weak matrices are more toward functional models where the project managers are meant more for coordinating or expediting rather than managing. Strong matrices are more toward the projectized models where the project managers are needed to have strong skills in managing.

Here, the employee has *two bosses*: One at the departmental level and the other at the project level. Most times, the departmental boss takes care of the training, appraisals, etc., while the project manager handles the day-to-day needs of the project.

The reader is advised to refer to the PMI PMBoK® for further understanding of this topic.

## 5.6   Demystifying the Structural Frame

**A Quick Hands-On**

We have been seeing a lot of discussion on the characteristics and potential problems of the structural frame. However, one needs to understand that there has to be a balance while deciding the project organization *assuming one can 'call' the shots!* Hence, the cardinal questions for the project manager are:

**What is the appropriate project organization for a given project?**
**How should one decide on the same?**

We try to list a few of them for the understanding of the reader. However, there are several other factors that are to be considered and these are discussed at length in subsequent chapters. What is more important is that large projects typically have functional groups; it is possible that smaller teams are more 'projectized' despite their 'functional' designation. The criteria are actually a combination of the following dimensions:

1. *Project Discussions*
   One aspect that is often interesting is the content of discussion that occurs in a typical project meeting. Usually, the functional organization is comfortable to discuss issues in technical depths. However, a projectized environment has more of 'management discussion' limiting the talk to scope, cost, schedule, etc. Hence, a high-end technology project would often require a functional organization to ensure that the issues are correctly addressed.
   Projects that have lot of open points in the technical issues of the scope, often do well with functional teams rather than projectized teams.
2. *Resource Pool Maintenance and Administration*
   In many organizations, the project manager is often good at driving the project, but is not interested in working out the details in the resource pool maintenance. In other words, he is not interested in the additional burden of recruiting and maintaining the team. This is essentially true when the projects are small. Even in larger projects, the project manager is, at times, overly bogged down with project activity and, therefore, prefers to keep it discrete if the pool already exists elsewhere in the organization. Then, he would opt for a matrix model.

3. *Project Interference*
   One of the potential problems in the matrix or the functional setup is potential interference from the 'line' managers of the employee. When such a situation takes the project organization to a 'gaming mode', where a power and posture tussle become active, it is time to reconfigure and buy better commitment or alternatively go for the projectized environment.

4. *Recruitment Issues/Competence Issues*
   At times, the project manager is exceptionally good at managing a project. However, he ends up riding the wave of a 'failed' project. This is because he has team members that are not very strong. Often times, in recruitment, there are team members who are less competent simply due to budgetary constraints or issues arising from the HR who want to 'close' the position without 'exposing' their inadequacies. One sees many organizations where the project manager tells the HR team that a particular team member isn't performing. Oftentimes, HR managers in such cases ask the manager to 'adjust' or 'compromise' or even 'manage better'! So, the problem reflects later in the project success. Such situations are dangerous and a projectized organization might be able to maneuver this situation better.

5. *Fitment to Project Issues/Team*
   One might find employees being good performers. Yet, they are not actually performing in the particular team. Team dynamics is a critical component that needs to be addressed in the project organization. In such situations, a matrix organization helps quick resolution of such issues by quickly shunting-in or out the appropriate pair of employees.

6. *Duration of the Project Cycle*
   Longer project cycles should ideally have a projectized environment as compared with shorter project cycles. This parameter is, perhaps, *the most important of all in determining whether a project organization should be of the projectized form or not.*
   While there could be a lot of alternatives to this philosophy, a longer duration actually creates a psuedo-static project organization. Therefore, it is advisable to have a projectized environment in such cases.

7. *Time of Appointment of the Project Manager*
   Another tricky parameter in consideration is the time of appointment of the project manager. While there are norms laid down in PMBoK®, the fact remains that the real world does tend to deviate. If a project manager is appointed late, then the project structure should necessarily be a projectized organization. If that does not happen so, the project manager is often a victim of politics due to the 'pre-existing' organizational dynamics.

8. *Budgetary Freedom/Resource Freedom*
   If a project manager has 'limited' freedom from a budgetary perspective for his resource management, it is better to have a matrix-based structure to enable better handling of the resource productivity.

9. *Interdiscipline Coordination*

When the interdiscipline coordination is reasonably high, a matrix or a projectized organization works best as it drastically improves the speed of communication in the project.

10. *Team Member Attributes*

Typically disciplined and high-performing team members need a particular ecosystem that has to be reasonably 'static'. A functional organization often helps such team members.

11. *The Skill of the CEO*

All said, the *buck stops here!* The skill of the CEO to manage a project and his ideas of translating it into action often are the biggest determinant to adopting a particular organizational style. Hence, one has to ensure that this is adequately factored into the given situation.

Since project success is not dependent on the structural frame alone, these points are meant to help the project manager guide his way through any given project. The dimensions could act in favor of or against a given configuration and hence, it is difficult to give a generic conclusion that would satisfy all the given conditions. Therefore, in this work, we have tried to restrict ourselves to *significant pointers* that affect the decision of the structural frame.

What is important to understand is that the project organization is, oftentimes, not under the purview of the project manager. It is usually decided by some key stakeholder depending on his/her comfort level in assigning, monitoring, and controlling responsibilities. Most times, the feasibility report assumes a structure that is 'just drawn' on a piece of paper. Unfortunately, the binding of the organization assumed is realized much later, when the project is actually in execution.

The other aspect is that of the employee productivity and availability. We have assumed that the organization positions are 'filled-up' rather quickly and there is no significant 'waiting' period for the project manager. This is seldom the case. As we will see in subsequent sections, the project environments assumed in most business situations are way different from what they are supposed to be! Hence, in this work, we are focusing on such advanced topics to enable the project manager equip himself for better decision making.

## 5.7   The Biological Entrepreneurial Model

**An Innovative Approach to Project Management: Integrating Entrepreneurship Using a Biological Model**

A new model was observed and developed while I worked on a large project in the US. A teaser to the model is taken from my own blog that is reproduced here for the reader.

### 5.7.1 Understanding the Biological Component

The model is inspired by the concept of cellular *totipotency* that is found in biology. Totipotency is the ability of a single cell to divide and produce all of the differentiated cells in an organism. For instance, the entire movie of Jurassic Park was based on the 'totipotent' cells and genes of the Dinosaurs found in the fossil. So, common examples of totipotent cells are spores and zygotes. In the spectrum of cell potency, totipotency represents the cell with the greatest differentiation potential. In our project environment, especially a greenfield one, the chief executive is an entity analogous to such cells. 'Toti' comes from the Latin totus which means *entirely*.

While we speak of zygotes and spores, it is possible for a fully differentiated cell to return to a state of totipotency. In other words, a muscle cell could regenerate into a full fledged individual. This conversion to totipotency is complex, not fully understood, and is the subject of recent research. Research in 2011 has shown that cells may differentiate not into a fully totipotent cell, but instead into a complex cellular variation of totipotency. We use this fact in defining the entrepreneurial potential of individual project managers in an environment that fosters such growth.

This is the logic behind the stem cell technology. A stem cell is able to differentiate into many other possible cells. In the same way, an entrepreneurial project team could grow and differentiate and take on different areas within a project. Just as stem cells are divided into the following types: (a) totipotent embryonic stem cells, (b) pluripotent embryonic stem cells, and (c) multipotent stem cells; so could the project team work on the same principles. In other words, at the beginning of the project, the 'biological' nature of the team makes it a totipotent type of stem cell that is capable of differentiating and growing across all the functional areas and requirements in the project organization. Just as in a laboratory setting, it is possible to induce stem cells to differentiate into specialized cells by changing the physical and chemical conditions of growth, the same principle could be extended to professional project managers. I am yet to see a project manager who isn't (somewhere) a generalist. So, in smaller work groups, this differentiation is more rapid.

Again, going by the analogy, several sources of stem cells are used experimentally, and are classified according to their origin and potential for differentiation. Human embryonic stem cells (hESCs) are extracted from embryos and are pluripotent. The adult stem cells that are present in many organs and differentiated tissues, such as bone marrow and skin, are multipotent, being limited in differentiation to the types of cells found in those tissues. Similarly, in the project organization, we have a similar situation whereby the differentiation starts getting limited over a period of time (and as the project organization truly grows).

To develop on this further, in order for a cell to differentiate into its specialized form and function, it needs only manipulate those genes (and thus those proteins) that will be expressed, and not those that will remain silent. The primary mechanism by which genes are turned 'on' or 'off' is through transcription factors. A transcription factor is one of a class of proteins that bind to specific genes on the DNA molecule and either promote or inhibit their transcription. The preconditions in the project environment are similar to these transcription factors.

## 5.7.2   A Quick View of the Model

Do you think you can improve your project management style? Modern Project Management finally falls into three models: functional, matrix, or projectized. Each of these models has been tested with time and has really not stood the acid tests for project management. Seeing this room for innovation, I am currently looking at a model that actually defies the logic of conventional project management. I call it the Biological Model. Let us dive into the details!

Every model works with certain assumptions. The biological model is one that bases itself on intra/entrepreneurial skills of the employees in the organization. It starts off with a functional model, however, enabling growth in lateral scope and vertical scope as the project moves on. There is no definite structure of the team. Each team member works on his set of ideas to see how they fulfill the goals of the project. Its like cells growing in a culture medium…they grow into ideas like cells forming tissue. After the idea reaches a critical mass, the interrelationships between the various streams are evaluated and the strategic decision is arrived at. Such a model allows people to 'own and work their ideas through' on a project. There is also a good scope for innovation in the project. As the intrapreneur works his way through, he also develops his team to support the idea depending on what the idea requires. This enables unimpeded growth of the tissue and makes it gel well into the body of the project that is being cultured.

The success of this model, however, relies heavily on the management structure. Since every cell will tend to grow and become a big tissue, there is a greater need to coordinate the growth of these tissues. The control of direction needs to be clearly defined, identified, and communicated. This model requires the management to control each of the interfaces and ensure that each of the concerned tissues operates and grows smoothly.

While conventional project management talks of information flow and chain of commands, this model requires a flat structure; therefore, an ocean of information with appropriate information tools that enable people to sift through the required ones and discard the rest. Needless to say, intrapreneurial competence and subject expertise is a key to this model. The management also needs to be reactive enough to address the needs of the intrapreneurs. For this model to succeed, the management must ensure that the success of the intrapreneur must gain priority over the management of the organizational structure. This is a critical success factor for this model. This also means, in other words, that the underlying infrastructure of management processes like accounting, legal, staffing and hiring resources must be simple, flexible, and fast. Decision variables for decision making must be well defined and well communicated. They should be iterated with sufficient feedback to avoid wastage of time on 'wrong directions' of thought. Decision levels need to be delegated properly in order to avoid overburdening and thus, 'bottlenecking' the coordination effort.

While this model ensures good quality decisions and project management, it also runs high risks. The highest risk is that of scope. When a 'cell' is absent or grows extremely slowly in one area of the media, it means that the scope element is not being developed appropriately. In such a situation, the likelihood would be to wind up with 'fire-fighting measures' in that area. This would occur due to the simple fact that the growth of the other cells and tissues would take time till they span this part of the medium. Hence, the management needs to understand, in concrete terms, the scope of the project. It is extremely risky to prioritize areas as important as this could create the 'out-of-sight-is-out-of-mind' situation. Hence, the management must ensure that they understand every deliverable or the 'work breakdown structure' associated with the project. This point defies conventional logic in management sciences that insists on a simple priority action plan. The management, on the contrary in this case, needs to be prepared to accept flat information that helps them control the entire span of the project. This is easier said than done in most cases.

Having understood the program plan of the project, the next risk associated with this model is that of decision making. While on the one hand, the quality of decisions is expected to be better, on the other hand, decisions and information could become 'event driven' rather than being 'schedule driven'. Aligning these 'events' with the 'schedule' calls for a focus on the project time management. The intrinsic nature of working of this model calls for ownership. However, when it comes to coordination, there must always be a clear and single owner of delay for each train. Most projects, however, have transfer of ownerships. While most companies and professionals would argue that the ownership needs to be transferred due to the different nature of jobs (like civil works, mechanical works in industrial construction or requirement gathering and programming in software), I insist that the owner of the last deliverable must be the owner of the chain. This owner needs to check the alignment of the scope, cost, and time issues with the project.

As we see here, the key issues are the areas of growth and the rates of growth for the management. The critical success factor is the management style, structure, infrastructure, and the communication (I am listing this last point separately for obvious reasons!). As a tool, the management of uncertainty in large programs is the key. However, with the strategic options concerning the deliverables for the individual elements in the scope are variable, it makes sense to look at controlling these options from a schedule/time management perspective.

*This article is taken from the White Paper of the author at Consulting Connoisseurs. For more information, visit* http://www.consultingconnoisseurs.com.

## 5.8 Key Takeaways

While modeling tools are fairly rigorous, the conceptual tools used to discuss the structural frame are also extremely useful in providing the manager with business acumen and direction. However, the concept is further developed in subsequent portions of the book using modeling and simulation techniques. In the present chapter, a reasonable understanding of the concepts in the management context is, thus, provided.

The biological entrepreneurial model is a 'game-changing' approach to the traditional structural frames. In the traditional frames, it is quite possible that there are 'gaps' or 'loopholes' in the management methodology. The entrepreneurial component in this frame ensures that the gaps are plugged and the project manager gets a platform to grow and assume more and better responsibilities. It is, therefore, a win–win situation for the entire team.

A successful model however, requires the right kind of alignment with the top management and the CEO. Most times, structural frames are too rigid to consider such variances and lead to disastrous results. Again, it is upto the chief project executive to take on these issues and drive them across.

Another important aspect of the biological entrepreneurial model is that it incorporates the views and factors of the project team. In today's management literature, very little importance is given to the project team, although the team comprises 80 % of the workforce. So, it is a more inclusive paradigm that has greater relevance in today's business environment.

# Chapter 6
# Basic Mantras of Project Management

**Abstract** In this chapter, the focus is on the typical failures encountered in project situations. While a lot of these are known, most companies have issues because they tend to keep even the cardinal factors 'flexible'. This is nothing, but a recipe for disaster. We systematically touch upon six key mantras that are critical for any project's success. The first mantra is on stakeholders and stakeholder management; the mantra provides an improved framework of stakeholder management, involving the identification and the basic guidelines that need to be followed. The second mantra is on the scope management in a project. It shows the essence of scope management and control. The third is on schedules in the project context and how schedule management needs to be understood in the overall environment. The fourth mantra is on costing principles. These are not on accounting, rather touch upon the basics of how costing estimates are to be derived and incorporated in the project. In many cases, creative talents demand innovative execution options. While these are to be welcomed, the costing and the estimation are to be dealt with in a slightly different way. This is elaborated in the fourth mantra. The fifth and the sixth are on quality and communication. These key areas are often treated as support faculties and ignored in conventional literature.

**Keywords** Stakeholders · Stakeholder management framework · Scope management · Schedule management · Project costing · Quality · Communication · Mantras

## 6.1 A Serious Note to Begin with!

**A Quick Hands-On**

Before we move ahead, the reader is requested to do some preliminary reading on the reasons why projects fail. After all, our cardinal question is:

**Are readings truly relevant? If yes, why do projects fail?**

© Springer Science+Business Media Singapore 2017
N. Gurjar, *A Forward Looking Approach to Project Management*,
Lecture Notes in Management and Industrial Engineering,
DOI 10.1007/978-981-10-0782-8_6

One of the best compilations on project failures is published annually by the Standish Group. While the factors do not change as much, their rankings keep changing over the years. According to the Standish Group's report CHAOS 2001: A Recipe for Success, the following items help IT projects succeed, in order of importance:

1. Executive support
2. User involvement
3. Experienced project manager
4. Clear business objectives
5. Minimized scope
6. Standard software infrastructure
7. Firm basic requirements
8. Formal methodology
9. Reliable estimates

The reason for taking this report and not the latest is that the factors mentioned here and the ranks correspond well with cross-sector experiences in industry even today. Although these are the findings that have been reported, the reader must be aware that there are a lot of areas that could be 'camouflaged'; as the causal nature of the factors and their relationships are not mentioned here.

However, all said, this survey provides us some 'magical mantras' that could well have a strong *pareto effect* providing us with a good probability of project success.


## 6.2  Wits of Experience!

An experienced project manager once forwarded a message to me. It read as the *Laws of Project Management*:

1. Projects progress quickly until they are 90 % complete. Then, they remain at 90 % complete forever.
2. When things are going well, something will go wrong. When things just cannot get worse, they will.
3. When things appear to be going better, you have overlooked something.
4. If project content is allowed to change freely, the rate of change will exceed the rate of progress.
5. Project teams detest progress reporting because it manifests their lack of progress.

*Remember these universal truths are the ones that you are likely to see, once you are done reading the book*! A lot of such 'laws' seem to shape the ground for cynicism within the community and outside. So much so as to question competence, skills, and the methods of project management. Every manager needs to be cautious about this phenomenon.

## 6.3 Formal Project Management

We are now entering the realms of formal project processes. Before we actually 'plunge' into formal methods, we need to understand the benefits of going for formal tools and methods. This is relatively easy…for the entire PMBoK® actually speaks of these methods.

1. Better control of financial, physical, and human resources
2. Improved customer relations
3. Shorter development times
4. Lower costs
5. Higher quality and increased reliability
6. Higher profit margins
7. Improved productivity
8. Better internal coordination
9. Higher worker morale, etc.

Thus, any formal method must incorporate the nuances of basic project management concepts. Literature speaks of the basic project attributes and then goes on to develop tailored concepts keeping these attributes in mind. Projects are those that have *all of the following attributes*:

1. Unique purpose
2. Temporary
3. Require resources, often from various areas
4. Should have a primary sponsor and/or customer (internal or external)
5. Should involve uncertainty.

If any of these are missing, then the entity will not be called a project! As simple and as bluntly stated here, most researchers are keen on evolving project management to a scale where it could be used in 'other faculties'!!! In other words, they are trying to 'projectize' areas like marketing, operations, etc. In doing so, they tend to change the nature of the environment or the attributes. However, practically speaking, in doing so one is actually compromising the entire concept of project management.

In this book, therefore, we will ensure that we do not allow such a compromise of the project concept and we will take up issues accordingly.

At this stage, one needs to reiterate the problems in acceptable standards for project management practice. While the PMBoK® is one such standard, it fails miserably due to the fact that the breakdown of concepts given in the book is seldom seen in reality. That is to say, there is never a 'department' called Project Scope Management. Therefore, mapping the processes given in the book to reality is often times a challenge. Moreover, the processes are not always independent of each other. For instance, time and cost management have several tie-ins. So, a conceptual framework like the PMBoK® is nice to have, but at the same time, it *has lot of room for improvement due to the loopholes seen when it comes to implementation*. The several thousands of books on project management are only a testimonial to this aspect.

## 6.4  Project Stakeholders

Maley [7] in his book has given an excellent treatment on the project stakeholder. He has gone on to lay a fundamental definition of the concept of a stakeholder and says that stakeholder management is a *skill* of the project manager. While that may be true to some extent, it could simply 'overburden' the project manager with too much of responsibility. He defines stakeholders as individuals who represent specific interest groups served by the outcomes and performance of a project or program. In his definition, he also mentions that the interest could be a positive or a negative one with respect to the project. He goes on to define five basic parameters viz.

1. *3D Type*: Which basically speaks of whether the person is a Driver (sponsor), Doer (project team), or Deliverer (beneficiary of the project).
2. *Interests*: That is rated in a matrix as either low or high
3. *Importance*: How much the stakeholder's issues, expectations, needs, and interests are related to the aims of the project
4. *Influence*: Both formal and informal power held by the stakeholder
5. *Priority*: A score that is derived from these parameters

If one evaluates this framework closely, it has three basic limitations. First, what this framework lacks is the project perspective: that is, the framework is *too close to the stakeholder and too far from the project*! Therefore, a lot of managers would find it difficult to *translate* the points into action. Second, the entire view is extremely static. This is normally not the case in most projects. The stakeholder's interest and involvement could change over the lifecycle of the project. Third, the management of interests is defined by its intensity in the framework and not the 'content'. This is again not the best way of doing this kind of an analysis.

To overcome these issues, we, therefore, define just two parameters and insist on looking at the behaviour of these two parameters over the course of the project life cycle as shown in Fig. 6.1. The framework is the Consulting Connoisseurs Simplified Stakeholder Management Framework impact here is the potential of making an impact on the project course. This is different from 'influence' that actually looks more at the ability rather than the potential. An able person might not always want to engage in changing the course of a project. However, a potential change could always be made by any person (able or otherwise) so as to change the course of the project. Hence, identifying the level of impact is essential.

The interest, as defined here, can be both positive or negative. It could include a wider variety of situations than are covered in the previous framework. For instance, a jealous colleague could have a negative interest in the success of the project. So, identifying the intrinsic nature of the interest is more essential than the intensity of the interest. A person might have a negative interest of high intensity, but if the potential to impact would be low, it should be easier to deal with such people. On the other hand, even if a person has a positive interest of lower intensity, if the potential of impact would be high, it would be a more significant aspect for the project manager to factor in.

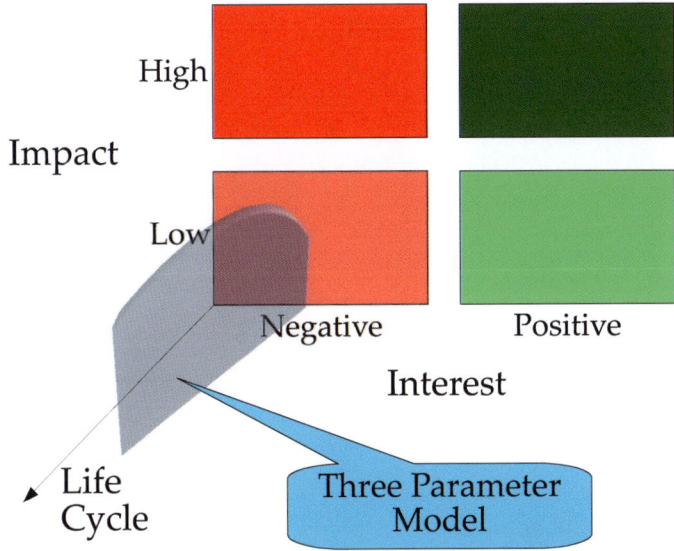

**Fig. 6.1**  Consulting Connoisseurs Simplified Stakeholder Management Framework

Third, these parameters do change over the lifecycle of the project. They say timing is everything! Well, couldn't agree more on that for understanding the stakeholders. A project manager, therefore, needs to understand how these parameters change with the entity over the period of the project lifecycle. That would help him maneuver through the challenges in the project environment.

## 6.5   The First Mantra

Always know the stakeholders of your projects. Every stakeholder can have *implicit or explicit requirements that could affect the 'scope' of the project.*

Be systematic while defining your scope. Use the systems thinking methodology to identify potential interfaces and the relationships between each of the interfaces and the stakeholders. If required, go for multiple layers of interfaces. (Ex: customer's customer, etc.). Identify the type of implicit requirements from each of these stake-holder groups.

Use this information in the framework discussed above to derive a solid approach to understand and manage your project stakeholders.

## 6.6   Project Scope Management

Scope refers to all the work involved in creating the *products* of the project and the *processes* used to create them.

Project scope management includes the processes involved in defining and controlling *what is or is not* included in the project.

The project team and stakeholders *must have the same understanding* of what products will be produced as a result of a project and what processes will be used in producing them.

The PMBoK® has beautifully elaborated the domains of scope management in terms of the following:

1. Scope Initiation
2. Scope Planning
3. Scope Definition
4. Scope Verification
5. Scope Change Control

It further lists out the tools and techniques for each of these blocks as follows:

1. Project Selection Methods
2. Expert Judgement
3. Product Analysis
4. Benefit-Cost Analysis
5. Alternatives Identification
6. Work Breakdown Structure Templates
7. Decomposition
8. Inspection
9. Scope Change Control System
10. Performance Measurement
11. Additional Planning

Undoubtedly, the most important technique is the one used to define the work breakdown structure (WBS). Most project managers would agree that the devil starts showing himself in the detail! A WBS is no exception! A Work Breakdown Structure is a deliverable-oriented grouping of project elements that organizes and defines the total scope of the project: work not in the WBS is outside the scope of the project. As with the scope statement, the WBS is often used to develop or confirm a common understanding of project scope

The WBS must be differentiated, in theory, from the other possible schema like:

1. Contractual Work Breakdown Structure: That uses the level of reporting the seller wants from the buyers
2. Organizational Breakdown Structure
3. Geographical Breakdown Structure
4. Bill of Materials
5. Project Breakdown Structure

**A Quick Test**

Take the sample project from your organization (that you just completed!)
Define the scope of the project. Detail it out as much as you can! Check the
scope for completeness. Try to DRAW a tree of the WBS. Discuss it with your
colleagues, reportees, and superiors.

All said, one should be cautious when one uses a WBS. In this context, the areas
where project management literature has serious limitations or gaps is that:

1. They are silent about WBS perspectives. One would see when one discusses the
   WBS with colleagues that their 'style of the WBS' is different from one's own. In
   other words, a WBS cannot be universally applicable. This has multiple fallouts,
   the major one being the need for a WBS Dictionary!
2. They tend to emphasize the 'difference' between the WBS and the other schema
   and *go overboard with it*! While differences are many, there are a lot of useful
   similarities that are to be significantly leveraged in an organization.
3. They do not actually state the ownership of the WBS. Who owns the WBS? For
   whom is the WBS? Why do we need to have a WBS in a particular way?
4. Most importantly, what is the *practical purpose of the WBS*? Is the discussion
   facilitating the practice positively???

At a rudimentary level, just as a car is to take you from one place to the other, the
WBS is meant to facilitate the management of a project. Everything else in it is
actually meant to satisfy this objective. Sometimes, this kind of a blackbox approach
helps the manager be more flexible and practical.

Once the reader has taken the quick test, he needs to ask himself the following
parameters to check his WBS:

**A Quick Hands-On**

**Is the WBS truly complete (NO Out-of-sight-is-out-of-mind!)?**
**Does it truly organize the Devil-in-the-Detail across multiple per-**
**spectives (Departments, Phases, Contracts/Procurement, Trades, Team**
**Members, Responsibilities, etc.)?**
**Does it help focus on the Project Plan?**
**Did the reader try to incorporate the areas of improvement that were**
**identified at the end of the project?**

This simple introspection will provide the reader with several answers...most
importantly, *the reason why WBS documents are not taken as seriously as they should*.
It is the job of every project manager to ensure that it is taken seriously.

## 6.7   The Second Mantra

Always know what you plan to do in as much detail as possible! Anything missed out will cause a snowball very soon…Remember procrastination of defining a detailed WBS is a recipe for DISASTER! Experience means you have seen it all and NOT that you are doing it better now!!!

Always organize the detail to enable the use of multiple perspectives (and levels)! Ensure it is communicated and sincerely used/followed!!! Most management literature fails to recognize and emphasize these two key points and these are the trigger points for scope issues in projects.

## 6.8   Project Plan and Schedule Management

I often fail to appreciate Project Managers who call their schedules *living documents*. Just as James Bond had the License-To-Kill, every project manager who tries to show 'life' in the plans actually believes that these are his 'Living Daylights' and he has the License-To-Kill (The Plan/Schedule) at Will:-). This is a dangerous situation.

Again here, the cardinal question for the manager is therefore:

**What are the points to be concerned with an Activity Plan?**

While this may sound simple, the answer is potentially fairly complex to implement. To understand this aspect in a little more detail, let us consider a simple example.

**A Sample Workout**

Let us consider a simple project with six activities as given in Table 6.1. Such a table is typically the starting point of any project planning process. These tables are made by the entire 'core' team of the project. However, in large programs, the practice is usually such that the tables are made by the planning team with their own experience. In any case, it is the starting point to the planning process.

The reader is advised to use a regular project planning software (like Oracle Primavera, Microsoft Project, etc.).

The next step is to verify this table. This is the *most difficult* part of the project planning process. In any case, it needs to be done.

In this example, we will show you variations of the table above to emphasize this point. The project plan now appears to be similar to the Gantt chart shown in Fig. 6.2. The duration of the project is clearly derived as 14 months.

On further introspection, the team realized that this same project had some finer aspects that could be used to 'optimize' the schedule. The relationship table was now changed to the one as shown in Table 6.2. Clearly, this exercise has changed the overall project scenario. The new project has now got a different timeline as

**Table 6.1** Activity table of a project

| Activity number | Duration (in months) | Predecessor | Relationship with lead/lag |
|---|---|---|---|
| 1 | 3 | – | – |
| 2 | 4 | 1 | FS |
| 3 | 3 | 2 | FS |
| 4 | 5 | 1 | FS |
| 5 | 4 | 4 | FS |
| 6 | 2 | 3,5 | FS |

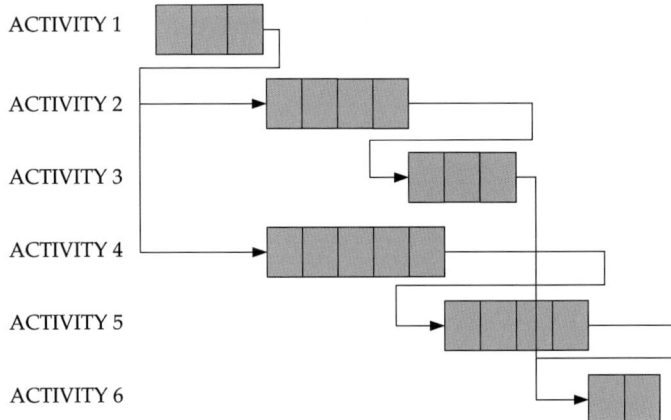

**Fig. 6.2** Project Gantt chart/timeline

**Table 6.2** Activity table of a project after introspection

| Activity number | Duration (in months) | Predecessor | Relationship with lead/lag |
|---|---|---|---|
| 1 | 3 | – | – |
| 2 | 4 | 1 | FS(−1 Month) |
| 3 | 3 | 2 | SS(3 Months) |
| 4 | 5 | 1 | SS(1 Month) |
| 5 | 4 | 4 | FS(−2 Months) |
| 6 | 2 | 3,5 | FS(−1 Month) |

represented in the Gantt chart in Fig. 6.3. The project duration is now 9 months instead of the previously planned 14 months.

When the Director reflected his views on the introspected schedule, he reduced 0.5 months across all activities in the project, claiming that there was too much of a buffer. And he added an adjustment activity at the end of the project of 0.5 months as a separate activity.

**Fig. 6.3** Revised project
Gantt chart/timeline

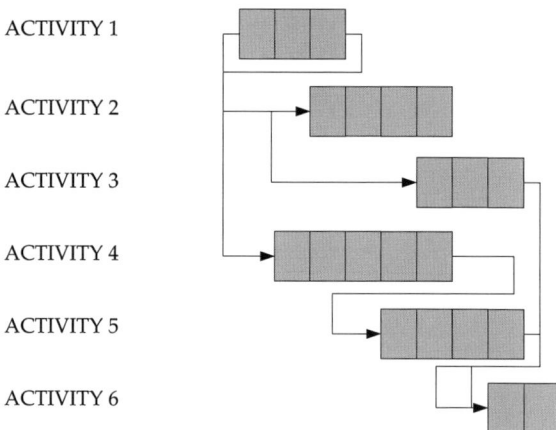

ACTIVITY 1

ACTIVITY 2

ACTIVITY 3

ACTIVITY 4

ACTIVITY 5

ACTIVITY 6

Clearly this further changes the project scenario. The reader can easily verify the outcome.

---

**A Quick Test**

Let us conduct two tests here.
　Define Activities in your project. Define their interrelationships. Estimate durations.

**Identify 'Hard' and 'Soft' relationships. A 'hard' relationship is one that is a must happen type of relationship. A 'soft' relationship is one that may not be required, but is preferred. However, at some point, a 'soft' relationship becomes a 'hard' one. Can you precisely define when a 'soft' relationship becomes 'hard'?**
**List possible causes of duration variations. What caused duration variations in your previous projects?**

---

## 6.8.1   Two Problems to Avoid

Most project managers think like the commercially available project planning software. They tend to have a lack of focus on the 'soft' relationships making every relationship 'hard'! This actually skews the critical parth and the critical chain in practice. Hence, it is necessary to be conversant about how one is to use the Critical Path and the Critical Chain the right way.

The second aspect is that of improper estimation! There seems to be a Knowledge Gap in Estimating that is plaguing the industry. Every project manager needs to know that 'realistic' estimates come at a premium over 'beautiful' estimates…And in doing so, the project manager must understand the dynamics of the management game called 'fine tuning'!!! I recall an incident in a refinery where there was a Director who just would not approve employee applications for leaves. Anytime an employee asked for leaves, he would just cut it short to half the requested duration. And each time he did that, he would get some 'sadistic' pleasure when the employee would 'bitch and moan and cry'! One fine day, a fresh trainee applied for leave and he cut the duration by half…the trainee was happy and he was about to leave. When the Director asked him why he was not sad, the trainee replied that I knew that you would cut it by half…and so, I requested for twice the duration in my applications:-).

In another experience, the Project CEO called me to his office one day and asked me to explain the concept of the buffer! He wanted to understand what exactly was being communicated in the schedule. Not many executives would express their ignorance and want to overcome it this way. I was glad at the end of the day that the CEO was keen on understanding how the projects were being managed and the tools and techniques used in planning.

## 6.9 The Third Mantra

Remember: The lesser the project schedule changes, the more the chances of success!!!

As a manager, the schedule is your true reflection…Its as realistic (and practical) as you are!!! The more mismanaged the project gets, the more 'dynamic' and 'living' the schedule will be!

If key project personnel do not understand schedules, train them on day one!

## 6.10 Project Costing

Another challenging area in modern project management is project costing. Just as planning, costing too is a critical issue. However, there is a slight difference when we consider costing. While time is purely an 'executing manager's' perspective, the costing has a lot of overlaps between the accountants and the financial experts. Consequently, there are practices that are derived from accounting operations and there are concepts that are derived from finance. All said, due to this fusion, there are a few tricky issues that are often left to the project manager.

From a project manager's perspective, it is important to have *enough money to complete the project at hand*. However, in many situations, one sees project managers having a tough time due to budget issues, fund availability issues, etc. These get further complicated when the pay-outs are due. In some organizations, the legal process takes over with court summons, lawyer meetings, etc., that further delay the project.

From my experience, though, most project managers are appointed after the project is approved. This always has serious repurcussions and the budgets, at times, seem inadequate. However, even when the project manager is appointed sufficiently in advance, the budgets *still seem inadequate* at times. What this essentially says is the following:

**How realistic are the budgets?**
**And how aligned are they with the execution plan?**

In most projects, overruns are common. Which only means that either the budgets are not realistic enough or that the choices made to execute the project are different from those used in deriving the budget.

---

**A Quick Test**

Develop a cost breakdown for your project. Work on it in as much detail as possible.

Next, develop a cost tracking method for the project. Use a 'bottom-up' approach to calculate the overall cost tracking curve in the project.

Discuss the gaps in the two with your colleagues, your subordinates, and your superiors. Try to 'baseline' both the exercises.

---

### 6.10.1  The Pseudo Management Problem

In many companies, the project manager just 'inherits' the project. What is worse is that the assignment/handover of responsibilities is not done in a professional manner. In other words, the project manager comes in and is loaded with several files without adequate explanation. The only attribute that is seriously considered is his availability for the work and his willingness to do the job. Capability is often an 'assumed constant':-) All the specifics are deemed to be outside the 'need-to-know' faculty! Or probably, things that he should 'discover' or even 'stumble upon' as he goes along! This is a resource-based concept. It is most times, therefore, not a management concept.

Second, there are many project managers who are given a more 'inclusive' role and are assigned the budgets and all the good things required. However, here the project managers are not *strong enough to understand how to leverage the situation*. The end result is that the Accounts Department and the CEO are forced to take on his role (or the 'brunt' of his incapability) and the resulting situation gets more and more chaotic as it starts 'bureacratizing' the project environment in an attempt to make it better managed. The resulting situation is that the project manager is totally out-of-sync with the project management process and hangs-on due to his past/present goodwill.

If one tells a project manager the following, what do you think his response would be?

**You are appointed as the PM to a USD 1 billion project. It involves the IT Systems Implementation and Transformation of the entire back-office at our bank. Your team is on the 4th Floor. Any questions?**

What do you think your own response would be? Given a budget and the nature of the role, most project managers think that is *all they need to know*! This is the irony of the situation. Most project management literature does not focus on what the project manager should do next!!! I am sure, you all agree:-)

Hence, two aspects that are important from a project manager's perspective on the costs and such a system are the *professional assignment and the handover*, which brings us back to two fundamental sets of questions:

**Can I 'own' the budgets?**
**How reliable are the estimates?**
**Do you verify them with the market?**
**Do you incorporate global changes like inflation, interest rates, insurance, etc.?**
**How are you proposing to control the costs and the schedule?**
**Budget versus Actual**
**Earned Value Method, etc.**
**What *exactly* does the project manager do?**

These are often important, yet ignored. A proper assignment/handover involves a clear understanding of all these issues. And more importantly, the project manager must be *resourceful enough* to get these things in place; and rather quickly at that. In other words, what is required is a smooth and sound bridging of the 'management'– 'resource' gap.

**A Quick Demonstration**

While much of the discussion appears theoretical, project managers who have burnt their fingers do understand the critical message that is being conveyed. However, it would clarify the situation further with a simple demonstration. A particular project has three ways of execution. Since this is a purely financial treatment, we only focus on the costs in the demonstration as given in Table 6.3. The first option is the conventional execution plan that has been used in the company. This plan is fairly well known. The second alternative is a new one that has come in the market. It is relatively less known at the time. The third too is another method.

**Table 6.3** Project costing for a sample project

| Execution option | Cost estimate in USD million | Reliability in % | Interval range for final costs in USD million | Maximum value of costs in USD million |
|---|---|---|---|---|
| 1 | 100 | ±10 | 90–110 | 110 |
| 2 | 90 | ±30 | 63–117 | 117 |
| 3 | 110 | ±15 | 94–127 | 127 |

Now, the choice of using a specific estimate is an important one. Which one would a project manager use? That is a tricky question for many, especially when the overall scenario is evaluated. However, there is a simple thumb rule here that is summarized as the fourth mantra.

## 6.11  The Fourth Mantra

Understand execution options in terms of their costs; always take the most reliable data for the execution option in the budget! Anything lower, through any other way, is a definite bonus, but shouldn't be used to make the calculation over-optimistic.

Always follow a combination procedure: Develop the scope 'top-down' and determine the costs using the 'bottom-up' approach; a fully informed 'top-down' helps estimate well and a 'bottom-up' approach helps in better cost control! Reconcile them eventually!!!

## 6.12  Project Quality

From the very tangible components of project scope, time and cost, we now move to the important, yet 'softer' issues in project management. One of the issues is that of quality management. When I talk to many Married But Alone(MBA) project managers, they say that they realize their wife's role only when she isn't around. So is Quality! Quality is important, but its importance is felt only when it is gone:-)

Most project managers are more concerned with 'results', and to such an extent that they openly admit that they *do not care how it is done as long as it works reasonably*. This thinking is more inclined toward the Quality as a Product Attribute concept. Project managers from the execution background tend to strongly go by this concept. However, it is important to understand that results are important on the one hand, but it is even more important, on the other hand, to have an integrated system. This is analogous to that of a car that is made of several engine parts. Each part is made and the car is assembled. However, if the parts do not have the desired smoothness, the resulting assembly would see a lot of losses. This is typically in the form of friction, abnormal wear, compromised performance, etc. In the UK, sometime in the 1970s, it was realized that about 40% of their energy requirements were actually for overcoming friction. This was when, Tribology as a subject, got serious attention. Similarly, in formal environments, Quality as a System is an extremely important concept. Its not the product alone that matters, the process that goes across the spectrum of project operations too needs to maintain quality. Many organizations are trying to take it further to the point where quality integrates as an ideology, making the faculty 'omnipresent'.

---

**A Quick Test**

Define the term 'Quality' as used in the context of Project Management in the sample project.

**How was the Quality measured? Who measured it? Who reported it? How was it controlled? What exactly did you do when you say that you 'controlled' it?**

List three measures through which the quality could have been improved in your previous projects.

---

## 6.12.1  Quality: An Ignored Concept

At the back of everybody's mind though, quality is understood as a 'support function'. And many people think Quality is finger-pointing, when actually it is meant to avoid that! So, when one asks a project manager the question:

**Who is responsible for Quality in your Project?**

Chances are that he would name some person XYZ in the organization. That is the irony of the situation. Try thinking of the parallel where one holds the traffic police responsible for a traffic violation that one just made! Interesting scenario, isn't it??? So, this is essentially an understanding issue.

Another critical parameter is that a lot depends on the commitment of the top management; hence, ensure you make the costs of quality clear at the outset. And that the top management is having the same 'understanding' of quality as the project manager has. In many companies, quality is seen to be synonymous with certifications like ISO, CMMi, JCI, etc. That is again a ridiculous way to start analyzing the situation. However, this is a specific aspect of quality philosophy that is best dealt with in the advanced books on quality. We will, therefore, not overdrive this discussion at this stage. Fundamentally, the project manager needs to understand that the quality management system is like a traffic rule book. If one follows it diligently, the journey would be smooth. For those who cannot appreciate this as much, I would suggest you drive a car in Switzerland and compare that experience with driving the same vehicle in India or Mexico!

If your project has multiple handovers, you are bound to see quality issues costing your system a lot of 'coordination' costs. Very few companies acknowledge these costs or even make an attempt to evaluate these costs in the right spirit or method.

## 6.13   The Fifth Mantra

The points made are simple: Ensure that your quality standards, practices, and procedures match the policy.

Make sure that the quality aspects are usable!!! Always keep the quality documentation simple, handy, and easy to use. I have seen many companies having several big books like the 'bibles' of quality. They are hard to go through in the time of need. What is worse is that they are not always accessible. Some copy might be available at the executive's office. Most people down the line do not have their own copies. Still complicated? Yes, because the documentation most times is written by the 'quality' guy who doesn't always know the problems or the questions of the user. So, the way it is organized is a major issue. The mantra, therefore, is self explanatory.

The other aspect that one needs to focus on are Quality Documents pertaining to Assurance and Control. Every project manager must ensure that the Assurance and Control Reports are *easy to search.* Many organizations just keep a pile of it without giving the details of what is searched and where and how! This inability to use the documents reduces their 'intrinsic' value!

## 6.14   Project Communication Management

Any project manager will surely agree that good communication is probably the biggest foundation block of the project's success. What is interesting, however, is that most companies do have a formal reporting plan. So, be it weekly reporting or monthly reporting or whichever, companies often have the details spelt out. In fact, many plans also have the structure of the report and the recipients of the report. However, they *do not have a formal communication plan*! A communication plan is different from a reporting plan. Although both are designed to meet the requirements of the stakeholders in any project, a communication plan is more decision-centric while a reporting plan is more of a post-mortem; if we look at the 'content' of information.

We will look at these issues in detail in subsequent chapters. Our attempt here is to stick to the 80 : 20 rule so as to help the forward looking manager decide and move forward quickly (remember, these are the *basic* mantras out here).

### 6.14.1 Pitfalls of Improper Communication Management

PMBoK® says that there has to be a formal communication plan. A formal communication plan needs to be diligently followed to avoid making the project a political minefield! Many times, the project manager is less clear about his decision variables and the way in which he needs to communicate across the project. The end result is that some stakeholders are often working on things that are not particularly aligned with the corporate objectives. This also starts creating an 'inner circle' that almost always has disastrous consequences. I recall interesting examples here. The project manager had four team leads and he would discuss issues over lunch with one of them. They would also 'decide' a few things during their discussions. The other two team leads were being left out and the decisions were leading to a lot of dissatisfaction. Over time, these decisions were not always conveyed on time and the whole system started getting chaotic! Hence, a formal communication plan is important.

Another important situation occurs when a project is delayed. Once a delay is reported in any project, it is like 'the devil gets on the manager'. Every senior manager starts getting concerned and is looking at multiple ways to work around the delay and ensure there is minimum damage to the company. However, most often than not, there is no transparence on the progress of such corrective/follow-up fraction. This is one of the reasons why, despite having good communication, the project fails to come back to the 'green status'. It is, therefore, imperative to use a system that is transparent on the progress of corrective/follow-up action.

**A Quick Test**

Take the sample project from your organization (that you just completed!)

**Define three major problems that you encountered. Specifically look at: Who were involved in the resolution? When did the problem occur? When was the problem reported?**
**What mode was used to report the problem?**

## 6.15 The Sixth Mantra

Always prefer to communicate in formal (record-able) methods only. Informal methods must be avoided as much as possible. Your communication plan (internal and external) must be formal!

If the communication plan is not followed by a project member, *replace him! This also applies to the senior management working on any project*!

Another important and powerful mantra is that one must ensure all decisions are formally communicated by the decision maker! Many times the CEO decides, the senior management conveys it to the project manager and the project manager writes to the team informing them. This system actually makes the management task more like that of a 'postman' and less like that of a 'manager'. Such situations are complex and waste a lot of resources under the cost head of 'coordination'.

## 6.16   Key Takeaways

This important chapter actually equips the Project Manager with critical mantras that solve over 80 % of common project problems. They are extremely powerful and globally tested and they work wonders in any project. As a project manager, one needs to understand how to implement them early-on in one's project. Since this book is all about forward-looking paradigms, every decision is based on the circumstances that a project manager finds himself in. It does not look at any kind of a postmortem analysis to explain what *went wrong*, rather an evaluatory framework is always presented to explain what *could go wrong*. At any point in time, therefore, only that potential information made available to the project manager is used to explain the situation.

The reader is expected to work out all the exercises in detail and present the same to his team and his superiors. Each exercise would typically take an hour of the project manager's time, but would save several potential hours and dollars in any project. Hence, we would encourage you to do it yourself, or with your team, typically with a facilitator, who understands what to do next, given a particular set of responses.

# Part III
# Management and Business Review

# Chapter 7
# Second Level Review-I

**Abstract** In this chapter, the focus shifts to the project tools. Often, the project tools are operationally treated 'larger than life' but from a management perspective, they are seen as 'operational enhancements'. The mantras touch upon how one needs to balance both the perspectives. We then move on to understanding a typical business situation involving an 'on-time' and, an 'on-budget' project requirement with changing boundary conditions. In doing so, we come across various concepts such as the critical path analysis, crashing strategies, and costing challenges in Schedule Optimization. We introduce the Quick Solutions three-parameter-framework of Consulting Connoisseurs for risk management. The chapter then demonstrates how modeling and simulation could also be used to evaluate risks. In the process, we touch upon how the reworking of budgets needs to be done and how risk management needs to be factored in the corporate context in terms of probability and costs using modeling and simulation. In doing so, we have also delved into probabilistic evaluation in a practical scenario and have cursorily touched upon the PERT methodology as well.

**Keywords** Project tools · Mantras · On-time · On-budget · Case study · Critical path analysis · Crashing strategies · Costing in crashing · Three-parameter risk management framework · Modeling framework for risk evaluation and management · Schedule optimization · Reworking project budgets · Risk management · Corporate risk norms · Probabilistic evaluation · PERT · Project post mortem analysis

After learning the six mantras of project management, the project manager needs to understand how to incorporate his learning from this book so far, into his own project ecosystem. This is not difficult, but does not take place directly.

Moreover, the overall understanding of the previous part was toward the 'process-centric' view presented in PMBoK. So, every project manager must understand that the implications of all the mantras were largely on the process areas.

© Springer Science+Business Media Singapore 2017                           103
N. Gurjar, *A Forward Looking Approach to Project Management*,
Lecture Notes in Management and Industrial Engineering,
DOI 10.1007/978-981-10-0782-8_7

## 7.1  Reengineering the Processes

One has to be cautious while discussing any changes or even re-engineering any process. This is essential for two simple reasons:

1. When it comes to changes, everybody loves to be asked! Well, let me qualify it further at this stage. Recommending change to someone else is like suggesting make up tips to make the other's wife look more beautiful...
2. Anyone can 'talk through their hats.' Therefore, everyone *thinks* they are half consultants!

It is only the project manager, who is going to *responsibly incorporate* the changes in the processes, who understands the actual problems in this situation. Nevertheless, change is an essential process of learning. Which is why, after we have completed the first two parts of the book, it is time to review and understand what can be changed.

**A Quick Hands-On**

List two changes that you propose to implement in your project processes.
How and when do you propose to implement them?
How did you evaluate the appropriateness?
How will you ensure the sustenance of the changes? How will you choose to control the outcome of the changes to your desired outcome?

The reader is advised to commit himself to do this as an *exercise that is to be undertaken over the next two days without reading further*. Usually, it is observed that people take a 'good night's sleep' before they can think clearly about what needs to be done. Therefore, we are advising our readers to do the same and come up with reasonably good and convincing answers before they move ahead into other concepts.

As we are recommending a 'break' at this stage, it would be great to receive your feedback on the book so far! You are invited to give your inputs to make subsequent editions more effective.

## 7.2  Project Management Tools

We spoke about recruitment practices earlier and, therefore, we need to touch upon Tools as well. Most recruiters, as we mentioned earlier, are keen on focusing on issues related with processes and tools. Hence, from a 'pseudo recruitment perspective', we need to briefly discuss about tools. We will mention a few ideas here to explain a few quick takes.

### 7.2.1   So, What Is a Tool?

I have spoken with many managers and team members and the one thing they fail to clearly identify is the concept of the project management tool.

1. *Tool*
   Any instrument or apparatus necessary for the efficient prosecution of one's profession or a trade.
2. *Technique*
   Working methods or manner of performance as in arts, sciences, etc.
3. *Method*
   A general or established way or order of doing anything

This differentiation is important and we will quickly see the same in our subsequent discussions in later chapters.

### 7.2.2   Some Popular Tools

The most popular tools used in project management are project planning tools. These tools help the project manager understand and model the scope, the time, the cost, the resources, and the organization. Their primary focus is on execution and reporting. Hence, they are also called scheduling software. The most popular of them are Oracle Primavera and MS Project…, although there are a host of others that are in the same domain. Today a lot of freeware is available. For instance, Linux has versions like the Planner Project Management. Each of these has certain advantages. However, as a brand, perhaps Oracle Primavera has a position that is way above the rest.

The second most popular set of tools used in project management are project accounting system tools. These tools are like the accounting ERP solutions that are used in project management. Like any ERP, these are 'mega-systems' and they bring in their own rigid constraints; mainly due to the fact that they are normally tying into the remainder organization's ERP solution. Nevertheless, they are extremely popular today. Oracle Apps, SAP…and many such solutions exist today. However, these tools are yet to gain the *populist votes* from modern day project managers, who work on a wide spectrum of projects. Most project managers even today, feel that accounting tools are meant for the accounts department rather than for the project manager, to help him in his function. This understanding (or the lack of it) has made the project managers shy away from using such powerful tools to assist their functions.

From a 'traditional' project manager's perspective, therefore, the second most set of tools are the project document management system tools. We saw that communication is an important issue in any project and it is *believed that* a good project document management system is like *half-the-battle won*!. Some of the popular tools in the market include EMC Documentum, Oracle Primavera DMS… Their main function is organization, storage, and retrieval of project documents in an effective way.

In the connected world, information technology solutions have improved a lot over the years. New generation project management tools are project collaboration software tools. These tools help the project manager to share across organizational entities and bring in a systematic control to the *flow of information across legal entities.* Some popular ones include Conject, MS Project Server, Oracle eProcurement… Each of these have their own functions and positioning. Ideally speaking, they should be further classified according to the areas they support. However, as this is just an overview, we are not going to delve deeper into this here and would reserve that treatment for later.

Moreover, IT based tools keep changing with newer versions coming each year. So, it won't be meaningful to include too much of content (in this book), that could get outdated rather quickly, from the reader's perspective.

## 7.3   Mantras for Project Management Tools

Many times the project manager is not particularly aware of the tools he should use or needs to be using in a particular project. In such situations, there is often an expectation mismatch as far as the output is concerned. Worse, the project manager does not understand the 'efforts' required to develop or maintain or even use a given system in such a situation. The 'field' project manager fails to even appreciate the need and the support of the tools at times. So, in line with our methodology, we are defining new mantras (for the next generation project yogis! :-) (Fig. 7.1). However, this is the last of the mantras, as we know that too many mantras would make it hard for the project manager to read, remember, and implement. Hence, it is extremely important to draw the right balance.

There are a few quick mantras that help project managers in such circumstances:

1. Define a formal strategy to use the tool (extent, functionalities, users, access, etc.).
   Usually such a document is made, but it is not read, managed, and meticulously

**Fig. 7.1** More rules!

*At this Rule of Mantras,*

*I could well be called a*

*'Project Yogi'*

controlled by the project manager. Therefore, the gap between the objectives and the execution gets conceived right at the beginning.

2. Never start using the tool without adequate preparation. Quick-and-dirty implementations cost the company a lot; usually a reimplementation.
3. Train your personnel for best management practices for the given tools.
4. While every software guru will preach workarounds, use an alternative system for functionalities that are not built-in! This is because workarounds hamper upgrades and patching and, most importantly, *are trying to use the software for what it is not intended or designed for.*
5. Getting the system in place requires more than just the software. The positive involvement of all the concerned team members is a prerequisite, just like the proficiency of the lead tool architect (resource).

Let the tool stay a tool! The project tool should reflect your thoughts and not vice-versa… Ensure that you use it to verify your decisions, unlike basing decisions on the system. These might be common knowledge, yet very uncommonly used…:-)

## 7.4  Understanding the Management Context

The most popular discussion in any project management situation is the following:

### How to complete the project On-Time and On-Budget?

Every practicing manager would agree that any mention of the word 'Delay' or the word 'Overspend' in any meeting would suddenly shift the focus from the general agenda to these two areas only! Yet, most project management literature tends to oversimplify these areas. We will try to get a feel for these areas now.

## 7.5  A Sample Case

Since we are dealing with a generic concept, we will refrain from naming the activities with real word mnemonics and would continue with a generic naming. Let us consider a typical project situation as given in Table 7.1. As a best practice, we will consider only Finish-to-Start Hard Relationships. The reader is requested to create this project.

The management directive is to complete this project in 31 units.

This is a typical kind of a schedule problem. I am sure most managers have already gone through this situation at least once in the project life cycle. Although one does not know what the activities A, B, C, D, E, and F really stand for (in this example), any real-life project can have some relation there. For instance, A could be the engineering phase of the project. B could be the actual development cycle or the manufacturing cycle for the equipment. D could be the site-related activity (foundations, etc. for construction or preparing the data center for a software project).

**Table 7.1** A sample project

| Activity | Predecessor | Duration in units |
| --- | --- | --- |
| A | – | 7 |
| B | A | 10 |
| C | B | 10 |
| D | C, F | 8 |
| E | A | 9 |
| F | E | 10 |

And so on. Therefore, even if we do not give the mnemonic name to an activity, we do know what this could represent in a 'typical project environment.'

Having said this, it is possible to think of strategies to modify the project plan to think of ways in which one can reduce the durations. In doing so, one has to question the assumptions made in the plan. In this example, we are not going to enter into the specifics of the assumptions, but would try to discuss the alternative strategies that the project manager can come up with.

### 7.5.1   A Look at Potential Maneuvers

Thinking in rather abstract ways, and, *given just the above information of the project*, one can come up with a variety of options. Let us discuss them here. Our aim, in doing so, is to give the project manager an idea of the kind of information he is looking at, and, sensitize him to the view through the 'eyes' of other stakeholders. These are not accurate in any way, but give the manager a logical thought process behind the action. Like we said, this is a book on *forward looking* concepts. Hence, at any point in time, one needs to understand what are the knowns and what are the unknowns and try to make the maximum out of it.

**Equal Reduction Across the Project**

One of the most popular 'management methods' is to go for a compression across the board; for all activities. This again has, the following advantages:

1. Since it is across the board, nobody is gaining any favoritism. It is a case where pressure is on …for everyone to perform.
2. It is simple to 'advocate.'
3. It is easy for the accounting team to make changes and monitor.

Disadvantages are many. I am sure every project manager can summarize his own version of the views.

**Equal Proportional Reduction Across the Project**

The next variant is again a popular one. In this, the reduction is again across the board, but proportional to the duration of all activities. Hence, if an activity takes 3 days and its successor takes 9 days and if we are to reduce 4 days from the total duration, one would reduce 1 day from the first activity and 3 days from the second activity. This again has, the following advantages:

1. Since it is across the board, nobody is gaining any favoritism. It is a case where pressure is on for everyone to perform. However, in practice, this might not be believed to be right.
2. It is simple to 'advocate'.

Disadvantages again are many and the reader can reflect on them as he goes ahead.

**Reductions Along the Critical Path**

Most project managers are aware of the concept of the critical path. Hence, as the theory says, it makes sense to go for a compression of the critical path. This approach too has many takers as it is believed to have several advantages:

1. It is not across the board, avoiding a total mess of the plan. Nobody is gaining any favoritism; yet the critical path that is truly critical will have 'focused management'.
2. It is believed to be easier to 'implement.'

Again, I would leave it to the reader to check the disadvantages.

**Squeeze the Initial Activities**

Most project managers who are from the deployment / construction background would agree with this strategy. The advantages of using this logic include:

1. Initial activities have lot of buffers. Most often than not, the initial project team uses up all the available buffer!
2. Activities at the beginning of the project have less amount of 'cross dependencies', thereby, making it easier to expedite the process in the project.

The reader is advised at this stage to verify from his own experiences, situations where the logic of each of these strategies was found to be true.

**Squeeze the Longest Activities**

Again in most projects, the longer activities are most times on the critical path, and they are often good 'targets' of schedule optimization. The general thinking is that:

1. These activities are usually later in the project, providing one with a better chance for planning a compression.
2. Bigger activities have bigger buffers (potentially) and can accommodate the compression desired.

The reader may want to verify these claims for his own projects.

**Squeeze the Last Activities**

This is probably the *most common way in which project managers operate today*. The logic is argued as the following:

1. As time passes in a project, there is a better handle of the unknowns. Therefore, a compression of the last activities is always free from any typical assumption that is done otherwise (in other strategies).
2. Following this strategy will allow the manager work in a less hectic environment in the beginning and will reduce the potential 'interference' from senior management and other stakeholders.

Again, the reader is asked to verify these claims with respect to his own experience.

**Refine the Activities**

A very popular move by any project manager is to resend the project to the planners and ask them to rework the entire plan with a much higher degree of detailing. The logic adopted here is as follows:

1. Try to trigger successors before the modeled Finish-to-Start relationships given in the current plan and schedule.
2. Make an attempt to recheck the dependencies as the manager feels there might be room for changing them (in other words, they maynot be true).
3. The project manager might have been a 'hero' in one of his past projects. Thereby, making him want to be the *super-hero* this time. (although the risks and the conditions in his previous project might have been far too different from the current one).

Essentially, these strategies have been defined by us *without knowing enough about the project*. Typically, a senior management project stakeholder could be one such person. He often knows little of the project, yet calls the shots. So, the project manager needs to understand that it is possible to use some kind of logic based on certain assumptions and provide directives for the project. How far each of these is applicable and appropriate is debatable, but it does happen in practice.

The second aspect is that the project manager himself…can have a strategy developed based on his past experiences. This again is an exercise for the manager himself at this stage. Learning good strategies from a unique set of poor and good experiences could be challenging!

---

**A Quick Test**

Identify your own unique strategy to reduce the duration to 31 units.

Explain the logic and describe the same. We will use this in our future discussions in subsequent chapters.

Also try to pin-point the *point-of-ignorance* in each of the strategies mentioned above.

## 7.5.2 More Information to Decide Better!

The reader has to understand the use of 'additional' information in the project context. This is easier said than done in most cases. In this case study, we try to develop this concept in a more rigorous way. At this stage, the reader is advised to take the base case and then, show the results of the quick-test as two separate scenarios *before he reads any further*. This is essential as it is meant to help the reader / project manager conceptualize better.

Now, we will see how superior knowledge starts working in the project context. The additional knowledge is now given in Table 7.2 and the maximum compression possible is restricted to 25 % of the remaining duration (Rounded: Greatest Integer). That means, if an activity is of 10 days, the maximum compression possible is [2.5 days] = 2 days. If an activity is of 12 days, the maximum compression possible is of 3 days.

Given this information, the project manager is now asked to strategize and 'optimize' the compression. The entire problem suddenly appears simple and is easily amenable to a simple schedule optimization. The original schedule looks like that given in Fig. 7.2. This schedule clearly has 35 units of time. The mandate is to bring it down to 31 units of time. The critical path is marked in red while the Non-critical activities are marked in yellow.

Given the constraints, we need to develop a new plan. The first step is to understand which of the activities could be candidates for our process and then we need to understand the costs. The activities of interest are given in the subsequent Table 7.3. Clearly, the first pass is to identify the critical path and set the non-critical path aside.

Next we take these 4 activities and try to understand how one could save 4 units from them. This is simply done by checking those activities that are the cheapest to

| Table 7.2 Cost of individual activity compression in the sample project | Activity | Cost of compression (per unit) |
|---|---|---|
| | A | 2 |
| | B | 1 |
| | C | 4 |
| | D | 4 |
| | E | 5 |
| | F | 5 |

**Fig. 7.2** Gantt chart of the
sample

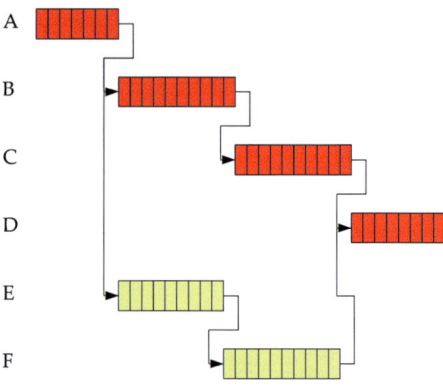

**Table 7.3** Identifying
candidate activities in the
sample project

| Activity | Maximum allowable compression in units | Of interest (i.e., on critical path)? |
|---|---|---|
| A | 1 | Yes |
| B | 2 | Yes |
| C | 2 | Yes |
| D | 2 | Yes |
| E | 2 | No |
| F | 2 | No |

compress. They are, obviously the priority. Therefore, we next try and rank them as
shown in Table 7.4.

Now in this ranking, the project manager needs to take the minimum rank and
assign it the maximum duration possible. This is shown in Table 7.5. We see that 8
units of costs are *additionally* required to compress the project. Since C and D were
at a tie, we have chosen D for the compression.

**Table 7.4** Ranking candidate
activities in the sample project

| Activity | Maximum allowable compression in units | Cost per compression | Rank |
|---|---|---|---|
| A | 1 | 2 | 2 |
| B | 2 | 1 | 1 |
| C | 2 | 4 | 3 |
| D | 2 | 4 | 3 |

**Table 7.5** Allocation of compression

| Activity | Allocation of compression in units | Cost per compression | Activity cost of compression |
|----------|-------------------------------------|----------------------|-------------------------------|
| A | 1 | 2 | 2 |
| B | 2 | 1 | 2 |
| C | 0 | 0 | 0 |
| D | 1 | 4 | 4 |

This is an interesting situation because the additional compression has come at some cost. The resulting schedule is shown in Table 7.6. This schedule looks like the Gantt shown in Fig. 7.3. Unfortunately, the duration of the project still exceeds the targetted 31 units! And there is now a switch in the critical path of the project as shown in the figure! The path AEFD is now longer and has 32 units. This means, we now have to repeat the process for the new critical path. The resulting procedure is again highlighted for the path AEFD. Note that the steps will now focus on how to minimize AEFD. Of the two activities A and D, A has used up its compressible duration while D still can spare 1 unit.

**Table 7.6** The revised schedule after the first pass of the sample project

| Activity | Predecessor | Duration in units |
|----------|-------------|-------------------|
| A | – | 6 |
| B | A | 8 |
| C | B | 10 |
| D | C, F | 7 |
| E | A | 9 |
| F | E | 10 |

**Fig. 7.3** Gantt chart of the sample after the first pass

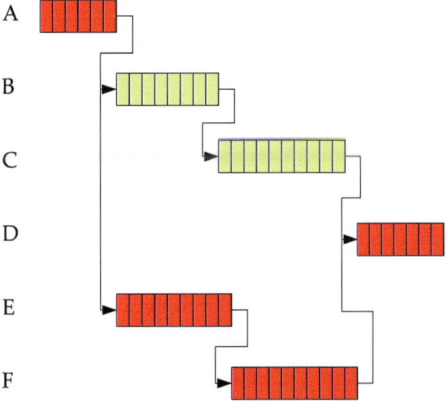

**Table 7.7** Ranking candidate activities in the sample project in the second iteration

| Activity | Maximum allowable compression in units | Cost per compression | Rank |
|---|---|---|---|
| D | 1 | 4 | 1 |
| E | 2 | 5 | 2 |
| F | 2 | 5 | 2 |

**Table 7.8** The revised schedule after the second pass of the sample project

| Activity | Predecessor | Duration in units | Additional costs by activity |
|---|---|---|---|
| A | – | 6 | 2 |
| B | A | 8 | 2 |
| C | B | 10 | 0 |
| D | C, F | 6 | 8 |
| E | A | 9 | 0 |
| F | E | 10 | 0 |

**Fig. 7.4** Gantt chart of the sample after the second pass

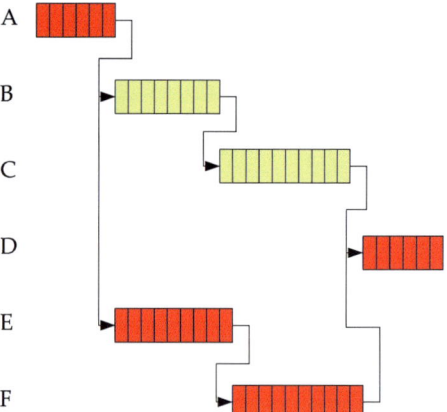

Hence, we need to look at E, F, and D only in order to gain the 1 unit. Again, a ranking needs to be done. This is shown in Table 7.7.

Again, D is the activity that can be used for the compression. The additional cost is another 4 units. The revised schedule is now shown in Table 7.8.

The *total additional cost for the project is now 12 units* and the duration is now reduced to 31 units. The Gantt chart is represented in Fig. 7.4. The duration is now reduced to 31 units. This is an *algebraic derivation*. Naturally, projects in real life are far more complex than the one described here.

But the critical aspect to note is that the project critical path has changed! Now this is a very important point to note. In any given project, there is a 'mathematically derived' critical path and there is a critical path 'by norm.' Any project manager goes for the normative path that is essentially reinforced through the years of experience he has had. Therefore, when such an optimization changes the norm, it leads to a lot of discomfort among the project circles.

Therefore, although this case is technically 'solved' at our end from a mathematical perspective, it is far from being solved from a 'practical' perspective.

In order to solve this issue, the project manager needs to retain the normative critical path as the final critical path! This means that the duration of B and C together must exceed that of E and F put together. Thus, the work-out is bound to be further different, if we consider this as an additional *implicit requirement* while compressing a project. This logic is true as far as we do not change the execution options (say manual coding versus CASE tools). However, once the execution options are changed, the discomfort should 'vanish' or at least reduce. Unfortunately, this is a situation where the *experience of a project manager goes against him*!

---

**A Quick Hands-On**

As a project manager, consider the following questions:

**What is the critical path before the application of the compression?**
**What is the critical path after the application of the compression?**
**As the project manager, how are you going to present your case to the management?**

---

If one is to keep this implicit requirement in mind, the solution would be similar to Table 7.9 that is reflected in the Gantt chart shown in Fig. 7.5. In other words, the additional cost required in this project is 16 units. At that price, the project *retains its original critical path*. This is an important criterion in many projects. The change of critical paths is a 'difficult' phenomenon to put across, and as stated earlier, does

**Table 7.9** The revision with retained implicit requirements

| Activity | Predecessor | Duration in units | Additional costs by activity |
|---|---|---|---|
| A | – | 6 | 2 |
| B | A | 9 | 1 |
| C | B | 10 | 0 |
| D | C, F | 6 | 8 |
| E | A | 8 | 5 |
| F | E | 10 | 0 |

**Fig. 7.5** Gantt chart after optimizing and retaining original critical path

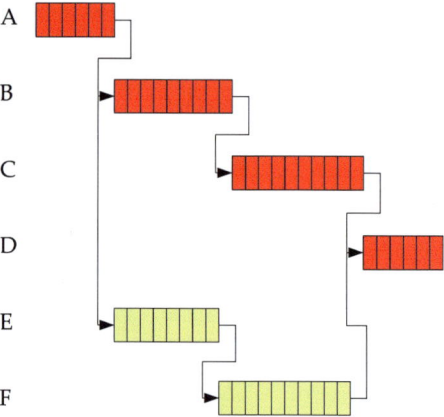

not normally occur unless there is a drastic change in the resources or the method by which a task is done.

Thus, we have seen in this case, as to how additional information on the compression costs changes the entire scenario. This is typical in projects where most contracts are sent out early or cases where we are asking for shorter lead times where the contractor is already loaded with other projects. It is, therefore, not a very uncommon situation and does occur frequently in real life.

## 7.6   Factoring Additional Information in the Sample Case

In the example above, we have not yet considered the base cost of the project. In other words, the allocated cost is not given to the reader. In reality, many companies don't share costing information among their *entire* project team. Therefore, individual resources are able to talk to the agencies only for the compression costs. But a project manager, most often, does get the overall cost information. This is important to understand.

So, as the next step, let us incorporate this information and see how it affects the way we think. The base costs (originally contracted costs) are given in Table 7.10 as additional information.

Given the picture of the overall average base cost, the reader is now requested to check the original contract cost of the project. This is relatively simple as one takes the individual costs and multiplies it with the durations. In other words, the original or the base cost of the project is found to be 119 units.

As a 'management' opinion, a situation that shows a huge variance between the 'base' cost and the 'compression' cost is one in which the *contractor is potentially taking the company for a ride*! Thus, there are three fallouts of this management decision: (a) Accept it directly, (b) Accept it with a renegotiation of risks, or (c) Renegotiate the rates. This kind of management opinion results as a 'tough' call for

**Table 7.10** The base costs in the sample project

| Activity | Average base cost per unit duration |
|----------|-------------------------------------|
| A | 1.5 |
| B | 0.5 |
| C | 3 |
| D | 2 |
| E | 2 |
| F | 4 |

the project manager. Usually, a lot many stakeholders start getting involved at this stage, as the project is *technically delayed and likely to have a cost overrun.*

Even if one considers this the start of the project, such calculations change the timelines as well as the costs of the project. Most times cost estimates are sampled from a few contractors to get an indicative value for the budgetary estimates. This is especially true in large projects. Hence, any such estimate needs to be verified again with existing market conditions and the resulting cost situation, with a differential as described here, is common, even at the feasibility stage. At times, this variance is ignored at the feasibility stage as the *issue is soon going to be the 'baby' of the procurement team and NOT the project manager:-)*

**A Quick Hands-On**

So, the cardinal question for the project manager at this stage is:

**What Line of Action is recommended after knowing the Base Costs?**

While the additional costs for schedule compression are not too high when compared with the total costs, they certainly are significantly higher on a piecewise relative scale. In other words, the increase in the overall costs is between 12 and 16 units (15%). However, in that process, certain activities seem to be having 'twice' the costs.

**A Quick Test**

**What course of action would be ideal in your corporate situation? Discuss how the three options would be used in your own company. What are the specific trigger points for each course of action?**

## *7.6.1  Devising a Good Framework for the Decision*

The above question is critical and, often times, not so trivial. Hence, a good decision framework needs to be suggested to ensure that there is sound logic and judgement for each of the options.

To keep matters simple, we propose to use the Consulting Connoisseurs three parameter cost evaluation portfolio framework for such a decision. The three parameters are shown in Fig. 7.6. The current practice, in most literature, only considers two of the parameters viz. (a) cost variance at activity level and (b) cost variance at the project level. We will now see how this framework is used in the conventional sense, using a few thumb rules. This is similar to the ones in the literature, although they are only for 2 parameters. At the same time, we will also show how this reflects in the modeling and simulations environment introduced later.

Simplistically speaking, conventional wisdom divides these elements as shown in the portfolios described in Fig. 7.7. Thus, it is recommended to go for an 'Accept-and-Move-Ahead' line of action for the case where the cost variances at both the project as well as the activity level are low and the agency (in all likelihood) agrees quickly to the change. At the same time, if the agency is taking long to arrive on an aligned agreement, it might be essential to 'Re-negotiate- the-Risks' as a course of action.

If the activity variances are high, and if the project variances are low, many companies tend to renegotiate the risks with the contractor if the expected time to 'agree' is reasonably low. However, if that isn't the case, the contract may need to be renegotiated completely.

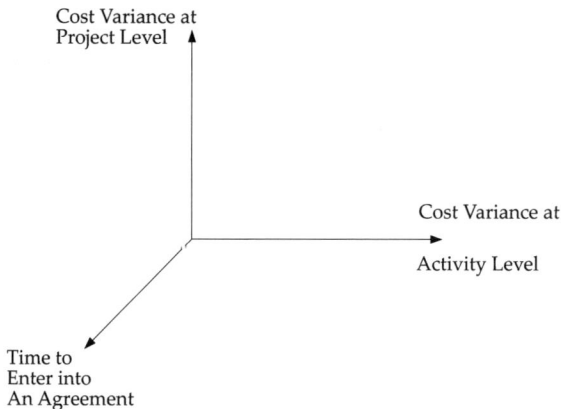

**Fig. 7.6**  Three parameter framework

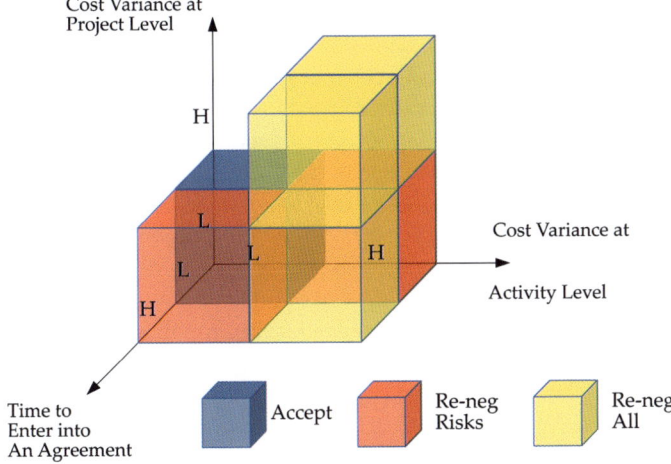

**Fig. 7.7**  Consulting Connoisseurs three parameter cost evaluation portfolio

**A Quick Test**

At this stage, the reader is advised to answer a simple question:

**How does one ascertain the intensity as Low or High in the portfolio? In other words, can we fix a definitive value for the same?**

The reader must also think of the acceptance of the 'definitive value' among the stakeholders of the project. This often becomes a point of debate and discussion. However, either way one would go, it is important to *rationally define the same and be able to justify the outcome with rigorous logic*. This is, often times, a big stumbling block in the conventional methodology.

## 7.6.2  The Modeling Approach

Our sample project will now be worked out using the modeling and simulation methodology. In doing so, we will bring in a more realistic picture in the analysis. Since we have an agency / contractor already, the discussions are going to have a slightly different nature than what is mentioned above.

Every contractor takes some time to review the revised proposal. Therefore, though we put the actual timelines at $t=0$, in reality, this is not the case. For instance, the first contractor would give his response after 3 days while the second might give his response after 7 days. This means that there is constant 'juggling' at the project manager's end. Despite everything done, there is a potential for the negotiations to fail. In other words, there is a greater risk than that we have thought. A lot of this depends on how the procurement team works and even on the directives of the top management. In any case, the situation can be characterized by Table 7.11.

In the table, we have a more realistic evaluation of the entire situation. Of course, the risk of disagreement is what the project manager *believes* to be and is not a particularly scientific number. Nevertheless, the gut feeling of a good project manager is, often, very strong and reflects close to the reality. It is possible to involve others and derive an *average* figure. But for the sake of this treatment, we firmly stand by the side of the project manager to take this treatment further.

While the project manager will be discussing the issues of getting a favourable agreement, there is always a risk component attached to this situation. If the project manager wants to minimize the risk to 0, he will need to pay an additional premium. Most contractors have this kind of a strategy, whereby they keep room for 'further negotiation' while quoting for a project. So, in order to make it a 'riskless' proposition, there would be an additional premium required. In negotiations, this is a battle between time, money, and acceptance. At a particular additional premium, the contractor is seen to be *more than willing to accept any contractual obligation*. That is, therefore, a different posture from the contractor. He quickly agrees at any condition above a particular price in the negotiation. Such a scenario is listed out in Table 7.12. This is a situation where the additional premium is higher along the critical path and lower elsewhere.

**Table 7.11** Characterizing negotiations in the project

| Activity | Time to decide | Risk of disagreement (%) |
|---|---|---|
| A | 3 | 20 |
| B | 5 | 30 |
| C | 10 | 10 |
| D | 15 | 40 |
| E | 6 | 50 |
| F | 10 | 15 |

**Table 7.12** Additional premium for minimizing risks

| Activity | Compression estimate | Additional premium for minimizing risks(to 0) |
|---|---|---|
| A | 2 | 2 |
| B | 1 | 1.5 |
| C | 4 | 1 |
| D | 4 | 2 |
| E | 5 | 1 |
| F | 5 | 0.5 |

Table 7.12 actually could be comparable to the 'asking rate' of the contractor while the previous table of the compression costs is comparable to the 'manager's opinion' on where the negotiated costs can lie. There are, therefore, multiple *real life scenarios* that are possible in this situation. The reader is expected to reflect on the various scenarios possible as we move ahead in the discussion. In other words, the discussion here is pretty generic and can characterize most project environments fairly accurately. This is important for the reader to understand and appreciate.

First, we factor in the risks that are given. For the project now, Table 7.13 represents the cases that are interesting and reasonably significant, which means that the distributions need to be calculated by bearing these cases in mind. The end result of these cases is described in Fig. 7.8.

**Table 7.13**  Risk profiling for the project at time $t = 0$ additional premium not incorporated

| Case number | Case characteristic/scenario | Probability of the case | Additional cost in the case | Duration in the given scenario |
|---|---|---|---|---|
| 1 | A agrees; B agrees and D agrees | 0.336 | 10 | 31 |
| 2 | A disagrees; B agrees; D agrees and E or F agrees | 0.0777 | 15 | 31 |
| 3 | A disagrees; B disagrees; D agrees; E agrees or F agrees; C agrees | 0.02997 | 21 | 31 |
| 4 | A agrees; B disagrees; C agrees; D agrees | 0.1296 | 14 | 31 |
| 5 | A agrees; B agrees; D disagrees; E or F agree; C agrees | 0.1865 | 18 | 31 |
| 6 | A agrees; B disagrees; D disagrees; C agrees; Either E or F agree | 0.07992 | 15 | 32 |
| 7 | A disagrees; B agrees; D disagrees; C agrees; E agrees; F agrees | 0.02142 | 25 | 31 |
| …and so on … | | | | |

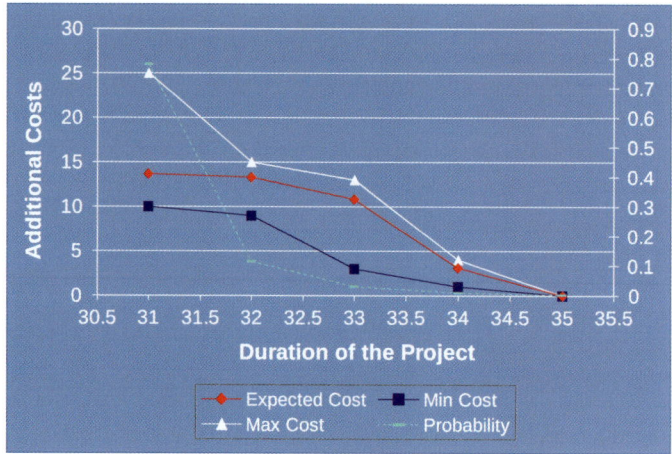

**Fig. 7.8** Risk distribution: duration versus additional costs and probabilities

### 7.6.3   A Few Points to Ponder

The project management context involves complex decisions that actually are made or are based on integrated information from various processes / faculties. This is important for every project manager to understand.

Wide varieties of outcome need a good handle of computer-assisted tools and techniques like modeling and simulation, for every outcome is important and any incorrect decision can become costly! In the present case, we are discussing these topics to give the reader a good overview of the concepts involved.

Typical compression risks are often characterized by the *risk funnels* as shown in Fig. 7.8. The tendency of the funnel is to open up inwards. That is, as the duration reduces, the funnel opens up. Even the current funnel will show a tendency to open up further as the project goes. Thus, the project manager must have a good handle of managing the funnel.

---

**A Quick Test**

The reader is advised to understand two questions at this stage:

**How does one evaluate a compression strategy?**
**Under what circumstances is a compression strategy required and justified?**

The reader is requested to get a definitive solution for the same.

**Fig. 7.9** Decision rule
involving risk norms

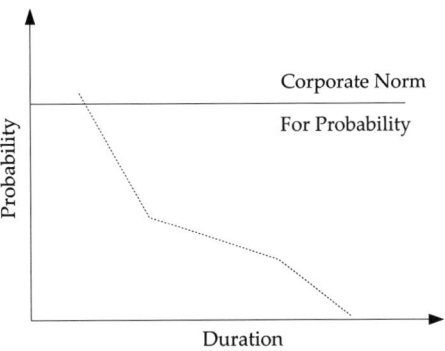

Coming back to our discussion now, we will look at two important criteria when premium needs to be considered for a project.

The first parameter is the corporate norm on probability. Most companies believe in having a typical 'risk norm' or 'risk appetite'. In other words, a project is *deemed to be in the green* if it covers 80 % of the success in this case. That is, when a project manager says his project is doing fine his organization believes that the project would be successful with a probability of 80 %. So, this norm is an important one as it depends on several factors. For instance, in space research, the norm is close to 95 %. However, in other sectors, it may have a significant variance. Typical construction projects hover around the 80 % norm. Software projects many times are a little lower. In any case, the representation of the probability curve given in Fig. 7.8 is again shown in Fig. 7.9. If the cumulative probability for 31 weeks is more than the corporate level of comfort, then one does not need to go in for the additional premium payments. In other words, the company will be happy going its way by negotiating with the vendors and trying to bring down the costs close to their targeted values. This seems to be the case in the current project. If, on the other hand, the probability of the duration would have been less than that of the corporate norm, it would be essential to secure the norm and ensure that the project is within the acceptable norm. In such cases, additional premium payments would be strongly considered (and recommended) to ensure that the project gets back on track.

The second parameter that is important for the project manager is the cost parameter. If one finds that the difference between the 'risk free' additional premium and the expected cost for the specific duration (31 units in this case) is within the organizational norm, it might well be meaningful to go for the option of paying the additional premiums (Fig. 7.10). Thus, the distribution is critical as shown in the figure. If the difference is just a couple of units, it might be meaningful to go for the same and avoiding an unpleasant situation later in the project. A non-committing management could make it a more challenging situation than required, in such cases.

These two norms are extremely important for the manager. The Risk Officer and the Project Manager are continuously working at the two ends of the spectrum:

1. The risk officer tries to define meaningful norms.
2. The project manager tries to control his project within the norms.

**Fig. 7.10** Decision rule
involving cost norms

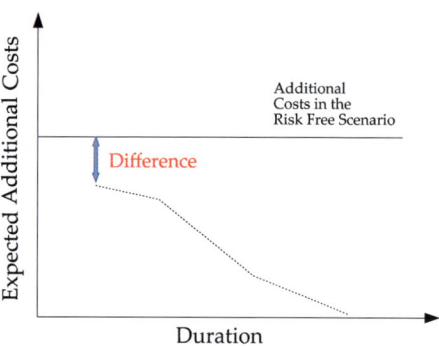

The leadership has to give them clear guidelines to operate efficiently in this frame-work. Unfortunately, in most organizations, this norm is not defined. In fact, the Corporate Risk Officer is, many times, clueless about his role and is usually busy doing some periodic reporting. To make matters worse, the project manager, who is the main stakeholder of the project, himself isn't clear as to how to interface with the corporate risk officer. This leads to a chaotic situation where decisions are random and not based on appropriate logic.

Complex set of probabilities and network outcomes have seen innovations in con-tracting like: (a) Lumpsum, (b)Time-and-Material with an option to become Lump-sum after a finite time, (c) Target pricing, etc. With improved techniques, one can determine appropriate strategies for these types of contracts. These are discussed in greater detail in subsequent chapters.

## 7.7   Understanding Critical Paths

A critical path in a project is a path that has zero float. In other words, all the activities *must start and must finish* at the same time as planned, so as to ensure that the overall project is not delayed.

The critical path refers to the way the diagram shows those activities that must be completed, and complete in a specific order, so that the project can be completed successfully and on time. Activities on any other path should have a few units of time available before they can start impacting the overall project schedule. For instance, in our sample project shown in Fig. 7.2, the E and F activities have 1 day less than B and C. Hence, even if E or F would delay by a day, it would not affect the end date of the project. However, the moment the delay exceeds 1 day, it would extend or delay the project. Thus, E and F are said to have a float of 1 day each. So, technically, the project's longest path should serve as the critical path.

There is a fundamental difference between the critical path and the critical activ-ities. All activities on the critical path are critical. That is, they do not have any float or free room from a schedule perspective. However, there are other activities that are often found to be critical as well. These activities are made critical from a 'software' perspective using different ways and means like:

1. Adding constraints to activities.
2. Increasing their durations to make them reflect as critical ones.
3. Imposing relationships that are often not the 'hard ones'.
4. Tightening schedules and contracts by forcing compression to reduce floats.

**A Quick Hands-On**

A schedule might indicate critical activities or short paths that are not along the critical path. This is typically a software issue and less of a project issue.

**As a project manager, what does a project with 'most' activities being critical mean to you?**

*A Project with more than one critical path is typically a planning disaster to start with!* Such a situation occurs in schedules which are said to be *network sensitive*. However, one often finds schedules with many critical paths and this is more of a project management issue rather than the skill of the planner. The reader must understand that in the real world, a project with all activities being critical is *inconceivable*. Though it reflects in the schedules, it is most times the failure of the project manager and the leadership, than the planner.

A project schedule is considered sensitive if the critical path is likely to change during project execution. The critical path is simply all the tasks that drive the end date of your project schedule. If your project schedule has multiple critical paths, then your project schedule is considered sensitive. If your project schedule has only one critical path and there is some slack among the various tasks, then your project schedule is considered insensitive.

Understanding the critical path is useful to know when one is tracking a project schedule's performance. It is said that a project manager needs to 'focus' on the critical path and the critical activities more than the others. While this is true, the word 'focus' gets synonymous with 'neglect' and the project manager ends up, many times, using floats on other paths. This, therefore, needs to be looked at, in the right perspective. Understanding how many critical paths are in the schedule and, the available slack in the schedule determines how much time the project manager needs to spend on specifically managing the critical path.

A less popular, but increasingly important process that is gaining popularity in modern day project management is the *Project Post Mortem Analysis*. One of the popular aspects of discussion is the specifics around As Built Critical Paths, i.e., the actual critical path that a project had. This step is useful for understanding assumptions and the actual performance issues that affect the project performance. However, in large programs, this is seldom done. This is mainly due to the fact that:

1. Team members often change (or are changed) during the course of the project.
2. Many team members have roles specific to certain phases of the project. Therefore, getting them back for the analysis becomes cumbersome.
3. Most times, at the end of the project, everybody is happy that the project is 'over.' So, most members do not want to go back and discuss these issues in detail.
4. Most managers do not have a scientific appraisal system to check the true performance of their team. So, they refrain from activities like these that are rigorous.
5. There is a gap in many projects between reporting and reality. That makes the truth practically 'untraceable.'
6. Finally, any such activity requires excellent project infrastructure. Hence, it is utmost essential to have the right kind of infrastructure to capture this information.

Yet, the *true and effective use of the critical path is visible only when one performs such a post mortem analysis*. The reader must realize this concept and perform it at least once to get a feel for how to manage it effectively. Certain aspects critical to reporting are given in subsequent chapters.

## 7.8   Three Point Estimation

A lot of project literature speaks of three point estimates. Let us quickly understand the concept at this stage. The most popular three point estimate is given by PERT. It is a simple way of determining the duration of an activity. Let us understand it using an example as shown in Table 7.14. The objective is now to determine the expected duration using PERT. Also, it would be interesting to determine the project buffer at this stage.

The reader must have already read about PERT in first-level textbooks. However, a quick recap is given here. PERT uses a simple statistical relationship to determine the duration estimates. In any duration estimate, the realistic duration is also called the most likely duration. So, going by our table, activity A has a realistic or a most-likely duration of 7 units. This is often denoted as MLD of an activity. Optimistically, one checks the minimum possible time for an activity. In our table, activity B can be done

**Table 7.14** Activity durations in a project

| Activity | Predecessor | Duration (in units) | Variance of duration (in units) |
|----------|-------------|---------------------|----------------------------------|
| A        |             | 7                   | +4/−2                            |
| B        | A           | 10                  | +1/−3                            |
| C        | B           | 10                  | +3/−1                            |
| D        | C, F        | 8                   | +1/−2                            |
| E        | A           | 9                   | +2/0                             |
| F        | E           | 10                  | +3/−2                            |

in a minimum time of 7 units. This is often denoted as the OD of an activity. Finally, potential 'worst case scenarios' are also called pessimistic estimates. They give the maximum duration of the activity or the PD of an activity. In our table, activity E has a PD of 11 units. The three point estimate of PERT states that:

$$\text{Expected Duration} = \frac{OD + 4 \times (MLD) + PD}{6}$$

The three point estimate is statistically derived and is said to be a reasonably accurate representation of the activity durations. The variance is said to be the square of the deviation and is given by:

$$\text{Variance} = (\text{Deviation})^2 = \frac{(PD - OD)^2}{36}$$

When it comes to the project buffer, there are different schools of thought. However, here we follow that the project buffer is the square root of the sum of all the project buffers *along the critical path*. Using these formulae, one can easily find the expected values and the project buffer.

Our aim here was not to describe or discuss the formulae that the reader has already learnt from first-level books. Our aim here was to discuss PERT from the 'eyes of the manager'. PERT is one of the most popular tools in Project Management theory. Although widely taught, it is probably the *least used concept in projects globally*! This is another glaring example where the knowledge of project management is not entirely used by the industry. There are reasons for the same:

1. Mathematically, PERT tries to map variances as a distribution of durations to a *deterministic value*. Therefore, a good understanding of the nature of distributions and the reliability of the deterministic values is required. Hence, it is best to use the concept selectively.
2. Since the estimates themselves are highly dependent on the skills and approaches used by the estimators, they tend to vary a great deal across teams that are all within the same organization! Hence, this aspect makes it difficult to 'standardize' within an organization.
3. The top management of any organization would like to 'impose' the optimistic durations on all activities. This actually causes a lot of problems at the execution level.
4. Lastly, it is always better to be realistic rather than look for three values. Imagine a project with 10,000 tasks, it becomes a nightmare!

Hence, all said, PERT is not as popular as it would be expected to be by many readers.

## 7.9   Key Takeaways

In this chapter we focused on certain tools and techniques used in the manager's decision-making framework. We also reviewed some of the commonly used techniques like PERT and CPM.

One of the most common problems faced by project managers is that of schedule compression. In this chapter, we have discussed and elucidated the way in which such decisions are to be practically taken by the project manager. Since real life projects can be large, an analysis at the first level WBS would normally suffice for a quick evaluation. Of course, with the advent of advanced computing software, it is possible to do a rigorous analysis of all the activities.

One key takeaway from this chapter was the difference between a modeling-based methodology and the conventional approach in deciding about project compression. While modeling- and simulation-based methods are more rigorous, they also give a far *clearer picture* of how the line of action for a given project situation is to be determined. Unfortunately, despite its inclusive nature and simplicity, there is a widespread lack of awareness on the methods and their uses; and this problem reflects itself at the operational fronts in most projects. Most projects fail due to faulty decisions and this is *mainly because the conventional framework used in projects is grossly inadequate and yet, overrated*!.

Another important takeaway from this chapter is the concept of 'conditioning with partial information'. While most argue that the project manager often knows too little, the context shows that he is not able to use his knowledge well in many circumstances. With each additional information, the manager must be able to make better quality decisions. Hence, the understanding of how this information gets incorporated in his final decision criteria is extremely important. We saw how information was changing the line of action in the sample project, and we also saw that the modeling and simulation approach gave us a definitive guideline to take the project further.

One other point that needs to be realized by the manager is that the roles and responsibilities of other stakeholders need to be clearly understood. Most organizations have designations and poorly defined roles and responsibilities. For instance, the reader might have interacted with many a 'risk managers', but knowing what is to be expected out of them and using them in a meaningful framework is often more important. This is something that is described well in the chapter.

The reader is advised to question the fundamentals and review his project situation each time. Modeling- and simulation-based approaches are far more effective in projects and are becoming increasingly popular in diverse areas like scope management, execution control, and claim management. Note that the modeling scenario makes our decision variables simpler to deal with.

# Chapter 8
# Second Level Review-II

**Abstract** This chapter focuses on reporting processes and conceptual challenges to defining reports in a project environment. It delves on the complexity of the choices that are available to the project manager. We have, then, described a new framework that assists the project manager to define the right metrics. This Consulting Connoisseurs Simple Metric Definition Framework is then elaborated. The chapter then moves on to the Earned Value Analysis. We look at the problems with the EVA methodology and despite them, the chief reasons for its popularity. An accounting-centric approach, the EVA has a lot of information that does not really help the project manager translate his thoughts into action. We also touch upon certain scheduling issues including the concepts of retained logic and progress override.

**Keywords** Reporting · Simple metric definition framework · Earned value analysis · Problems with earned value analysis · Scheduling and earned value analysis · Retained logic · Progress override

The reader, now, probably understands how we are penetrating into the conceptual domain from the process domain and vice-versa. As we began with this book, we clarified that the context of project management is not particularly well understood because a lot of literature focuses only on the *hows* without looking at the *whats* or the *whys*. As we go further into the concept driven arena, many more situations, tools, and techniques will be discussed.

## 8.1 Reporting

Just as it is said that any rule is just as good as it is implemented, an analogous project management view is that *any project management philosophy is just as good as it is 'reported'*. In other words, the world of reporting is a tricky one and there are a lot of areas that concern any project manager as far as reporting is concerned. To demonstrate this, we will first look at simple project reporting.

© Springer Science+Business Media Singapore 2017  129
N. Gurjar, *A Forward Looking Approach to Project Management*,
Lecture Notes in Management and Industrial Engineering,
DOI 10.1007/978-981-10-0782-8_8

One of the key aspects of reporting is the development of project metrics that reflect the health of the project. Just like temperature, blood pressure, BMI, etc. are used in medicine to reflect the 'status' of health, reporting metrics are used in project management.

The definition of appropriate metrics is a complex one in project management. However, it is oftentimes an ignored area as well. Many project managers go for the 'usual' tools like the S-Curves to define their project progress. However, before one actually plots the S-Curves, it is essential to understand the weightages given to various activities. We will show this through an example and substantiate it with the necessary logic.

There is *no single prescription to define the weightages to be used in project reporting*. However, there are reasonable characteristics that each of the methodologies must have.

**A Quick Test**

At this stage, the reader is advised to review the following questions critically.

**List the information you would expect from a project status report.**
**What management decision process is associated with the information?**
**How do you track projects in your organization?**
**Do you think your tracking is adequately done? Do you get 'surprises' in your reports? When and why (give 3 reasons)?**

While most times, reporting is done to 'align' with the boss's directives, there is more to it in the project manager's court than is believed. An honest answer of the questions above would help the project manager to independently delve into some of these specific areas.

### 8.1.1  A Sample Case Demonstration

To make matters simple, let us see a sample project. For the sake of our discussion, let us define a sample project of five components, viz., A, B, C, D, and E. Note that we are speaking here of components, although this method can be used for any entity (activity, etc.)

**Little Information!**

Given just this much of information, the only possible line of action that a project manager could use to define weightages is to give each component the same weightage. That is, each component gets the weightage as:

**Table 8.1** Component cost information of the sample project

| Component | Cost of each in USD |
|-----------|---------------------|
| A | 10,000 |
| B | 30,000 |
| C | 500 |
| D | 50 |
| E | 10 |

$$\text{Weightage} = \frac{1}{n}$$

where $n$ is the number of components.

The reader is advised to analyze, evaluate, and spell the advantages and the disadvantages of this methodology.

**Component Cost Information**

Now, let us imagine that the components of the deliverable are given as per Table 8.1. Since each activity also could have a physical component associated with it, we are using that as a metric. If we attach components to activities, since the later part of the project is more on assembly, testing, and commissioning, very few components might be made at that time. Hence, this idea gels well with the component description given.

The reader is advised to analyze, evaluate, and spell an appropriate methodology for the given information. In other words, define metrics using the information in Table 8.1. What are the potential advantages and disadvantages of using this methodology? How is this different from the previous metric?

For instance, should one choose the cost directly or the square of the costs or the square roots of the costs? Should one go for the logarithmic values of the costs? How and why should one decide on the same? These are some of the tricky questions that often 'bug' project managers when they define the metric. I am sure that the reader too will have many more ideas and will be able to substantiate his choice well.

**Component Effort Information**

The next perspective could be to incorporate the Level of In-House Effort for each activity. Let us assume that this is given *in addition to the previous set of information*. How does this affect the definition of the weightage? The data are given in Table 8.2. Again, the reader is advised to analyze, evaluate, and spell an appropriate methodology for the given information. What are the potential advantages and disadvantages of using this methodology? How is this different from the previous metric? These are a common set of questions that need to be asked after each iteration. The reader is expected to give a strong logic that can be used to *convince the stakeholders* for each of his methodologies.

The cardinal question remains the same for each iteration:

**Table 8.2** Component effort information (for inhouse efforts) of the sample project

| Component | Level of effort in man days |
|---|---|
| A | 50,000 |
| B | 5,000 |
| C | 7,000 |
| D | 1,000 |
| E | 50 |

**What should be the weightage for the reporting schema?**
**How does it help the project manager better manage the project?**

Again the reader here has an infinite possibility range to choose from. Should one look at the cost × effort, or the square of the product, or the square root of the product? Should one look at the sum of the two? And so on…We leave it to the wisdom of the reader to think and try out the various possibilities that only become increasingly complex as we try to incorporate more things. The reader can keep looking at several possible options to take this forward.

### Modified Component Effort Information

In addition to all the information given above, if the project manager also knows the Level of In-House Effort at Site, there is a strong potential for the metric to change. This additional information is given in Table 8.3. The reader is now asked to derive a metric incorporating the information given. With each iteration, the cardinal questions remain the same. Therefore, again, the reader is advised to analyze, evaluate, and spell an appropriate methodology for the given information. What are the potential advantages and disadvantages of using this methodology? How is this different from the previous metric?

### Schedule Information

All along, our analysis was more oriented toward costs and efforts. However, now we would try to incorporate some schedule information as given in Table 8.4. As we go along, we now have many more factors to choose from.

**Table 8.3** Component effort information (on-site detail) of the sample project

| Component | Level of effort in man days on-site |
|---|---|
| A | 100 |
| B | 300 |
| C | 500 |
| D | 100 |
| E | 5 |

**Table 8.4** Additional schedule information of the sample project

| Component | Latest date of completion (O is the operational due date) |
|---|---|
| A | O |
| B | O-10 |
| C | O-30 |
| D | O-45 |
| E | O-15 |

This schedule information gives us a different perspective of how to re-allocate the weightages. Many project managers show progress early on, during the project, and then, start reporting delays. This is often a reporting issue, though many things go out of control later as well.

## 8.1.2 Important Aspects of Reporting

If you talk to any project manager, he would say that reporting is a double edged sword…and hence, it could both 'kill for' or 'cut' the project manager. Reporting, therefore, is very tricky. For *result-oriented organizations*, a fair report is always important. However, there are many challenges to reporting:

1. It is usually a *static view*. This is probably the biggest area of concern in reporting. In a journey, if one knows the current location and speed, does it give one the time he would take to reach the destination? The answer is pretty straightforward! Hence, most reporting exercises see a characteristic phenomenon here as shown in Fig. 8.1.
   The activity in the project peaks during the reporting meeting period and then, tends to 'cool' down till the next reporting meeting period in the project. The more the duration between the reporting periods, the greater could be this tendency. It

**Fig. 8.1** Reporting dynamics

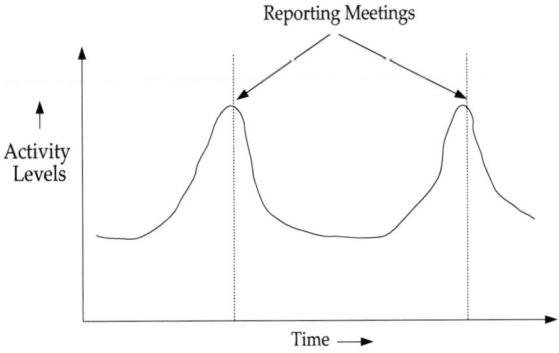

is like students giving exams. (where they actually 'study' more when they are closer to their 'exams'.)
2. Predictive reporting often ends up as a subject of political decision-making. The whole concept of predictive working gets compromised most times when the approach toward reporting is different on the two (opposite) sides of the table.

### Impression Management at Work

The reporting shifts from a *fact-based reporting model* to something that is an *impression based reporting model* in many projects. A good treatment of this phenomenon is given by Harris and others [8] where they have dealt with different aspects of this phenomenon. A quick overview is given below for the reader.

The objective of impression management is to create a desired image in the mind of others, which can be achieved using a variety of tactics. In this context, the work actually talks of political skills that provide an individual with the ability to understand others and use that knowledge to effectively influence situations. While reporting project progress, several impression management tactics are used. The most popular ones are those given by Jones and Pittman, which cover five tactics:

1. *Self Promotion:* Which involves exaggerating or highlighting one's achievements and abilities so as to be seen as competent. This is often used in project review discussions where the project manager tends to focus and keep highlighting what is done or completed. The reports keep reiterating and reflecting this stand.
2. *Ingratiation:* Which involves doing favors or giving flattery in the hope of being seen as likeable. This is often used in project reviews where the managers quickly change sides to agree with the views of the top management, thereby, often leaving his team in the lurch!
3. *Intimidation:* Which involves acting threateningly or intimidatingly to colleagues so that they will view the manager as forceful or dangerous. This is often done by the stakeholder who *wants results and doesn't buy excuses* and occurs many times during project reporting reviews. The unfortunate situation being that the stakeholder is, many times, fully aware that the project is in an infeasible situation.
4. *Supplication:* Which involves broadcasting one's shortcomings in an attempt to be viewed as a needy person. This is a tactic used by many project managers who *inherit* projects that are in the later stages of execution. They use this tactic to get better resources and reset expectations of the stakeholders.
5. *Exemplification:* Which involves making others perceive that their actions are exemplary and worthy of serving as a role model. For instance, during project reviews, many project managers state that they 'worked late' on many days. This is to give them the mileage of instances where individuals go above and beyond the call of duty to appear dedicated.

The project manager should be aware of these tactics because he too has subordinates reporting to him (just as he reports to his superiors). All said, this is a tricky aspect that sometimes *camouflages the overall reporting schema!*

People indulging in impression management are categorized as self monitors. There are, fundamentally, two types of self monitors, viz., Low and High Self Monitors. In comparison with low self monitors, high self monitors

1. pay more attention to the behavior of others in social situations;
2. prefer to enter situations that provide clear guidelines for behavior; and
3. are more attracted to careers that emphasize the importance of public behavior, such as acting, sales, and public relations. High self monitors also
4. are more adept at reading other people's facial expressions, and
5. are better at communicating a wider variety of emotions than are low self monitors.

A project manager must understand these essential differences while evaluating the 'truth' in the reporting. While this is easier said than done, it leads to interesting and important results that one needs to be aware of.

### 8.1.3 The Most Common Mistake in Reporting

The most common mistake in reporting is that *Middle Managers define the Reports!* In other words, most reports are designed and prepared by middle managers. Thus, there is always a gap between the picture the manager wants to portray and the image the top manager wants to see.

After having interacted with several top managers over the years, the one interesting thing I observed is that they are *unable to articulate what they want to see in their project reports*. This is the genesis of the dichotomy that brings in a 'gap' between the manager's views and the stakeholder requirements to start with. To make matters worse, most top managers *don't believe they need to dedicate time* to define and control the reporting process. It is analogous to a patient going to a medical diagnostic center and asking them to perform 'whatever' tests they think 'important' to ascertain the health of the individual. Unfortunately, when there is no 'doctor' who would prescribe the tests, it would be upto the 'skills' of the diagnostic technician to understand and decipher what is in the mind of the doctor. The patient in our case is the project, where the diagnostic team is the project manager. The doctor is the top management/stakeholder. Yet, this practice is fairly rampant. So, a conservative technician would do some preliminary investigations, something less than desirable. What needs to be done is that the reports must be defined by the Top Management and the Top Management only! This helps contain and eliminate many of the problems with the reporting process.

### 8.1.4 Defining Metrics

In the earlier subsections, we tried to define appropriate project reporting metrics. There were two things that were of significance there. The amount of potential information that could be incorporated into the metric was enormous. Actually speaking, the project manager, most times, has *too much information* to incorporate in his metric. It is like the world of astronomy, where there is too much of information for the scientist to decipher and understand. Similarly, even in project reporting, the difficulty of the manager is in *choosing the parameters that should go into the metric*. The second aspect that is of significance is the way in which the metric is to be defined. We saw several possible variants and each of these variants brings in unique sets of advantages and disadvantages. We will now focus on how one should define project metrics for reporting. We introduce the Consulting Connoisseurs Simple Metric Definition Framework. This is based on two basic parameters. These two parameters are shown in Fig. 8.2. Let us assume a project that made substantial progress and needs to be reported. Consider the project completed 80 % of its work in terms of manhours, but is only 75 % complete in terms of the proportional scope. The metric used should be representative of the quantum of work as well as the difference between what was planned and what was achieved. Thus, defining the metric is tricky. In this case, if we consider two parameters, i.e., manhours and scope, if we add the two, we get 155 where the planned figure should have been 160 if we add the two. If we multiply the two, we will get 0.6 instead of 0.64! Now, if we look at the sensitivity, the first is less sensitive giving a 3 % variance, while the second is more sensitive, giving a 6.67 % variance. How representative is each, is a matter of debate, and, what is the understanding of the term 'accuracy', is a matter of the corporate culture. However, while defining the metric, a lot of understanding must go into these two critical aspects.

The framework helps the manager understand how to design and use the metric. It also tells the reader about the potential pit-falls that could be encountered while defining the metric. The project manager must be able to strike a balance *jointly with his superior (the recipient of the report)* to define the right level of sensitivity and accuracy required for the project metric. This is an important aspect. Most project managers usually end up fighting a 'lone battle' to define the right kind of metric for their project. The end result is that they are quickly engaged in the battle for survival through impression management or politicizing the reporting process.

The next step in the definition is to enable a meaningful connection of the metric with two critical dimensions or criteria, viz.,

1. the ability to trace the problems through a report and
2. the ability to trace a potential line of action for 'management focus'.

**Fig. 8.2** Consulting Connoisseurs simple metric definition framework

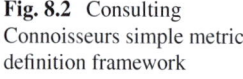
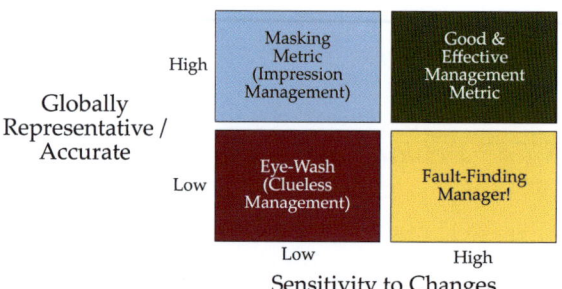

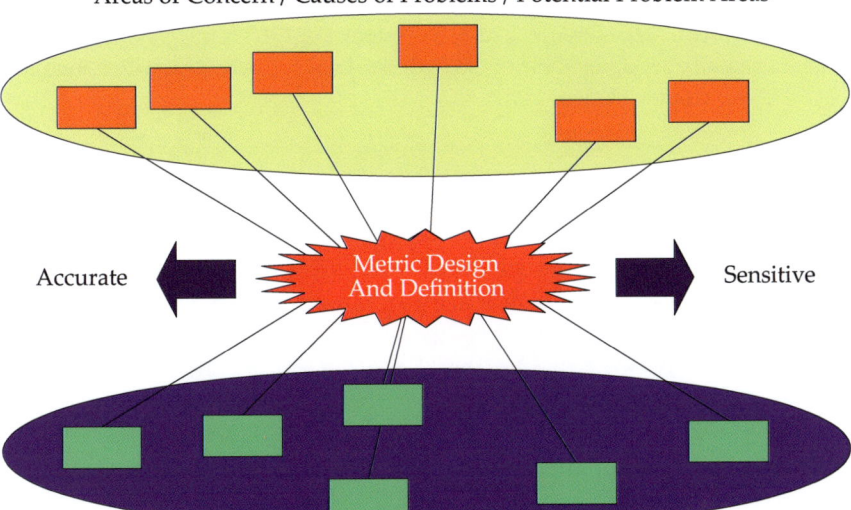

Areas of Concern / Causes of Problems / Potential Problem Areas

Fig. 8.3   Consulting Connoisseurs detailed framework for metrics

Just like the diagnostic test, the metric should help understand both these aspects when one designs reporting processes. Hence, a quick prescriptive procedure would involve the development of a map, as shown in Fig. 8.3, between the causes and the potential lines of action. Thereafter, the metric would be designed and optimized for the accuracy and the sensitivity. This is shown in Fig. 8.3.

A more detailed discussion on the metric will follow in the next few sections, where some of these aspects are again covered in greater detail.

## 8.2   Earned Value Analysis

One of the first systematic attempts to combine the cost, scope, and time elements of a project environment was the Earned Value Analysis. This has led to the EVA, as it is called, being extremely popular in project circles. Essentially, EVA stems from an accounting perspective of the project.

Carnegie Mellon University, in their online Project Management Resource [10] by Chris Henderickson, have covered many important and interesting issues in project management. The reader can review some of these details to get a fair understanding of how the entire concept is structured from a 'higher education perspective'. Essentially, the resource covers the fundamental definitions and the progression of the concept very well. We will, however, go by an alternative view while presenting the Earned Value Analysis. The idea is to *uncomplicate the concept and avoid excessive jargon-centric treatment*.

To begin with, we will try to understand what earned value means. Since we have already given a brief reference and some interesting literature pertaining to this method, we will now cover some of the intricate details. But to start with a summary statement, we would say that:

**A Quick Hands-On**

**Earned Value Analysis is an Integrated Form of Reporting.**

**The Reader is, therefore, advised to review the previous sections and see if he already has a good reporting method available. The cardinal question is: Is Earned Value Analysis the best way to evaluate the project at hand? What could be the problems? What could be the alternatives?**

So, with these cardinal questions in mind, we would slowly tread over this path. The EVA methodology involves the understanding of the following abbreviations, listed here for the reader to get familiar with the concept:

1. EV = Earned Value
2. BAC = Budget at Completion
3. CV = Cost Variance
4. AC = Actual Cost
5. CPI = Cost Performance Index
6. EAC = Estimate at Completion
7. ETC = Estimate to Completion
8. VAC = Variance at Completion

We will now focus on how this is calculated using a simple example.
The entire concept revolves around three definitions. The first is the basic formula for Earned Value given by

$$EV \text{ or Budgeted Cost of Work Performed} = \% \text{ Complete} \times (BAC)$$

PV, or Planned Value is also called Budgeted Cost of Work Scheduled is the cost in the budget for the work that is scheduled to be complete at the particular time. AC, also called the Actual Cost of Work Performed is the actual cost incurred in a project. These three are the basic definitions and terms that set the ball rolling!

## 8.2.1  A Sample Case

Consider a project with the budget and the status as given in Table 8.5. Each of the activities has a budget that is given in the second column. The sum total of all the budgets is the Budget At Completion of the project or the BAC. If we multiply the

**Table 8.5**  A simple project budget and EVA

| Activity | Budget | Percentage complete | Earned value |
|---|---|---|---|
| A | 10,000 | 95 | 9,500 |
| B | 5,000 | 50 | 2,500 |
| C | 8,000 | 35 | 2,800 |
| D | 15,000 | 0 | 0 |
| E | 20,000 | 0 | 0 |
| F | 4,000 | 0 | 0 |
| | 62,000 (budget at completion ) | | 14,800 (current earned value ) |

**Table 8.6**  A detailed EVA spreadsheet

| Activity | Budget | Earned value (EV) | Actual cost (AC) | Cost variance (CV) |
|---|---|---|---|---|
| A | 10,000 | 9,500 | 9,700 | (200) |
| B | 5,000 | 2,500 | 2,000 | 500 |
| C | 8,000 | 2,800 | 3,800 | (1,000) |
| D | 15,000 | 0 | 0 | 0 |
| E | 20,000 | 0 | 0 | 0 |
| F | 4,000 | 0 | 0 | 0 |
| | 62,000 (budget at completion) | 14,800 (current earned value) | 15,500 (actual cost) | (700) (net cost variance) |

percentage completion with the budget, we get the value of the Earned Value for each activity. The sum total of all these values is the Current Earned Value of the Project.

This is fairly straightforward in the analysis. It shows that the project is rated on the basis of its budget and the progress of the project. This, however, is the first analysis that is done in the EVA methodology. Once we get the EVs, we go to the second set of analysis that is given in Table 8.6. Here, the comparison of the Earned Value is done with the Actual Cost, and the net cost variance is obtained for the entire project.

Now the important relationships according to the EVA are:

$$\text{CPI} = \frac{\text{EV}}{\text{AC}} = \frac{14,800}{15,500} = 0.955$$

and

$$\text{EAC} = \frac{\text{BAC}}{\text{CPI}} = \frac{62,000}{0.955} = 64,921 \ (\text{approx})$$

Other formulae include:

$$\text{ETC} = \text{EAC} - \text{AC}$$

**Table 8.7**  A detailed EVA spreadsheet for the schedule

| Activity | Budget | % Complete | Earned value (EV) | % Planned | Planned value (PV) |
|---|---|---|---|---|---|
| A | 10,000 | 95 | 9,500 | 100 | 10,000 |
| B | 5,000 | 50 | 2,500 | 45 | 2,250 |
| C | 8,000 | 35 | 2,800 | 50 | 4,000 |
| D | 15,000 | 0 | 0 | 0 | 0 |
| E | 20,000 | 0 | 0 | 0 | 0 |
| F | 4,000 | 0 | 0 | 0 | 0 |
|   | 62,000 |   | 14,800 |   | 16,250 |
|   | Budget at completion |   | Current earned value |   | Current planned value |

and
$$VAC = BAC - EAC$$

The arithmetic is fairly simple for the reader to understand and calculate. We now focus on the schedule in the given project. The EVA spreadsheet is given in Table 8.7. While the earlier table was intended to identify the cost parameters, the current one is intended to evaluate the schedule parameter. Therefore, the next step is to understand how the schedule was planned for the project. This is done by understanding what the percentage of the planned completion for the schedule was. The planned value is then derived from the planned completion expected and the budget.

Two important relationships that the EVA now defines are given as:

$$\text{Schedule Performance Index} = \frac{EV}{PV} = \frac{14,800}{16,250} = 0.911$$

And
$$\text{Schedule Variance} = EV - PV = 1,450\,(\text{Cost})$$

Again the arithmetic is fairly straightforward and the reader can use it in his projects with relative ease... The reader needs to note that the Schedule Variance in the EVA approach is actually a cost, rather than a time parameter. This is an important observation, as the EVA is a methodology that *tries to translate aspects in dollar terms*.

## 8.2.2  Problems with EVA

Although this is one of the most popular metric used in modern project management, it is *chiefly used for monitoring rather than control*. Most times the EVA is used to raise flags in a project, but its efficacy gets compromised by the fact that it does not direct the user to a definitive set of options as far as the Line of Action (described in the previous section) is concerned. In other words, it tells us the health of the project,

**Fig. 8.4** EVA correlating with % completion

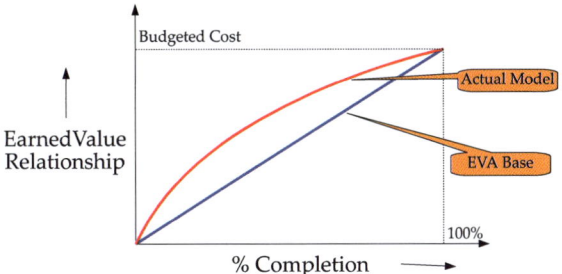

but does not actually map the metric to adequately define a workable plan of action.

Second, it is very *budget centric* in its approach. If a project starts with an 'optimistic' budget, the EVA would just 'keep ringing bells' throughout the project. By conventional wisdom, the project manager would start getting engulfed into meaningless meetings that would, manytimes, be a part of a 'fault-finding exercise'. This is a very serious problem with the EVA methodology.

As we get to the methodology, there are a few interesting aspects that every project manager needs to know. The EVA method assumes that the progress has a *linear* relationship with the cost of the project as shown in Fig. 8.4. From a practical standpoint, this is an oversimplification and a serious limiting parameter at times. For instance, preparatory work for a heavy lift in a project involves the transport and assembly of a crane at the site. While the actual activity costs less (assembly at site), the transport and the assembly of the crane could cost quite a bit! Hence, the relationship doesn't always hold good. In such cases, over 50% of the cost might be spent without any progress on the project front! Including the crane assembly in the progress calculation would, therefore, skew the assembly activity!!! Hence, even if a project is *as planned*, its EVA might reflect some deviations. The reader can easily verify this with a sample case study, the essential point being that the duration and the cost are not directly proportion. Hence, where most projects actually get updated by remaining duration methodology, the EVA goes chaotic. The EVA expects the manager to track the project progress on the basis of costs and *not* on the basis of schedules.

To illustrate this aspect, let us consider the progress of one activity as given in Table 8.8. Let us consider that the activity has made progress as planned for the first three days. However, there is a delay in the next 4 days. It appears that the project will take one additional day to complete. There isn't an extra cost associated at this stage. So, let us perform this update on the project. The two different scenarios are given by the two perpsectives of time and cost.

From a time perspective, if we use remaining duration, the progress of the project comes out to be:

$$\text{Percentage Complete} = \frac{\text{Duration Spent}}{\text{Duration Spent} + \text{Remaining Duration}}$$

which actually gives us a value of $\frac{3}{8}$.

**Table 8.8** Varying results using EVA for cost and schedule of activity of 7 days

| Duration complete of activity | Cost budgeted on $n$th day |
|---|---|
| 1 | 10,000 |
| 2 | 10,000 |
| 3 | 10,000 |
| 4 | 5,000 |
| 5 | 5,000 |
| 6 | 5,000 |
| 7 | 5,000 |
|   | 50,000 (budget at completion) |

From a cost perspective, the progress of the project comes out to be:

$$\text{Percentage Complete} = \frac{30,000}{50,000}$$

This gives us the value of $\frac{3}{5}$.

Note that these two will give us very different results when we populate the overall spreadsheet. The reader may verify this for himself and see how the EVA results get skewed.

Yet, the popularity of the EVA is driven by the clout of the finance manager and, therefore, the finance manager's perspective. In the accounting world, people are very fond of using a linear relationship just because *it is the easiest to use!* Hence, despite its shortcomings, the EVA is 'hailed' by many as one of the best tools to evaluate project performance. These two aspects pertain to the popularity of the method and the basic assumptions involved.

As far as the methodology itself is concerned, the implementation is a challenge in large projects. The major issue in this exercise is *to determine the percent complete!* This is again a dicey situation. Most times the percentage complete is a *guessed value* that is given by the project manager or his team member. This skews the whole reporting scenario as it leaves for sufficient room to 'doctor' reports. We have seen how the project reports get skewed in the earlier table. Yet, one can still apply the EVA. And, as we see that the cost data are closer to the 'current reality', we need to leverage this knowledge in defining the right methodology. Hence, there are two basic rules that every project manager must follow:

1. Where possible, the project manager *must* go for physical progress using a proper *measurement-based system.*
2. In case the project manager is not able to get the actual physical progress using a reasonable measurement-based system, he should use the 50–50 rule.
   The *50–50 Rule* states that any activity has just three states. The first state is 0 % where the activity has not begun. If the activity is currently in progress, its

progress can be reported as 50 % *regardless of what the actual progress on the field is, since it cannot be reasonably ascertained.* The third state is 100 % complete, where the activity has been completed.

This method is very powerful in large programs or in schedules that show a lot of parallel activities. In schedules that do not have the details built in, or the parallel activities in execution, this method might yield a pretty different picture from the reality. Moreover, this method is useful if the *average activity duration is less than the reporting cycle frequency.*

The other issue with the methodology is in updating costs…as not every cost gets uniquely allocated to an activity in a clear and unique way. This again, is similar to our example of the crane, whereby, allocations can skew the overall progress and the EVA.

### Understanding Schedules for EVA

One of the important aspects about EVA is to understand *how the schedule has been defined.* People often get confused with the usage of the term '*Out-of-Sequence*', and use them in inappropriate ways. While we talk about *Out-of-Sequence*, we also need to discuss the difference between *Retained Logic* and *Progress Override.*

The term 'Out-of-Sequence' from a schedule perspective simply means that the successor activity has reported progress even though the predecessor activity was not complete and their relationship was a 'hard' Finish-to-Start one. In legal terms, however, it essentially means a disruption of the Contractor's schedule. For example, the crews lost time in hopscotch from one floor to another floor, as the Contractor is held up by the owner's interference. However, to quantify the labor inefficiency due to such a situation is very difficult.

The earned value disregards this issue totally and, therefore, could mask the legal exposure that a company could have. Worse, it could also generate a *feel good emotion* about a schedule that might just be welcoming claims and litigation.

The biggest assumption is that the EVA expects you to make the same mistakes all over the project and in proportional magnitudes!!! So, if there is a cost variance in the EVA, it only *increases proportionally* when one uses the EAC for the project. This is, in my humble opinion a ridiculous stand taken by the EVA advocates. My only question here to provoke your understanding is:

### Are project managers dumb as the EVA puts it???

Yet, it has been christened the best way to track projects… While one of the problems with the EVA analysis was that of the Out-of-Sequence acivities that were mentioned earlier, the other problem is that most project plans are made using a *Remaining Duration* logic. Thus, activity completion in terms of duration could be very different from that in terms of the cost. This poses additional practical difficulties when one tries to track both the parameters using the same system.

The biggest problem is when one has multiple baselines for a project. Usually, the schedule baselines are more *dynamically* changed as compared with the cost baselines. Hence, here too, the budgets and costs become cumbersome metrics to define accurately.

All said, the Earned Value Analysis is a wonderful tool for project managers to monitor their projects. However, it is not a great decision tool and this aspect needs to be understood by every manager.

---

**A Quick Test**

Given the backdrop of our discussions, the reader needs to look at newer tools now, to combine the strengths of EVA with management practices like:

1. Incentives and Penalties
2. Correcting issues like management philosophy used in the EVA for proportional calculations (as Project Managers are definitely NOT as dumb as the method assumes)

Keeping this in mind, list two possible variants to the EVA approach. Explain the advantages of the two variants.

---

## 8.3  Key Takeaways

The current chapter has identified some of the critical aspects of reporting and has also made the reader sufficiently aware of the complexities involved in defining meaningful reports. At the same time, it also needs to be put into perspective that a lot of reporting issues are dependent on the organizational culture and the qualities of the owner-stakeholder. While this is true for most management principles, failure to acknowledge these aspects could only lead to the 'Traps in the Lion's Den!'. Hence, every project manager needs to understand the dynamics of his tools from a management Decision-making perspective, as well as, from a local environment perspective.

The Earned Value Analysis was the next section in the chapter that discussed the general methodology and the need to be cautious while applying the EVA method. The EVA, by design, can quickly distort the reality and give false signals! However, its simplicity is its biggest advantage and the *closeness to the Financial Managers due to its 'Accounting-centric' Assumptions* has made it very popular. There are several books on how planning and scheduling processes and practices need to be changed in order to ensure the EVA's efficacy. However, from our perspective, this is like forcing a round peg into a square hole and we do not, therefore, support the thought process of going overboard to make the accountants happy…certainly not at this stage:-)

# Chapter 9
# Second Level Review-III

**Abstract** Despite all the knowledge a project manager might have, he might still find himself misaligned with the corporate objective. In other words, the project manager needs to broaden his frame of reference. In this chapter, we begin by understanding this concept by touching on a broader phenomenon that involves the concept of Truth and its Forms at various levels of the organization. We then move on to help the project manager understand this concept in a practical scenario. A simple case study is used to identify the impact of cost and time and the role of conventional analyses on the same. This is taken further by assisting the reader integrate these concepts with two business-level issues viz. the Strategic Time Window and the Opportunity Costs. The exercise helps the manager understand the difference in perspectives at the corporate and the business strategy level while defining the future course of action in many projects.

**Keywords** Many forms of truth · Project culture · Project cost versus time debate · Strategic time window · Opportunity costs · Corporate strategy · Project crashing

We start by looking at some of the critical aspects on the concept of truth in the management context. We begin by a brief treatment of the concept.

We then move on to help the manager understand this in terms of the business strategy.

## 9.1 The Many Forms of Truth

Since we were discussing the aspects of reporting in the previous chapter, it is important for the reader to put things in perspective. The article included here is a well-rounded reflection of the environment that the project manager has to live in.

## 9.1.1  Introduction

The understanding of truth is a fundamental concern for most executives. It reflects itself in the most fundamental needs of management: data, information, knowledge, and wisdom. It is important to understand that truth(s) have different forms and levels. This fundamental understanding actually has two direct fallouts viz., the phenomenon of abstraction due to the levels of hierarchy and the phenomenon of applicability due to the forms. Just as companies operate across the continuum from operations at the one end (or lower level) to the strategy at the other end (or higher level), one could visualize a parallel continuum of truths across the different 'levels' and the application areas.

## 9.1.2  Approaches to Understanding the Truth

Talking of levels in this context, one could go by the 'old school' and talk of two approaches to understanding the truth. The first approach is the top-down approach while the second approach is the bottom-up approach. Although strategists always like the line of thought involved in the 'top-down' approach for various reasons, the basic corporate model in the neo-renaissance era has been a 'bottom-up' model for truth like the data, information, knowledge, and wisdom that form the progression of the needs of management. Thus, the modeling and acceptance of truth has always been bottom-up! Corporate recognition has always been on 'mastering the detail as one moves up the ladder' rather than 'working the metaphysical into the detail'! However, business success is more involved and intricate, revealing a totally different story. If one closely analyzes business success, one understands the 'top-down' approach. However, one needs to understand the conventional school before 'breaking the convention itself'. Hence, let us try to understand how the conventional 'business' or the 'bottom-up' approach is, and the issues associated with this approach in a little more detail.

For a business, from a practical standpoint, as there are different possibilities to explore the forms and levels, it is essential to note that not all truth is interpreted the 'right' way, thereby distorting the concept of truth itself! Further, not all truth(s) are put in the global context of their relevance. The contrasting varieties of possible configurations make it a challenge for companies to define and develop the right levels of understanding on each of these issues. From a common man's perspective, variances in any of these components, forms, and levels causes issues that lead to huge operational and strategic inefficiencies. Such inefficiencies usually tax not just the system, but also have a profound impact on the external stakeholders of the system. It is, therefore, important to align these elements in order to establish the required informational cohesion for the company to operate effectively. However, all said, there are several challenges to realizing this goal.

### 9.1.3 Parameter 1: Culture

From a management perspective of the business, there are three fundamental parameters that affect the understanding and the expression of truth: culture, the dilemma of choices, and the corporate direction. The first parameter is culture. In the broad sense, this is one of the strongest parameter that affects the definition of truth. Simplistically speaking, culture is defined as the tastes in art and manners that are favored by a social group. The social group extends over and includes:

1. The environment,
2. The management and
3. The organization itself.

The environment includes issues like the geography, economy, competition, status of the company, sector in which the company operates, the status of the truth informer, suppliers, and the customers. For instance, European geographies have a 'negative' tone to facts while the American geography has an 'overly positive' tone to express truth. On the same token, the Orient is relatively 'neutral'. Every large conglomerate looks at sensitizing itself to cultural factors, especially when they extend over multiple geographies. However, culture itself brings with it an element of ambiguity and this needs to be carefully worked out.

For instance, taking the element of management, many times the management might not be able to appreciate the truth in the communication due to limitations stemming out of the knowledge, the management style, or the 'unwritten' norms of communication, to name a few. Each of these actually add to the ambiguity and could be a potential ally to distort the truth.

The final factor, the organization itself, goes a long way in defining the culture. The knowledge levels and the business acumen of the people in the company do have a profound implication on the culture of a company. While it is relatively easy to define the knowledge levels of people, the business acumen is hard to determine. It is the role of HR professionals to actually try to map the business acumen and knowledge levels of potential candidates to the people around the roles they are expected to be in. There are many challenges to do this and our aim is not to delve into them now.

### 9.1.4 Parameter 2: The Dilemma of Choices

The next challenge toward understanding the truth is the dilemma regarding the choices. Culture works as a boundary condition. They are like the shore of a beach. The dilemma of choices or the tone, as I call it, is comparable to the colour of the water.

The general consensus in theory and practice is to do away with 'political correctness'. However, in real life, it is seldom going to be the case. 'Political Correctness' is like 'Inequality'. A situation where they don't exist in a society is unrealistic

(Utopia?). At the operational level, political correctness transforms itself into shades, like the tendency to focus on the component of optimism or pessimism in the 'truth'. In the software sector, for instance, the 'ugliest' statement is 'this is not logical'! In the construction and manufacturing sector, a similar statement is 'this is not true'! These are just the beginnings of scathing attacks…

This brings us to the other shades; viz. shades of reality and shades of perceived truth (imagination?). Discussions regarding these moods or the 'colour of the water' actually affect the understanding of the message it carries. The negativity of choices often has its bearing not just on the truth, but more so, on the informer of the truth. Have you heard of 'Shoot the Messenger'? A typical indication regarding dilemmas and the dynamics of the understanding of the truth are in the level of discussions that one sees in the organization. For example, it would be comparable to discussing the estimation on the accuracy of the length of the telescope, an activity that hardly has any 'business significance' to extrapolate the error to the remainder of the distance between the earth and the stars (the activity with greater 'business significance')! Do you see situations where discussions of 'accuracy' or 'logic' are used to camouflage the core message in a communication? Do you see situations where choices like the optimism/pessimism affect the acceptance of the information? These are some of the indications of the level of coherence in thought across the organization.

## 9.1.5  Parameter 3: The Corporate Direction

One of the challenges that plays a pivotal role in understanding truth is that of corporate direction. This is different from culture as it is more specific to the way the situation is approached. In our beach example, where we had the beach shoreline as the culture and the colour of water as the tone, imagine a person standing in the water. If the person is looking at or toward the shore or the sand, he might possibly have many 'surprises' from the waves. If the person is looking into the sea or away from the shore, he has a different view. If he stands perpendicular to the direction of the waves, he sees still more different things. For a typical management function, the corporate direction can be compared with the direction the person faces when he is standing in the water.

Corporate direction is often based on three factors:

1. Assumptions in the Management Philosophy,
2. Evolution of Focus Parameters, and
3. The Nature of Growth the Company has had.

At the rudimentary level, the assumptions in the management philosophy are the result and the basis for the organizational structure and the communication/reporting plan.

Do you often get wind of problems from 'an unexpected corner'? Is the management philosophy and the strategy firmly understood by all concerned? Do you hear words like 'I was not aware' in your meetings often? These are potential issues of disconnect in the corporate direction. Do you segregate your organization into management levels and groups a little too often? These are indications of a poorly implemented management philosophy.

The second factor is the evolution of focus parameters. The 'once bitten twice shy' paradigm works negative in this case. Many times, companies have a difficult time in focusing on the right things. The fact that someone got 'bitten' makes people change priorities to the less important, just because the management direction asked them to do so. Like I said in the article on the biological approach to project management, this translates, in practice, to an 'event-driven' approach rather than a conceptually sound approach. The result is that one has lesser reaction time to plan and prepare for attack. These issues again create huge inefficiencies. However, on the greater risk, they also bring in shades to the truth!

The next factor is the nature of growth of the corporate. Companies with organic growth have a greater tendency to become the 'white elephants'. It is like a huge empire where everyone has a fiefdom that stops promoting talent and focuses on retaining power. Let us defer that to a political philosopher. However, on the other hand, companies that have recently been a party to a merger or acquisition tend to perform better in understanding such factors. Companies that are 'young' with 'disruptive' technologies also tend to perform better in optimizing across these factors.

## 9.1.6 Integrating the Parameters

In our beach example, if we assume that one has the right understanding of the shoreline, the colour of the water and one is facing the right direction, we are then left with the sensitization of the decision making process to all these factors.

This actually is common to both the approaches. A sensitization is more dependent on the intellect of the person and how the person processes the various inputs provided by the different factors like waves, fish, wind, birds, etc., that one sees in the beach example. In this context, therefore, aligning of business knowledge with all forms of communication is required: data, information, knowledge, or wisdom. This calls for a strong systematic approach that enables one to work with different levels of abstraction. It is often a challenge in itself. In the real world, however, the key lies in working with a strong qualitative and quantitative model to understand the implications of various information 'bytes' and translating that to business.

Amidst all these factors, there is a strong feeling among various executives to actually align themselves and balance several factors, especially those of Truth, Ethics, Professional Responsibilities, and Business Needs. This is especially important for middle managers having difficulties in orienting themselves to the changing roles from managers to leaders and the dynamic needs of today's businesses. It is also useful for intrapreneurial skill-sets within organizations that are looking at ambitious targets for growth and that are wanting a dynamic work culture. A good handle of these factors also helps entrepreneurs to position their personal strengths and sift through the right opportunities in a more fruitful way.

## 9.1.7   The Way Ahead

Fortunately, these challenges are well addressed in the 'Top-Down' approach towards understanding truth. In the top-down approach, the understanding of these factors is relatively simple when one starts off with a solid foundation of the exact definition of the highest level of truth. This truth is characterized by universal applicability and has the highest level of abstraction. From practical considerations, it also needs to have a simple form, making it easy to understand and clear to interpret. If one fails in determining the right starting point, the 'Top-Down' approach fails. In this light, we have now developed a model of truth based on the metaphysical and spiritual levels of knowledge. We are currently looking at this scientific model that is an application of established models in religion. This model has its significance due to the fact that religion provides us with pointers to the highest levels of abstraction. A direct fallout of this fact is that religious principles are fairly universal in nature. Our research actually took this aspect into understanding the universal nature of the principles. If this assumption holds truth, they should, therefore, be applicable to ANY or EVERY situation at hand. Ethics and conduct, part of culture and environment, need to be revisited from the regulatory perspective, often codes that are instilled by law. These fall under the umbrella of the definition of the society.

And my recent research into the Bhagvad Gita has revealed an amazing wealth of knowledge in mapping business requirements with the orientation towards defining and understanding truth. The development program designed at Consulting Connoisseurs on the basis of the Bhagvad Gita is, thus, a warehouse of immense knowledge that enables us to understand factors in the broader context. This training program helps one work around the problems of 'conventional business thinking' by enabling its participants to become more effective managers and leaders. Get trained in understanding businesses through the Bhagvad Gita and jumpstart on your next career move.

*This is taken from an article of Consulting Connoisseurs and further details are available at the website* www.consultingconnoisseurs.com.

## 9.2   The Race Between Cost and Time

Before we go further, we will just revisit the definition of projects for the benefit of the reader.

> A project is a temporary endeavor, having a defined beginning and end (usually constrained by date, but can be by funding or deliverables), undertaken to meet unique goals and objectives, usually to bring about beneficial change or added value.

The business context looks at the purpose. Hence, any treatment in project management is incomplete if we don't understand the concepts used to analyze these issues. In this section, we will attempt to touch upon these aspects.

We have seen in previous chapters as to how project feasibility criteria are used. In doing so, the business context integrates different aspects of business:

1. Financial
2. Operational
3. Legal
4. Strategic
5. Market and Competition
6. Government and Policy, etc.

Tools need to integrate these issues with the project *effectively*. The current section will describe how this is done in practice.

**A Quick Hands-On**

So, to take it forward, there are two cardinal questions for the Project Manager at this stage:

**How does one perceive the race between cost and time?**
**How can one integrate the project feasibility and business criteria with the project issues effectively?**

We will first discuss the perception regarding the 'race between cost and time'.

## 9.2.1  A Sample Case

Let us try to understand how project issues are calculated in practice. While we would like to give it some amount of flexibility, it is also essential to drive the concept in the right direction. So, we would ask the reader to get a feel for his own project and business acumen, by doing certain tasks at his end, but would use a sample to demonstrate how we would go ahead.

**A Quick Test**

A Project with an initial investment of USD 10 million is expected to give a Rate of Return of 20% p.a. over 10 years (2 years project duration). Create a cash flow stream for the same.

While there are hundreds of possibilities here, the purpose of giving this exercise is to help the reader understand the *dynamics and the sensitivities* of the cash flow streams for the overall Rate of Returns.

**Table 9.1**  Cash flow stream for a USD 10 million project

| Year | Cash flow in USD million (outflow is negative) |
|------|-----------------------------------------------|
| 1    | −5 |
| 2    | −5 |
| 3    | 3  |
| 4    | 3  |
| 5    | 3  |
| 6    | 3  |
| 7    | 3  |
| 8    | 3  |
| 9    | 2  |
| 10   | 2  |

IRR is 20% p.a. for the project

To consider our sample case, however, we will use the cash flow stream given in Table 9.1 as the basis. Note that we have tried to keep it as simplistic as possible.

Next, we calculate the Rate of Return for four scenarios viz.

1. There is an overspend of USD 2 million.
2. There is an overspend of USD 5 million.
3. The project gets delayed by 1 year.
4. The project has an overspend of USD 2 million and a delay of 1 year!

And we try to analyze the results. For the sake of simplicity, we assume that the overspends are more in the second year when it comes to the second case.

The resulting scenarios are given in Table 9.2.

Simplistically put, the table shows that an overspend hurts a project *more severely* than a delay. Hence, if we go by this table alone, the general impression can be ascertained as cost is a more important factor to consider.

**Table 9.2**  Cash flow streams for scenarios in the USD 10 million project cash flows in USD million (outflow is negative)

| Year | Base case | Scenario 1 | Scenario 2 | Scenario 3 | Scenario 4 |
|------|-----------|------------|------------|------------|------------|
| 1    | −5         | −6          | −7          | −4          | −5          |
| 2    | −5         | −6          | −8          | −4          | −5          |
| 3    | 3          | 3           | 3           | −2          | −2          |
| 4    | 3          | 3           | 3           | 3           | 3           |
| 5    | 3          | 3           | 3           | 3           | 3           |
| 6    | 3          | 3           | 3           | 3           | 3           |
| 7    | 3          | 3           | 3           | 3           | 3           |
| 8    | 3          | 3           | 3           | 3           | 3           |
| 9    | 2          | 2           | 2           | 3           | 3           |
| 10   | 2          | 2           | 2           | 2           | 2           |
| Rate of return | 20% p.a. | 14.69% p.a. | 8.88% p.a. | 15.45% p.a. | 10.91% p.a. |

So, if a project manager is asked: Is Time more Important or the Cost?; he would be tempted to answer that the cost is more important. This is not very far from the truth, as this thought process is well aligned with the financial and accounting experts, in the company. However, in doing so, the manager needs to be cautious on how he has to evaluate the situation. We will cover these areas in the subsequent sections.

In many management meetings, however, there will be situations where the owner-stakeholders would tend to control the costs and allow the time to 'fly'. Fundamentally, this behaviour is well represented and explained using the concept of the IRR and the variance of IRR with variances of costs and time. However, this is just the project perspective. One needs to also look at the business persective.

## 9.3   Important Concepts on Integration

In our attempt to integrate, we need to look at two important concepts.

### 9.3.1   The Strategic Time-Window

The Strategic Time-Window is the expected/exact Time Period when the implementation of a Strategy yields maximum returns. So, if a project is to provide a company with returns, the returns are maximum when it is launched at some specified time. For instance, if one is launching a product that is meant for summer, the return would be maximum only if the project is completed just before the onset of summer. If the project delays beyond the summer, the revenues would shift to the next summer. Thus, the timing is the key.

Similarly, when De-Hydro-De-Sulphurization (DHDS) was announced as a legal requirement for fuels, the project implementation to set-up the plant was governed by the legal framework. Delays would, thus, affect the bottomline significantly.

The business perspective involves a detailed understanding of the Strategic Time Window. Simplistically put, it tells us: Why the Stakeholders want it NOW!

If we go back to our case described in Table 7.1, we might want to look at the strategic time-window in a more concrete way. The total project cost was at USD 119.5 million in that case. One would normally need to represent the strategic time-window in an easily understandable manner. We are describing the changed business conditions in Table 9.3. Such changes are further modeled into the overall strategy plan of the company to give the actual 'numbers' concerning the strategic options and scenarios. Notice that each week is a separate scenario and there will be *a detailed backup available with the Chief Financial Officer at the company that describes the derivation of the table*. The project manager also gives his inputs while calculating the various scenarios. This has been done to a larger part in the earlier chapters. To put the aspects into perspective, all this integration is represented in the table, and that same curve is expressed graphically in Fig. 9.1. In the Figure, the time is along the X-Axis and one can see the 'Loss of Benefit' using the Benefit-Cost Ratio on the project. This is more in line with the 'readability' requirement of most managers.

**Table 9.3** Revision of the strategic time window

| Week | Loss of benefit over 3 years in USD million (defined as *intended strategic benefit–benefit if launched in the week*) |
|------|------|
| 26 | 40 |
| 27 | 20 |
| 28 | 12 |
| 29 | 10 |
| 30 | 2 |
| 31 | 0 |
| 32 | 30 |
| 33 | 40 |
| 34 | 50 |
| 35 | 100 |
| 36 | 150 |
| 37 | 150 |
| 38 | 150 |
| 39 | 150 |

*Details of revision*

Info on week 35: Competitor (Lead Market Rival) has advanced his project to week 35 and has an intensive marketing campaign

Info on week 32–35: Potential loss of sales due to shorter time to market. Alternatives are infeasible

Info on week 31: Revised target due to competitor launch. Week identified based on market intelligence, media campaign, etc.

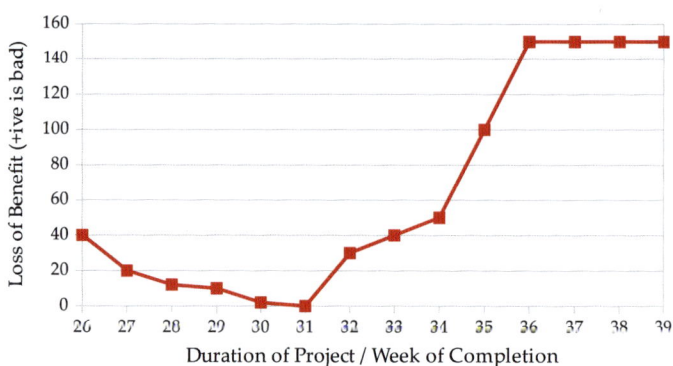

**Fig. 9.1** Graphical representation of the strategic time window

## A Quick Test

Given this information, and incorporating the same in the decision-making process, what would the recommendations of the project manager be? In this case, certain operational risks are factored in, through our earlier discussions. With the new information, most managers will get a better picture of how their project situation is.

The cardinal question to ask at this stage is:

**Given the Strategic Time Window, how does the decision on compression, in the previous chapter, change?**

The reader can work out ways to maximize the benefit for the 3-year decision criteria. Is this different from the earlier norms? If one looks at most projects, this additional information could (and most often does) change the results of a compression exercise!

## 9.3.2  The Opportunity Cost

In the previous discussion, we touched upon the concept of the strategic time windows. But that is not all that a project manager needs to see! Surprised? Unfortunately, it is true and we will demonstrate how…

To make matters easier to understand, let us consider our relationship with a bank. Let us say that the bank gives us 6 % interest. Now, this might be a lucrative figure in some countries, while in others, it might not be as good as the competition. We are not debating the value. However, as a customer, I would expect a bank to give me USD 6 for every USD 100 I have with them, *at the end of the year and regardless of how they use my money to generate the amount / surplus*. This is an important pre-condition. So, it is up to the bank to ensure that they are able to pay me the interest rate and are also able to generate some surplus as their own profits. If the bank fails to deliver my 6 % interest, it would be a breach of the conditions of the relationship! Consequences could be disastrous.

Any project scenario is no different. It is like any allocation of funds, and the simple definition states that

In any allocation of funds, the opportunity cost is the economic value of alternatives that could be candidates for the funds.

So, if one takes two alternatives, the one with lower return is the opportunity cost of the alternative with higher returns and vice versa.

For a business to be successful, the opportunity cost of an allocation must be *lower than the return from the allocation*. Please note that this is a very important concept. The opportunity costs can cause people to be fired and share prices to go down! One often hears about change of leadership in companies due to poor stock performance. This is nothing but the understanding of the opportunity cost. In other words, the opportunity cost tells us: Why the Stakeholders chose THIS Project/company over another!

Since a lot of stock holders and banks and other agencies provide funds to businesses, the business itself will in-turn look at ways to maximize the utilization of the funds. The best way to do this is using the opportunity cost analysis. In the opportunity cost analysis, candidate projects are taken and analyzed. The Opportunity Cost is, thus, based on a portfolio of possible allocations. The Opportunity Cost is

**Table 9.4** Evaluating portfolio of opportunities/candidate projects

| Project | Investment in USD million | NPV of returns in 3 years (strategic benefit) in USD million | Returns on investment using NPV in % |
|---|---|---|---|
| Current project | 119.5 | 180 | 50.6 |
| Alternative option 1 | 130 | 160 | 23.1 |
| Alternative option 2 | 100 | 145 | 45.0 |

important to ensure that the investment remains lucrative for the shareholders. This is shown in Table 9.4.

The table gives us an interesting picture.

$$\text{Available Total Buffer} = (50.6 - 45.0) \times \frac{119.5}{100} \tag{9.1}$$

It states that the maximum 'over spend' or 'additional investment' permissible in the current project is only USD 6.7 million. Anything beyond that would mean that option 2 would be a *better option to invest* than the current project! This is a potentially dangerous situation now!

---

**A Quick Test**

As a project manager, given this additional information:

**What do you think would be a suitable strategy to work in this situation? Explain why.**
**Are the Strategic Time Window and the Opportunity Cost published in your company? (Does the PM know it and understand the criteria?)**

These are some of the basic issues that one needs to look at.

---

A common problem that I have seen after interacting with numerous project managers is that most companies do not explain their strategic time windows to their managers. What is worse, even the opportunity cost is not known to most managers. This becomes a rather uncomfortable situation for most managers. On the one hand, they are under pressure to perform and are continuously driven 'against the wall' by their owner stakeholders, and, on the other hand, any alternative they come up with, doesn't seem to be 'getting a timely approval'. So, there is often a perception that their efforts are running futile. Many managers often express frustration when their efforts to maintain the schedule are not adequately supported by the management.

Interestingly, the moment I discussed the issues (mentioned here) regarding the strategic time windows and the opportunity costs with most managers, they realized that they were not having the complete picture and were mis-reading the decisions of their own management. Thus, this simple exercise helped them understand and appreciate the broader business context associated with project management.

**A Better Way?**

Companies need to realize that professional project managers today can understand how the opportunity cost is to be used in their project scenarios. They also understand that their project managers understand the strategic time-windows better. However, without appropriate tools for integration, the two seem to be going *in different directions, if not different planes*. This costs the company in terms of the synergy.

To solve this problem, the company should reveal these figures when requested by the project manager. A good manager, using the tools and techniques described so far, is in a fairly good position to evaluate these parameters and make better and more informed decisions.

Often times, these parameters are revealed much later to the project manager. Unfortunately, this strategy neither helps the project manager nor the stakeholder in achieving their objective. It is essential for companies to develop a broader perspective and work towards more integrated solutions. Empowering and training their project managers to understand these aspects and tools is the first step towards making the organization more dynamic and profitable. In any case, a customized framework for a company needs to be *jointly developed* by the project manager, the owner-stakeholder, the Chief Risk Officer, and the Chief Financial Officer.

## 9.4 Key Takeaways

Understanding the truth is probably an important aspect of any management philosophy. In the present work too, we have touched upon this aspect in the very beginning. Distortions in the corporate environment are highly subjective and sensitive. While most people would readily agree to this view, they have difficulties in estimating the volumes and the numbers behind such a phenomenon. The first part of this chapter focused on the concept of Truth in terms of its many forms.

Subsequently, we also wanted to provide the manager a feel for how the knowledge and information gaps between the strategic and the operations levels reflect in decision making. The evaluation of projects from their business or strategic perspective is, often times, very different from that at the operational levels. In most literature, these aspects are just mentioned in the project selection criteria; they actually play a far more significant role when it comes to actual implementation and controling. Most project managers end up spending a lot of their time, effort, and energy in 'planning' areas to maintain schedules and costs that quickly get 'scrapped' by the top management. One of the fundamental problems here is that the project manager *isn't equipped enough to handle the information effectively*. Fortunately, using the modeling- and simulation-based approach that is demonstrated here, the reader would now be able to appreciate how to use them in practice. Therefore, with our demonstration through the example in the chapter, the complexity of the decision is made clear and a suitable framework can be worked out by the company/project manager.

# Part IV
# Active Project Risk Management

# Chapter 10
# Basics Mantras of Risk Management

**Abstract** In this chapter, the basic mantras of risk management are covered. Like the previous chapter on mantras in general project management, this is an advanced treatment on risks. Six cardinal mantras of Risk Management are defined here. The first mantra involves the understanding of a Risk Prioritizing Framework that incorporates the three basic parameters in defining priorities. The second mantra introduces a new concept called the Golden Rule of Risk Focus, which helps the project manager to understand how to take on priorities in the Risk Management Scenario. The third mantra touches upon focus and delegation of risks along the organizational hierarchy and taking it further from there. The fourth mantra introduces the reader to a new concept called the Risk Web that is used to define, understand, and control the risks in the organization. The fifth mantra touches upon the sensitive aspect of risk evaluation cycles. The last mantra touches on the alignment and leveraging of the Risk Breakdown Structures along the Work Breakdown Structures and the Organizational Breakdown Structures.

**Keywords** Risk management · Mantras · Risk prioritizing framework · Golden rule of risk focus · Focus and delegation of risks · Risk webs · Synchronizing risk definitions along risk webs · Risk evaluation cycles · Risk breakdown structure · Work breakdown structure · Organizational breakdown structure

We had looked at probabilistic distributions earlier, but one needs to differentiate between the distribution and risks. Most times, the practising manager considers the modeling, by means of a distribution, as a model for risk. This is far from true. A risk is something that is *adverse to the business objective*. Hence, anything that isn't 'adverse', despite being a distribution, will not qualify to be a risk. As we go along, we will clarify this concept and explain how modeling and simulation are used in the risk management faculty.

While we do that, one needs to also understand that many times project management theory tries to create a 'deterministic view' to a 'probabilistic parameter'. This is done for two basic purposes:

© Springer Science+Business Media Singapore 2017     161
N. Gurjar, *A Forward Looking Approach to Project Management*,
Lecture Notes in Management and Industrial Engineering,
DOI 10.1007/978-981-10-0782-8_10

1. To give it operational simplicity. For instance, the duration of an activity is 7 days. This statement is simpler to understand than the alternative one: The duration of the activity ranges from 4 days to 10 days and more than 80 % of the time, it is expected to complete either on Day 6 or Day 7…
2. To enable decision making. For instance, the project manager needs to choose between alternatives that have a wide variety of possibilities. It is, therefore, necessary to develop a 'deterministic view' to enable decision making.

However, in the broader scheme of things, both these can link up with the concept of risk management as long as *they are meant to be directed to the reduction of the probability of the adverse outcomes or the level of adversity of the outcomes.*

## 10.1  A Taste of Practice!

Let us understand the basics with a simple hands-on.

> **A Quick Hands-On**
>
> Every practical project manager appreciates the risks involved in any project. However, from an institutionalization perspective, processes of risk management aren't always clear and standardized. Hence, take the past 3 projects that have been completed by you and answer the following:
>
> **What were the risks identified at the beginning of the sample projects?**
> **In your risk management, what were the alternative plans to workaround the risks?**
> **Did any of the risks occur? When was it confirmed? What was done to workaround the risk?**
> **Whose responsibility is/was to track the risks?**

One of the main problems in project management is Risk Management. Like project quality management, risk management is also perceived as a 'negative' concept. We have already discussed what implications this has on the 'truth' elements. However, the project manager needs to understand this concept and use it to his advantage.

In most projects, risk definition is probably the only aspect one actually sees about risks management. However, that too is grossly inadequate. We will touch upon these aspects as we go along. *Active Risk Tracking and Control is by far the most ignored area of Project Processes; but is the most talked of!* Hence, it is an ironical situation, yet true with most organizations. The other important aspect is that Risk Management is a cumbersome exercise. It doesn't typically have any quickies in management. Hence, there is a dearth of skill levels and most people end up relying more on their experience rather than any formal tool to manage the risks. This concept only *internalizes the risks*. It is like the gymnast, who is on the walking beam, who might believe that it isn't risky. However, to anyone watching the show, it is definitely a very risky affair!

A similar case in point is the Leaning Tower of Pisa. While thousands visit the tower everyday, the 'leaning' nature does create a scare in the minds of those who climb up. I recall our visit there in 2004 and, it was definitely a feeling that one was doing something risky! However, many people find it amusing and not risky at all! In other words, the perception plays an important role in such situations. Nobody applies laws of physics or engineering or management for that matter, when they try to go up the Leaning Tower!

### 10.1.1 The First Mantra of Risk Management

There are risk-based scenarios for everything: Cost, Time, Scope, and Quality. Most companies stop at Costs!!! However, have *alternative plans* for the aspects: Time, Scope, and Quality issues as well. This is the most important aspect for any project manager to consider.

Update the triggers to each alternative plan regularly. Publish the criteria for using each alternative plan. List the plan implications in your Risk Reporting!

With this quick mantra, we trust that your risk management system will suddenly become effective in the range of 60 % or more! Of course, the remainder 40 % is a little more elaborate and needs a little more clarification at this stage.

## 10.2 Some Basic Checks to Start

We have already covered a few basic checks when we began our treatment on project management. We will just hold the same 'thread' and move ahead to describe a few basic checks on project situations.

### A Few Caselets

*Caselet I*
Great, now we are able to look at overseas markets. This is a big order coming our way…We decided to quote the price of the project in USD, but the currency fluctuation has been too high lately. We should have probably used the Euro instead. *Do you know what impacts your profitability?*

*Caselet II*

"You have got to reach the office this evening. Take the flight at 6..."

"Fine..."

"Where are you?"

"I am sorry, I missed the flight!!! I was late there…"

"You should have started at 4 from the hotel…"

"I did Sir, but there was a huge traffic jam and the vehicle tyre punctured mid-way on the deserted highway!!! Never expected both to happen…"

*Do you know what impacts your operational feasibility?*

*Caselet III*

The boss looks like an 'academician' in risks… Each time there is an issue, he wants to understand the situation, the workarounds, the Plan B, and Plan C. I told him that if I have to focus on defining my Plan B and Plan C, I cannot work on the project!!!

*Do you have the right people who understand risk management?*

---

**A More Targeted Hands-On**

Our earlier hands-on was more oriented to the definition and the timing. We now look at the broader question of improving project skills.

   **Give two instances to demonstrate Risks in your recent projects. How were they contained? Did they occur? How were they managed (once they occurred)?**
   **Can you give a better way in which the situation could be managed?**

---

This is the key question to understand. It triggers a whole new approach towards risk management. We are looking exclusively at *forward looking management frameworks*. Hence, in a forward looking framework, there is no correction possible from hind-sight. Everything has to be done based on information available or potentially available at any given instant of time.

## 10.3  Prioritizing Risks

The first strategy toward de-risking is the risk mitigation strategy. Simplistically put, risk mitigation involves the lowering of the *probability of the risk* or the *adversity expected of the risk*. In other words, the risk is 'reduced' using some maneuvers.

**A Quick Practical Hands-On**

List out many (at least 20) typical risks in your project scenario. Pick any two that you want to mitigate. Explain the criteria used to pick the two candidates.

**Explain your mitigation methodology in detail. What is the cost of the mitigation?**
**How 'safe' are you with your mitigation strategy?**

Explain the method you used to determine the 'safety'.

Any practicing manager will understand the complexity involved in risk decisions once he tries to logically answer these questions. Most times, project managers just identify risks and put them in the 'Risk Master Table'. They then keep reporting the risks. The action plan needs to be drawn in advance! However, most managers do not believe they need to work out the painful details. The results are evident as overruns! No amount of certifications in Project Management can save one from the gory details of working out a mitigation strategy :-)

## 10.3.1 The First Fundamental Analysis

On the one hand, risk management requires a meticulous treatment of the details. On the other hand, it is relatively easy to maneuver once we know where to focus. There are many ways of doing this. We will define a simple framework here. It is called the Consulting Connoisseurs Risk Prioritization Framework. One can easily identify the parameters required for risk classification.

They are (as shown in Fig. 10.1):

1. *Impact of the Risk*
   Whereby, the impact is taken for the risk, when it occurs. For instance, if a key team member leaves midway of a software project, the impact on the project could be high. At the same time, a delay of a few hours in the arrival of a long-lead shipment may not have a critical impact on the project (although several operations might be affected).
   So, a project manager needs to assess these elements carefully. Again, there are several metrics possible here like the costs, the manhours, the waiting times, etc.

2. *Probability of the Risk*
   While the impact is definitely the first significant parameter, the probability of the risk is another major one as well. For instance, a nuclear explosion in the US tomorrow might have very serious impacts on the project. But the probability of the risk is extremely low. Hence, it may not be of any consequence to draw alternative plans for the same.

**Fig. 10.1**  Parameters of
risks

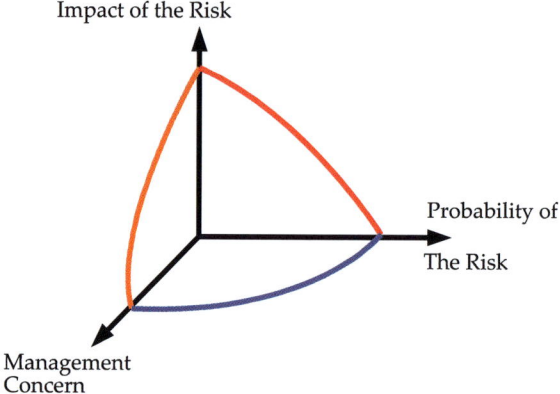

A project manager, while evaluating the probabilities, must consider his own experience, the track record of his company and his own *gut feeling*. Each of these would give him different values and he needs to understand how to use them judiciously.

3. *Management Concern*

   Most frameworks in the literature actually ignore this aspect. According to our experience and observed practices, however, this is *the most important parameter in prioritizing the risks*. Any practicing manager would subscribe to this view, many times management directives 'force' the manager to work on different types of risks that are, according to the manager, not necessarily the top priority!

Again, like the reporting metric, it is possible to combine these parameters and define a kind of a *ranking system* for risks. Since this is again dependent on the organization, we leave it to the reader to define and identify a combined metric that helps identify these factors. In any case, the project manager needs to:

**Give ranks/categories to each of the risks listed out in the previous exercise. Define the criteria used for ranking explicitly.**

## 10.3.2   The Second Mantra of Risk Management

This mantra is also called *The Golden Rule of Risk Focus* in the Consulting Connoisseurs Risk Prioritization Framework, and is stated as follows:

1. Set cut-offs for Ranks.
2. Select those risks with important ranks.
3. However, ensure that you have modeled the three parameters correctly for every risk.

4. Since risks change along the Project Life Cycle, you will need to *repeat* this procedure periodically.

As simple as it sounds, this mantra is pretty difficult to implement in practice. It needs an astute project manager to ensure the focus is right and in the direction desired.

## 10.4   Controlling Risks

In the previous section, we saw how the risks can be defined and ranked. The Golden Rule then spoke of how one needs to focus on the risks. However, this method is silent about a few issues. In other words, the Chief Risk Officer has given directives for the risk ranking framework. The manager has identified the risks, ranked them and prioritized them. So far, so good. Yet the cardinal question to the manager is the following:

**A Quick Hands-On**

**We have determined the key risks to focus on. What happens to the rest of them?**

Again, the literature is largely silent on these aspects.

In attempting to answer this question, we need to understand about the working of the organization.

**A Quick Hands-On**

List out the project organogram in your organization. Explain the authority to approve organizational spending in your respective organizations with respect to the organogram.

**Identify the number of approvals made at each (final) level in the past 5 years (you may need to dig into SAP???). Are all the approvals linked with the limit? Explain your answer.**

Many organizations have an *operational setup* different from the *formal setup*. This is an important aspect to understand for any manager and he should understand how to use this information to his advantage! Just like your decision trends, risk management trends have a tendency to skew… In other words, bigger risks might be managed by lower rungs of the organization! This is a dangerous situation, but does happen many times in projects.

### 10.4.1  The Third Mantra of Risk Management

Always ensure that risk management is comprehensive in your organizations. Most organizations start with comprehensive lists, but then, they fail to use their organizational hierarchy to leverage their efforts.

*Focusing on one set of risks doesn't mean ignoring the others.* On the contrary, it means that these risks have to be delegated to the lower levels for tracking... Understanding the dynamics of the organization is important while delegating the authority to the lower levels.

## 10.5  Classifying Risks

While listing out the risks and ranking them and mapping them to the organization, one also needs to understand and identify the types of risks and a better way to classify them.

---

**A Quick Hands-On**

Exchange the risks you identified with someone in your peer group (in your organization). Now discuss with him/her to try to understand which of the risks are:

1. Independent: Whereby one actually sees that the risk is not 'related' by means of any *cause–effect relationship or a complimentary relationship* with any other risk.
2. Dependent: Whereby one risk gets triggered, *partially or wholly*, by another risk.

Explain the dependence using a cause–effect relationship (and a flow model). Discuss your answer with your partner.

To conclude, define *independent 'clusters' of risks.*

---

### 10.5.1  The Fourth Mantra of Risk Management

In order to ensure good project management practice, always cross verify your risk assessments with your peer groups. While determining risks, be careful about the following:

**1. The Double Counting of Risks.**
**2. The Missing Out of Important Risks.**
**3. The Dilemma of Including or Excluding Rare Events in your list.**

Have a clear policy for each of these!

We now introduce you to the Consulting Connoisseurs Risk Web Concept. In most project scenarios, there is a web of 'root-causes', 'relationship or causal handles', 'risks' as we have defined them, and the 'impact'. This is shown in Fig. 10.2. For instance, a root cause of a key executive leaving the organization can have a wide variety of impacts ranging from overall project chaos, cost overruns, time overruns, etc. However, this isn't operating in isolation. If the systems and processes are inadequately formalized, the impact is greater than otherwise. So, the risk of poor systems and process implementation got triggered by the leaving of the executive (the causal relationship!). Hence, to summarize, the root causes of the impact are more than just this one. Hence, one needs to understand these issues and adequately factor in each of the elements so as to ensure a robust model to define the risks.

While this is a common problem, it depends on the elements that are *listed in the Risk Master*. If the elements are the risk definitions, as we have defined, then there is a huge scope for error due to phenomenon like double counting, etc. At the same time, if one were to look at the root-causes, the problems are more *controlled*, as the project manager knows the *exact element to control*. This method is better for projects where the differential probability of occurrence is relatively high for some causes than the others. In case the differential probability of occurrence is relatively equal, then the project has to probably be modeled through the causal trigger perspective.

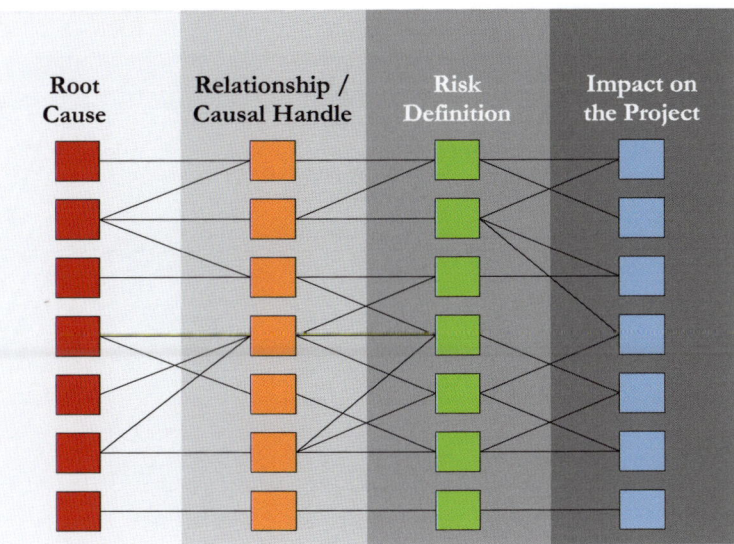

**Fig. 10.2**  Typical risk webs

A project manager must remember that:

**It is recommended to use the perspective that gives the minimum number of controllable factors. Hence, any of the four perspectives could be interesting depending on the project situation and the ability of the project manager to 'cluster' them.**

One of the important aspects that the reader might notice in this book is that there is a lot of emphasis on *hands-on exercises and discussions*. This is because a lot of project management concepts are based not only on the rigor of tools, but also on the situation and, more importantly, on the context. Therefore, in order to bring all the elements under the same 'scanner', we have attempted to ensure that there is a balance in the concept and the reader *stays connected with his reality at all times*. This also helps us maximize the efficacy of the content in this book, without *risking* the project manager/reader from going overboard, in an attempt to implement something new.

## 10.6   Re-visiting Rankings

When consultants come on-board, they bring with them a set of advanced methodologies, tools, and techniques. These may be understood by many managers, but they often don't have either the manpower or the management support to implement and evaluate their situation. This is probably true with most management disciplines. However, there are a few areas that do require these tools and techniques. One of them is the Risk Ranking faculty.

**A Quick Hands-On of a Different Kind**

A project has four areas of significant importance:

1. Schedule
2. Cost
3. Scope
4. Quality.

Assign weightages to each of these areas on the basis of your last project. Discuss it with your peers. Next compare them with values obtained by using the methodology of AHP. The reader is advised to read on AHP from any management textbook like Saaty [11].

### 10.6.1   The Fifth Mantra of Risk Management

Remember that you have to order risks for *every project you undertake*. It is not sufficient to do the exercise for one project and apply it to the others.

For long-duration projects (>18 months), you might want to repeat the procedure after reasonable time (9 months). The revision will help align the risk tracking better.

Most importantly, *the advantage with AHP is that it allows 'imperfect' perceptions to be modeled; hence, being more useful* (as it is easy to use as well).

## 10.7   Risk Hierarchy

We have spoken of the definition and the classification of risks. Often times, however, Risk is also expressed as an RBS, i.e., a Risk Breakdown Structure. While this is not very popular with *every industry*, there are some merits in defining risks in a hierarchial manner. This helps the project manager to combine various perspectives.

Thus, two important practices in Risk Management include:

1. The Maintenance of the Risk Register.
2. The Organizing of Risks.

Like scope, risk also is often depicted in levels. In other words, there is a concept of Risk Breakdown Structure (RBS) in many companies.

A typical RBS helps managers organize risks, create a hierarchy, and track them later in a more efficient way.

---

**Another Hands-On**

Develop an RBS for the project you just completed. Discuss the hierarchies with your peers and superiors. Check for inconsistencies, overlaps, and impreciseness during the discussion.

Map your RBS on your current Organogram. Explain the criteria used. Would you like to change the Organogram? Give reasons.

Compare this with the Job Descriptions in your company. Propose at least two changes in the Job Descriptions based on your findings here.

---

Needless to say, this is a complex exercise and yet, a very important one. Just as we mentioned earlier that quality needs to 'seep in' at every level, we will find that RBS mapping is an important exercise to help the concept seep-in at every level. While delegating the responsibility, it is also important to ensure that the organization re-inforces the delegation. In other words, changes in the job descriptions are many times required.

### 10.7.1   The Sixth Mantra of Risk Management

Always try to leverage the presence of your WBS in managing your RBS. Since WBS's are mapped to the OBS (Organizational Breakdown Structure), it is easier to manage the RBS along the lines of the WBS.

Never try to redefine the organizational structure on the basis of the RBS. The OBS must align itself with the WBS and the RBS must leverage on the WBS and the OBS to create meaningful delivery. The role of the Chief Risk Officer is instrumental in these cases.

## 10.8   Key Takeaways

In this chapter, some basic mantras are defined and developed. The reader might find a bit of conceptual 'criss-crossing' as we covered the mantras. The reason for sequencing the mantras that way was to allow a 'reasonable train of thought' to enable the reader to take it to a minimal conclusion. Once the conclusion was obtained, the concept was progressively revised.

The other aspect was that the *extent of explicit risk management processes* in practice, in any given company, varies significantly. Therefore, it is all the more important to understand the various aspects of risk management from the perspective of the extent of implementation.

Nevertheless, these mantras are extremely useful for the reader and the project manager can implement and start seeing results almost instantaneously.

# Chapter 11
# How Risky Is Your Risk Management?

**Abstract** Contrary to books that cover conventional risk management, this chapter focuses exclusively on the practical failure points in the Risk Management Process. It is, therefore, a unique compilation of critical 'audit' points when one is evaluating the risk processes. A total of nine such failure points are described in this framework.

**Keywords** Risk management · Risk assessment · Risk processing · Risk transition · Risk perspectives · Risk capture · Risk control system

## 11.1 Introduction

Risk Management is a highly specialized science today. Its widespread application has helped radically improve the management of businesses. Every company, therefore, strives to control its risks through robust systems for risk management that look like a typical 'risk race' (Fig. 11.1). However, many times fundamental issues in the business processes revolving around the aspect of risk management actually put the whole concept of risk management to risk! What are these issues? Let us take a closer look.

## 11.2 What Is Risk?

In simplistic terms, a risk is a probable situation that is adverse to the business, the key words by definition being probable and adverse. The key parameters in risk management, therefore, revolve around two cardinal questions:

---

This chapter is taken from two white paper of the author on www.consultingconnoisseurs.com. You may visit the website for more details.

N. Gurjar, *A Forward Looking Approach to Project Management*,
Lecture Notes in Management and Industrial Engineering,
DOI 10.1007/978-981-10-0782-8_11

**Fig. 11.1** Overcoming risks

1. How probable and
2. How adverse the situation could be.

From a management perspective, there are three philosophies used to handle risks. These are

1. *Risk Mitigation:* where one talks of reducing either the probability or the adversity of the risk,
2. *Risk Sharing:* where the attempt is to share risks between businesses and
3. *Risk Transfer:* where a complete transfer of the risk from one business to another is carried out.

Some people also include Acceptance as a philosophy, but we do not subscribe to the view in this book! All risk philosophies 'accept,' but in different ways. These philosophies need to be carefully understood in the context of the business, the stakeholders, and the business partners. Depending on the philosophy chosen, the company can optimize its risk exposure to meet the corporate strategy. While many companies operate on the risk sharing and risk transfer philosophies, the risk mitigation philosophy actually positions the business way ahead of the others! This is because a risk mitigation philosophy works at reducing the intrinsic probability and adversity of the risk. The other philosophies may not always achieve such an intrinsic reduction (depending on the way it is worked out). Is your company a believer of risk mitigation?

## 11.3  Phases in Risk Management

The risks in any risk management system actually stem out of the risk management process. Broadly speaking, they occur in different stages:

1. Risk assessment,
2. Risk processing, and
3. Risk transition.

The risk assessment stage focuses on the definition of the risk and its implications to the business. The risk processing stage actually looks at the integration of the risk management processes with the overall strategy processes. It also looks at the interface between the strategy and the operations processes from a risk standpoint. The risk transition stage looks at the issues involved when a particular risk moves from the 'probable' domain to the 'real world'. Each of these stages can be looked at from the process elements, process characterizations, and process perspectives. We will elucidate to help understand how these work.

## 11.4  Risk I: The Numerical Perspective

The idea of risk assessment is fairly easy to understand. However, determining the final deliverable of the exercise is often a big challenge. In the more rudimentary forms, one often encounters a qualitative treatment to assess the risk. While this treatment is good as an initial study, it is often one of limited potentials for beneficial use. A quantitative assessment of risks gives a more robust understanding of the risks.

**A Quick Test**

**Do you use a qualitative system for risk assessment? If yes, how do you ascertain the extent of the risks and the business impact each of these could have on the business environment pertinent to the sector.**

## 11.5  Risk II: The Extremes Perspective

The Extremes Perspective is also called the Extreme Axes that Balance Risk Philosophies.

Some companies use the Best Case-Worst Case scenario to assess risks. The challenge in these cases is the realistic definition of the Best and the Worst cases. How frequently do you find your 'reality' within the best and worst cases? Are all the parameters in your risk assessment system 'external' to the system or are they mostly determined from 'within' the system? If you are facing issues with defining the scenarios from this perspective, you might want to consider re-looking into your system. Do you have situations where your worst case is worsening progressively over time? Are there cases that are 'worse' than the 'worst'? If your answer is yes, you might want to optimize your scenario definitions.

## 11.6   Risk III: The Measurement Perspective

If you are trying to freshly design and implement your risk assessment system, you might want to align your risk management philosophy with the processes within the company. While there are different options used by companies to express the assessed risks, the most popular ones are those translated into costs. Expressions in costs are often preferred because they are easy to 'plug-into' an accounting framework. However, these often fail the test of business logic due to the limitations of business applicability of such accounting instruments. It is similar to painting pictures that create an eyewash, something that has to be avoided. It is, therefore, advisable to express risks in multiple parameters like time, money, goodwill, corporate image, operational reliability, environmental impact, etc. Does your company balance different parameters while assessing risks? If not, you might want to seriously reconsider your system. I have developed a tool by which one could look at risks from multiple perspectives, thereby enabling the understanding of the interplay of different elements and constituents of the risk. However, that being said, it is also essential to understand how combinations of risks could affect the business. Most corporates find this to be a major challenge. Does your system see multiple levels of risks or does it deliver a relatively flat structure of risks?

## 11.7   Risk IV: Method of Capture

The most interesting aspect in risk assessment is the concept of *the hidden risk…* A good system actually gives astounding results due to the fact that it helps one 'look beyond what one sees.' A well-implemented system also helps in the aggregation of these elements to enable this 'unique' feature. While posturing to risk management is a top management prerogative, the capture of risks becomes most effective when it starts 'low' in the organization. For it is often the operational level that actually sees the risks of the system the best.

**A Quick Test**

**Do you have a bottom-up orientation towards risk assessment? Are middle managers and line managers actively contributing to your risk identification process?**

If not, you might want to reengineer your risk assessment system completely.

## 11.8   Risk V: Periodicity

It is extremely important to maintain the right periodicity to assess the risks. A good system revisits the risks assessed and ensures that they are within the right limits for ensuring successful business. This helps in addressing the transient nature of the risks. Many corporates fail to understand this concept. Further, in addition to the transient nature of risks, it is often important to understand the dependence of a particular sequence of events. Systems tend to behave differently to different sequences of the same set of events, a concept that is beyond most accounting treatment of issues. One must, therefore, understand *the transient nature and the temporal nature of the factors that drive the risks in the business.*

**A Quick Test**

**Does your system incorporate the transient and the temporal nature of risks?**

My team has developed a unique method to identify and incorporate these issues in your overall risk management system. We will demonstrate this in subsequent chapters.

## 11.9   Risk VI: Control System

The next issue, that is probably the most challenging of them all, is that of the control mechanism. While it is a very fundamental issue that one must control some parameters to get desired results in other parameters, most companies have difficulty in defining their control mechanism. This actually leads to several issues. It calls for a good understanding of the controlling parameter and its relationship with the controlled parameter. Most often, the control mechanism is a 'hard to define' and a 'harder to implement' system. There are many simple tests to identify the robustness of the control system.

**A Quick Test**

**Has your company increased sales or made profits during periods of recession?**
**Ask any employee if he knows which of his actions help contain which kind of risks and to what degree?**
**How much have you changed in your working style and orientation between the risk profile of the company last year and now?**
**Can the changes be explained and justified fully?**

If you are having issues answering these questions, you might want to improve your controlling methodology.

## 11.10   Risk VII: Quality of Information

The skill in any management system lies in the process of decision-making. Management science revolves around the art and science of making sound business decisions based on incomplete information. As information forms the basis of good decisions, it is of utmost importance to understand the quality of the information that one is dealing with. Many factors decide the quality of information provided to the risk management system. However, having said that, it is essential to understand if the information is actually of good quality. Do you see many situations where there are 'big surprises' causing huge risks over short time horizons? Short strategic time windows or getting informed 'after the fact' are typical phenomena encountered in poor quality situations.

**A Quick Test**

**Is pessimism undervalued in your organization, thereby providing an overly optimistic picture? Does the management often plead ignorance while escalating risk factors? Do you see significant 'movements' in your targets over time (even if they occur in small incremental steps)?**

If your answer is yes, you might want to seriously consider an improved reporting methodology involving greater objectivity.

## 11.11  Risk VIII: Spread of Analysis

Most management systems are extremely limited to the knowledge pool of the management. However, learning organizations have now become increasingly popular, especially in technology companies, as they provide for more efficient ways to go beyond the knowledge pool of their management into the talent pool of the organization. A good spread of the risk analysis actually helps improve the understanding of the key elements and provides adequate insight to effectively manage the process.

## 11.12  Risk IX: The Transition

The Transition: The biggest challenge from a business perspective is when a risk actually becomes a reality for the business! While many companies have good risk assessment and risk management processes, they fail at identifying the rules of the game when the risks transform into reality. A risk transformed into reality most times, gets converted into another 'fire fighting measure' or just another 'challenge in the real world' and is often dealt with in that way. This actually defeats the purpose of having a risk management system. A good risk management system helps one identify the drivers and have a good handle on the drivers of the risk thereby, steering the process and maneuvering the business to success, even when it becomes a reality. It ensures that there are adequate feasibility checks to understand how this issue would be tackled in reality.

## 11.13  Key Takeaways

While there is a lot of sensitization to issues like the culture of the corporate, the strategic postures, and the outlay of the management process, it is essential to understand that risks in most cases start 'low' in the business. The term 'low' has two-fold meaning in this context: (a) low from the organizational hierarchy and (b) low in magnitude or value. Management must realize this fact and encash on this aspect when they design their risk management system.

While insurances, service providers, and lawyers are some of those agencies that actively assess risks in monetary terms, a corporate must understand how its risks are characterized from not just monetary or financial terms, but also from the elements of time, image, goodwill, etc. A simple test of the efficacy of the risk management system would be the situation where one would have 'no surprises' at the operational level!

# Chapter 12
# Project Management in R&D

## Misconceptions Galore!

**Abstract**  This chapter touches upon a less researched area of project management viz. R&D projects. It begins with the characterization of R&D projects that are typically misunderstood in their fundamental premises. Most people associate such projects with significant breakthrough innovation, while most R&D projects are not in the objective breakthrough innovation to that extent. Hence, given this backdrop the chapter touches upon the key challenges in such projects. Notable among them are those of defining exit gates, getting the right team sizes, and having a robust estimation model.

**Keywords**  R&D projects · Breakthrough innovation projects · Exit gates · Planning · Estimation · Exit gates · Teams

## 12.1  Introduction

From my experiences across various corporate functions including sales, operations, IT, consulting, projects, and research and development, I have seen that the research and development function is the toughest of all. We take a closer look at the challenges facing this area and the implications of the same.

## 12.2  Misconceptions in R&D

Most R&D endeavors are essentially efforts to create a 'new' element for the society or mankind. Hence, it is often argued that *the objective is not clear.* This is the first misconception that most people have of R&D. In most projects, the objective is

---

This is an article written by the author while he was in the US. It is taken from his blog. This article explains the dynamics of risk management and the challenges in intrinsically high-risk environments like R&D. It points out the exact nature of misconceptions and management issues that one faces in the R&D sectors. One can see more of this in subsequent chapters.

© Springer Science+Business Media Singapore 2017                                    181
N. Gurjar, *A Forward Looking Approach to Project Management*,
Lecture Notes in Management and Industrial Engineering,
DOI 10.1007/978-981-10-0782-8_12

very clear. However, the final outcome is not clearly known. However, the possible approaches to the objective are fairly well known.

The development of these approaches is the subject matter of 'conceptual tool-makers.' These people try to focus on the thought process that generates new concepts. Creativity, for instance, is one such faculty that they focus on. Most people think that *creativity is an inherent attribute in R&D projects.* However, this is another misconception that most people have. Most R&D projects run on applied methods and are not 'very creative' in the way they go. Creative talent, normally, is restricted in application to R&D projects involving breakthroughs or in emerging sciences. For instance, if one is interested in developing a transportation device, creative instinct might bring in solutions like 'Jet-Belts of Flash Gordon,' 'Pushpak Vimanam of Ravana,' 'Cryogenic Levitation,' etc. However, such projects are breakthrough projects and often involve a very high risk. Most companies are, therefore, reluctant to finance such projects. Most such projects speak of low success rates like 5 % or so for the development of new microchips and involve an incredibly long time. All said, the creativity lasts only as long as the project scope is defined. Once the project scope is well defined, the room for creativity reduces.

## 12.3   Challenges and Market Responses

Having said this, we now enter the domain of management. It is well known that R&D projects normally need iteration. That is, in other words, the project steps might have to be repeated. The key driver is normally a change in the initial approach. Most times, iterations need to do things in a different way. This brings in an important factor that manifests itself as a major challenge to management science. *We need here to deliver something without knowing the success rate, the number of iterations, and the appropriateness of the approaches used.* In short, we have a floating time frame, a floating budget, and a hazy picture of the outcome to start with! Corporate executives are very uncomfortable with such a situation. They would rather prefer someone to give some concrete time frame, a fixed budget, and a very clear picture as to what they can expect.

It is interesting to try and evaluate the risks in typical R and D projects. It is estimated by certain experts that most companies had much less than 2 % of their turnover dedicated to R and D. They have, thus, run on traditional models of investments, where one knows exactly what one buys, and tries to see how to implement it. This approach is often justified by 'old timers' as one that helps factor out the lower productivity levels in the workforce. However, if one looks at it more closely, it is not so. Anyway, the reason I am pointing this out is that one often avoids R and D. What happens then? We pay for the R and D efforts of another company. Needless to say, this makes the technology extremely expensive. If paying dearly for a technology is more acceptable than learning to manage the intricacies of such projects, then we are on the right track. This is significant because the investment cost goes up, but the running costs are much lower. Hence, the return on investment is assured to be high.

In the war of intellectual properties, however, this is probably *the worst strategy for a country, despite the economic relevance.* The problem is aggravated due to the fact that technology transfer is normally done for 'older' technologies. Hence, it is always a second best option.

**The trick in R and D projects is, therefore, efficient planning and controlling.** The R&D projects typically need a technocrat and a manager as their leaders. If either one of them is missing, we could have big problems. It is to address these specific niches that we have techno-management programs in most B-Schools around the world (US, Europe, and even developing countries like India). The techno-manager has his base both in technology as well as management. He is ideally in a position to evaluate the management risks associated with technological development.

Thus, **the determination of exit gates, development strategy, and a good estimate for the planning is often a key to R&D project management.** An interesting feature about innovation is that of team work. If one takes a look at most of the products that have been launched, one finds that the initial 'working' prototype was often the result of one or two persons, for example, Ford and his car, Edison and his inventions, the Russian developers of guns, etc. In very rare cases, one finds it to be more than two. However, the initial prototype is later developed further by teams (sometimes very large teams) and most corporate R&D projects stress on team work for the initial development! The success rates can well be justified!!!

It is interesting to see the financial implications of such models. Typically, large teams are expensive, and the chances of success are lower! So, each additional iteration implies a huge cost and makes this kind of teamwork nonviable. Hence, in most companies, accretion-based R&D models are used for optimizing the existing designs or prototypes.

## 12.4   Key Takeaways

From a management perspective, it is important to identify the direction of development and ensure concurrence of these with the objective of the project (something like the balanced scorecard). To this date, there are no reliable management tools to assess this factor. The second issue is to understand the iterative nature of the developmental work. This requires a practical approach rather than a perfectionist's approach. To understand this, think of a bug-free software code. Now think of a software code that is released in…say 60% of the time with 20% noncritical bugs. *Setting these standards is often a matter of corporate strategy and operational strategy.* Hence, the manager must focus on the same. Depending on the strategy and the decision criteria, the iteration could be partial or total. It is, in most cases, difficult to predict, nevertheless important to develop plans/alternatives for both partial and total iterations. This allows better management control. However, the existing tools in this area are highly mathematical, and the applications are, therefore, restricted to the technocrat managers. This is one more challenge for project management to take care of. What's in store for the future? Well…we wait and see!

# Chapter 13
# Tools in Risk Management

**Abstract** This chapter looks at established tools and techniques that have come from the Operations Research domain. Popular tools like the Decision Tree are first understood. We then modify the objective function using the utility function concept. In doing so, the chapter also touches upon how this could be used against the procrastinator manager. We provide with a few mantras on how these tools could be successfully applied in practical situations. The chapter then covers some of the common strategies that one could derive based on the analyses of decision trees like the Risk-Averse strategy, the Ready-to-Jump strategy, and the Balanced strategy. We then move on to understand another important concept called Game Theory and the applications of Game theory in reporting. The phenomenon of Schedule Chicken is touched upon and the common 'CYA' strategy is also touched upon using this tool.

**Keywords** Operations research · Decision tree · Utility functions · Procrastinator manager · Mantras for utility functions · Strategies for decision tree analysis · Game theory · Payoff matrices · Schedule chicken · CYA strategy

We now shift our focus to understand some of the tools in risk management. While there aren't tools available for every problem in risk management, the project manager is definitely in a position to develop and document relevant tools for his/her company. Companies that have developed such tools and techniques have been many; notable among them being DuPont, ICI, Phillips, etc. These companies have also developed their own powerful methodologies for managing projects and integrating them with their management systems.

Our focus, in this chapter, is restricted to tools that have a numerical flavor rather than those having a generic qualitative treatise. Hence, we delve into the specifics of certain developments in the project management arena that are specifically related to the dimensions of risks.

© Springer Science+Business Media Singapore 2017
N. Gurjar, *A Forward Looking Approach to Project Management*,
Lecture Notes in Management and Industrial Engineering,
DOI 10.1007/978-981-10-0782-8_13

## 13.1   A Brief History

Risks were formally treated in the 1980s when Chapman and Cooper gave a reasonably rigorous approach with PERT, Decision Trees, and Probability Decisions. They called it *Risk Engineering.*

Cooper went on to define what is known today as the *Risk Breakdown Structure,* something in the line of the Work Breakdown Structure! This was the **Risk Cost Approach** which is very common even today.

In the late 1980s and early 1990s, Kangari and Riggs introduced *Fuzzy Sets Theory* to risk management and used it to model risks and it demonstrated amazing results. In the 1990s, Yeo spoke of *Contingency Engineering,* to help people understand risks from the view to define the contingency. In 1991, Mustafa and Al-Bahar applied AHP which was a breakthrough to the traditional risk cost approach.

In the mid-90s, the idea was to model risks using AHP to combine both *subjective and objective elements.* Riggs went on to define quantitative and integrating tools for technical, cost, and schedule risks using *utility functions.* In 2007, Zhang and Zou used the combination of AHP and FST called the *Fuzzy-AHP approach.* In 2008, Hans gave a new method involving *Significance-Probability-Impact Assessment* of risks.

## 13.2   Critical Issues

While we have seen several critical issues in the earlier chapter, the important aspects for the reader, from a *tools and techniques perspective*, are to be understood in greater detail.

---

**A Quick Hands-On**

Risk management is based on the probability of occurrence, and any complex risk assessment is meaningful only if the probabilities and impacts are reliable. So, the cardinal question for the reader is to understand if

**One's risk statistic is good enough or if one needs to invest in further investigation to ascertain the quantum of risk involved.**
**In other words, is it worth investing in finding out a more accurate statistic?**

---

This is a tricky question and a lot depends on the *sensitivity of the decision to the parameters involved.*

**A Quick Hands-On**

After the last project review meeting, the following was the situation:

1. It is 60% likely that the project got delayed by a week, considering the current plan. The total project cost is USD 3000 million and the penalty is 0.5% per week (assuming the only effect).
2. It is 70% likely that the project can be completed on-time if we change the current plan. However, the cost of this change could be USD 5 million.

   **What would be your plan of action?**

## 13.3  The Decision Tree

The decision tree is among the most popular tools to capture risks. Yet, it is seldom used in projects *in an explicit way.*

While *Expected Monetary Value is a very popular measure in the accounting circles, it is not used in practice.* One would rather include a *utility function* instead of simple monetary value. It is believed that the utility functions give a better handle on the decision-making process than just cost.

One of the most concise treatments on the decision tree was done by Olivas in his primer for professionals [12]. This in conjunction with the elementary concepts given by Hillier and Lieberman [13] gives an excellent understanding of the theory behind decision analysis.

### 13.3.1  The Conventions

The decision tree is relatively simple to construct. Each decision is represented by a square. The alternatives then branch out of the *root decision node*, as it is called. These branches of options could further split based on the probabilities available. The outcomes can then be expressed and the tree can be closed. In our current problem, the decision tree would, therefore, look like Fig. 13.1. The tree ends with a triangle.

So now, if the project manager evaluates, he finds that he is better off by doing nothing. The expected value of doing nothing as such is USD $-6$ million. However, if he spends money, he brings it up to USD $-5 + (-4.5)$ million or USD $-9.5$ million.

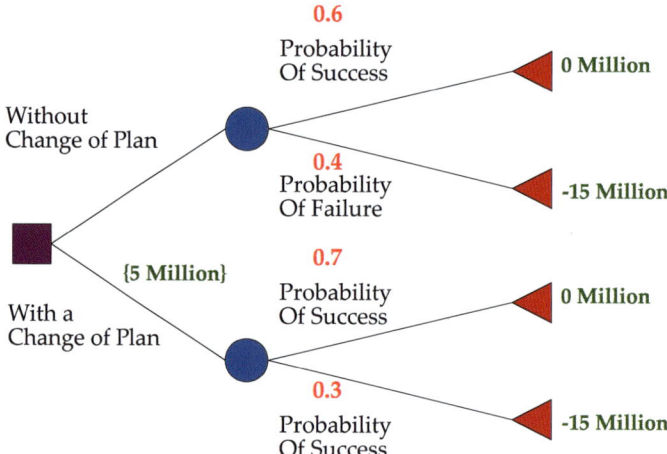

**Fig. 13.1** Decision tree for the hands-on

## 13.3.2   *More Information*

To incorporate more information, let us return to the hands-on and try to solve it again.

> **A Quick Hands-On**
>
> As the project manager, you are asked to come up with your choice. You know that rigorous expediting (to give a more accurate statistic/controllability) along with all existing measures for the current plan could help in 70 % of the cases. However, you know that this is an expensive proposition.
>
> Management wants to understand your position. They ask you for a budget estimate on the rigorous expediting. Your opinion is USD 1 million.
>
> **What would be your plan of action given this additional information?**
>
> With this additional information, one could draw a third decision option to the existing ones and find out the expected values that are important. One finds that this option is a better one in the *overall scheme of possibilities given*. However, we already know that this is not the only perspective one needs to look at.

## 13.4 The Utility Function

Very often, the cost perspective neglects the other management aspects. For instance, it might be possible to do a task in 8 h, but that would mean a great many things like ensuring that the people cut down their breaks, the lunch is brought to the table, the clarifications are immediate, etc. In other words, there are certain *uncommon conditions* that are used in such cases. These are *conveniently neglected* in the cost functions. Hence, the next model that was made popular was that of the utility function. The advantage of the utility function is that it enables the combination of multiple parameters into one. This is important many times, as the utility function *can incorporate* the other areas as well. However, just like what we have seen in the reporting context, these are complex to define and get in a representative and usable form. However, there are ways of doing this. Operations Research theory speaks of various tools that can be used, AHP being one of them.

### *13.4.1 An Application*

Now let us take the utility as follows:

$$\text{Total Utility } U = \frac{(-2)(\text{Total Cost in Million})^2}{15} - 10 \times (\text{Delay in Weeks})$$

What would be your plan of action for both the earlier problems given this info?

Given this information, the value is going to change significantly. Instead of the traditional cost, we also have a delay parameter that will affect the utility function. Considering this, each of the branches is going to appear different.

Without a change in plan, the cost incurred is nil for the no-delay case (60 %). However, there is a potential delay that could affect 40 % of the cases. This has an implication of USD 15 million. Therefore, the utility function for the first option of not making any changes now translates to

$$U(\text{Option 1}) = 0.4 \left\{ \frac{-2 \times 15 \times 15}{15} - 10 \times 1 \right\}$$

which yields a result of $-16$ units.

With a change in plan, the cost incurred is USD 5 million as a minimum. Therefore, the utility function for the first option of making the changes now translates to

$$U(\text{Option 2}) = 0.7 \left\{ \frac{-2 \times 5 \times 5}{15} \right\} + 0.3 \left\{ \frac{-2 \times 20 \times 20}{15} - 10 \times 1 \right\}$$

which yields a result of $-19$ units.

Thus, looking at expected values of the utility functions, we find that the first option is chosen. One can keep doing such an exercise to find out the better decision.

Note that most literature actually *includes the cost of the action while determining the cost of the payoff.* Some books also show them as separate entities to enable a *cost–benefit analysis* of the overall situation. We have digressed from the convention, therefore, in order to enable the reader have a flexibility of understanding that these are two separate entities and that *they can be optimized separately.* The idea, however, is not to confuse the reader, but to help him understand and evaluate his own detailing process.

Further, there is a lot of detailing that goes in any analysis. The management usually sees an *end result* which is often very *heavy with assumptions.* Hence, this allows the reader to understand the 'layers of optimization' that are involved in the overall process.

The one aspect that every project manager needs to understand is *how should a utility function* be defined in the first place. There are multiple perspectives and we will touch upon these in later sections. However, the right point of start is to determine the most appropriate utility function.

## 13.5   Incorporating the Procrastinator Manager!

The advantage with the utility function is that it is possible to incorporate the delay in decision making *more readily* than the pure cost approach. Let us understand this using a quick hands-on.

### A Quick Hands-On

If one wants to include predictive methods in the decision tree, one could do it as follows:
1. Delaying the decision by a week could increase the probability of delays of up to 2 weeks in Plan A by 10% and reduce the same from on-time completion
2. Delaying the decision of plan B would reduce the possibility of effective completion to 50% due to lesser time to plan the remaining activities.

   **What would be your plan of action for both the earlier problems given this info?**

Most senior managers would be very uncomfortable with such a model, and in practice, *it has the potential to explode into a blame game.* Moreover, the other important aspect is that one could see the victimization of the project manager through the potential long and winding 'interrogative' discussions that could try to prove the theory to be a false one!

The reader is requested to check the results of this analysis at his end and see what the numbers look like. Further, a quick sensitivity analysis based on both the cost-based evaluation as well as the utility-based evaluation is requested.

### 13.5.1 Two Quick Mantras

While using decision trees in a project, always have a good estimate of the cost of indecision before you start. To make this successful,

**Ensure that a reasonable Utility Function is approved right at the start of the project by the owner-stakeholder as well as the Chief Risk Officer.**

Procrastination is one of the most significant phenomenon found in project managers, which renders the exercise of risk management futile. This is an unfortunate observation, but true, in most organizations. The end result is beautifully summarized as two populist strategies used by managers:

1. *Keep-Subordinates-Busy* Syndrome and
2. *The-Ready-to-Firefight* Mentality.

The reader also has to pay attention to the next mantra:

**Avoid using explicit decision trees if you feel the discussions will focus on sensitivities rather than action-oriented plans.**

I believe the reader would easily understand these aspects and is able to use it in the right perspective.

## 13.6 Some Strategies for Decision Trees

Hillier and Lieberman [13] have beautifully explained some of the strategies that are popular and the context in which they need to be used. From a project perspective, these are not very complex to adapt.

### 13.6.1 Risk-Averse Organizations

A general prescriptive approach is given below:

1. Take the minimum along each path. Then, find the maximum value in those minimum values. Take this as the decision path. This is called *the Maximin criterion*.
2. The Maximin criterion is used by pessimistic managers. The focus is to base decisions on some kind of a *Minimum Guarantee*.

3. It is used by risk-averse organizations where *risk minimization is achieved using low-risk exposure.*

A common stereotype among financial experts is to try and minimize the risk exposure. When the organization is more 'driven' by their financial gurus, the tendency is to have the best possible 'worst case.' That being said, this approach is also true with other organizations that generally have a risk-averse approach.

### 13.6.2   Ready-to-Jump!

Many managers, however, have a different tendency. They try to look at the probabilities rather closely. Here is how this thought process operates:

1. Take the alternative that has the *Highest Probability.* For this decision, take the situation with the *maximum payoff.* This is the decision path. The criterion is called the *Maximum Probability criterion.*
2. The Maximum Probability criterion increases the risk exposure and is used by *Overly Optimistic Managers.* The focus is to base decisions on *the most optimistic scenario of the most likely case, ignoring inputs of other cases.*

### 13.6.3   A Better Balance…

With the rise of the 'techno-managers,' risk management has readily imbibed some of the better-known mathematical tools of probability theory. One such rule is given below:

1. Take the expected value of the payoffs. Take the decision with the *highest value of the expected payoffs.* The criterion is called the *Baye's Decision criteria.*
2. It is supposed to be robust as it calculates all the possibilities and, thus, includes all the information.

   Baye's Decision Criteria works well even with utility functions in most situations. It is the most popular approach in strategic risk management today.

   In the discussions on the decision tree, the one aspect that the manager must understand (a mantra) is that

   **The choice of the strategy is important. However, once chosen, it is even more important for the manager to understand how his control parameters look like, rather than revisiting the decision.**

This is one of the most critical aspects in the management of risks.

## 13.7  Game Theory

The decision tree is a very sound way of evaluating options and making decisions. It is extremely popular in strategic project management globally. However, *the decision tree sometimes fails in an interactive scenario.* When two organizations come together and have diverse interests in a project situation, they tend to decide on options that may not align well with each other.

So, one might need to go for something that is significantly different in this case. Game theory is being used in modern project management to understand these issues. The manager must realize that

> **A decision tree looks at your best posture, but can be considered a game against 'nature' (something passive). An active approach would be to use Game Theory and try to understand the different perspectives involved.**

Game theory starts off with answering the question *What's In It For Me...*

It then builds up a typical *payoff table* and tries to model different strategies to look at the outcomes from the payoff table. In short, *a simulation is done in these cases.*

Game theory has *successfully explained several interesting phenomena in project management, especially in negotiations and reporting.*

## 13.8  Basic Game Theory

A lot of discussion here is based on Hillier and Libermann [13] and their treatment done to explain the concept at the undergraduate level. To explain the concept of Game theory, there is a need to understand the nature of interactions. The simplest type of game is the *two-person-zero-sum* game. The important qualifier to this type is that there are two players and the gain of one is *exactly* the loss of the other. Therefore, the sum total is always zero. This is an important assumption.

So, the simple characterization of the Game theory is the definition of a *payoff matrix* that actually is a result of combinations of strategies. The important assumption here is that *each player knows the strategies she or he has available, the ones of the opponents and the payoff table.* The second important condition in this model is that each player simultaneously chooses a strategy without knowing the choice of the opponent.

Note that the sum of the payoffs on any row or any column is zero as shown in Table 13.1. This is how a typical game would look like. In determining the payoffs, most times managers need to consider utility functions rather than pure costs. For instance, a promotion means more than just a few dollars per month. So, it is essential to incorporate the right values before moving on. Again, since there are other players as well, 'similar' dimensions would exist and could be defined for the others involved in the scenario. This is difficult, because different people could have different priorities for the same dimensions of the utility function or even different objectives for

**Table 13.1** Payoffs in a zero sum game

|          | Strategy | Player 2 | |
|----------|----------|-----|-----|
|          |          | 1   | 2   |
| Player 1 | 1        | 1   | −1  |
|          | 2        | −1  | 1   |

the given utility function. Therefore, the definition of the utility function needs to be *simplistic and reasonably have a short-term time frame to enable speedy application.* One could go on listing out strategies for larger problems, but that would make it very difficult to incorporate and model.

Game theory believes in the development of rational criteria for selecting strategies. Hence, both players are rational and both choose their strategies to solely promote their own interests. This is important for the reader to understand. In case the reader wants to practically use the concept of Game theory, he needs to understand the assumptions and develop a *utility function* that appropriately models the behavior of the other player and his value system.

We will now apply this concept to reporting and demonstrate how this can be used to model a common phenomenon called the schedule chicken.

## 13.9   The Schedule Chicken

The term 'Schedule Chicken' is used in project management and software development circles. The condition occurs when two or more areas of a product team claim they can deliver features at an unrealistically early date because each assumes the other teams are stretching the predictions even more than they are. This pretense continually moves forward past one project checkpoint to the next, until feature integration begins or just before the functionality is actually due. The practice of schedule chicken often results in *contagious schedule slips due to the inter-team dependencies and is difficult to identify and resolve, as it is in the best interest of each team not to be the first bearer of bad news.* The psychological drivers underlining the 'Schedule Chicken' behavior in many ways mimic the Hawk-Dove or Snowdrift model of conflict.

Thus, this situation is actually dependent on the type of reviews that are done. Generally speaking, there are various types of project reviews that occur and every project manager must be experiencing the same. In most project reviews, any small indication of 'bad' news is viewed seriously and can often result in a difficult situation for the project manager. Hence, every person facing the review has three options that need to be clearly identified:

1. Reveal a Problem or a Potential Issue
2. Hide the Problem or the Potential Issue
3. Distract the Attention (by revealing somebody else's problem).

**Table 13.2**  Potential outcomes

| | | Other members | | |
|---|---|---|---|---|
| | Strategy | Reveal | Hide | Distract |
| Project manager | Reveal | Problems to solve! | Project manager is the chicken | Other members promoted (aka project manager fired) |
| | Hide | Other members chicken | Live and let live | Exposed (Inefficient project manager) |
| | Distract | Possible promotion | Exposed (other members inefficient) | Political problems |

These are fairly straightforward postures and could have different connotations in different contexts. Over the several projects that I have seen, the key to such a problem is *the nature and behavior of the project manager*. The way a project manager handles information, understands actions, and is able to comprehend what is expected from him and his team to ensure a successful project...ultimately defines the kind of reporting one tends to see.

Let us assume a 'charged' project environment where there is a 'political mine-field.' Here, the typical payoff matrix would be as shown in the table below.

**Table 13.3**  Payoff utilities of the schedule chicken situation (for the project manager)

| | | Other members | | |
|---|---|---|---|---|
| | Strategy | Reveal | Hide | Distract |
| Project manager | Reveal | +5 | −2 | −20 |
| | Hide | +2 | +8 | −6 |
| | Distract | +10 | +6 | +1 |

In Table 13.2, it is a challenge for the project manager to choose a strategy. Now, let us develop a payoff matrix for the situation as shown in Table 13.3. When given such a utility table, the first thing to check is whether there is a *dominated strategy* for the project manager.

To reiterate, when a project manager is confronted with such a matrix, the first point to check is if there is a dominated strategy visible. A dominated strategy is one of those strategies where, regardless of what the other members choose, the project manager always stands to gain less than another strategy. In other words, all outcomes of some other horizontal row *must be at least as good as or better than the dominated strategy*. In the current situation, the Distraction strategy gives higher outcomes for any condition chosen by the other members than the Reveal strategy. Here, we see that $+10 > +5$, $+6 > -2$, and $+1 > -20$.

Having this information, everyone going for the meeting knows that the project manager is not going to 'Reveal' much during the meeting. In other words, he is going to either Hide or Distract (that is, expose somebody else) in the meeting.

Similarly, the Distract (i.e., expose the project manager) strategy pays the maximum dividend for the other members as the other strategies are dominated. That is, $+20 > +2$ or $-5$, $+6 > -8$ or $-2$, $-1 > -6$ or $-10$. In other words, both the other strategies do not yield as much dividend as the Distract strategy for the other members. Hence, the situation moves toward a Distract–Distract posture of the project manager as well as the team members.

This is an extremely simple case to consider for any project manager.

### 13.9.1  Another Variation of Game Theory

Now that one is a little more comfortable about the applications of the utility functions in Game theory, let us move to a little more complex case for the project manager. Suppose there isn't a dominated strategy to consider, eliminate, and make life simple. What would the project manager and the other members do? We will demonstrate this using a 'CYA' strategy.

In a typical CYA strategy, the reviewer at the other end is a formidable personality who comes down heavily on every party. At the same time, he does not allow the other members to get undue advantages by means of promotions, etc. In such a situation, both the project managers and the other members are not interested in risking larger losses than necessary. For the sake of simplicity and demonstration, an example of such a matrix is shown in Table 13.4. In such a situation, by selecting the Reveal strategy, the project manager actually could win as much as 12, but could lose as much as 6. Moreover, the other members would try to seek a strategy that would avoid the project manager from making a large payoff. In the Distract strategy, the project manager could win as much as 10, but the other members could also cause him a loss of 8.

Therefore, the best possible choice for the project manager is when they actually look at the *minimization of the maximum losses*. In other words, this is also an example of a Minimax criterion. The criterion is proposed by Game theory because it holds even if the strategy is announced. Thus, the second strategy is the one that the project manager would adopt.

Now the other members need to make an intelligent choice. They could lose up to 10 or 12 if they were to choose the Reveal or the Distract strategies. Hence, they might want to settle for the Hide strategy. This is, therefore, a Minimax criterion for the other members *when we consider the same payoff table*. Thus, both of them would choose to Hide and be happy:-)

**Table 13.4** Payoff utilities of the variant (for the project manager)

|  | Strategy | Other members Reveal | Hide | Distract |
|---|---|---|---|---|
| Project Manager | Reveal | −6 | −4 | +12 |
|  | Hide | +4 | 0 | +4 |
|  | Distract | +10 | −4 | −8 |

### 13.9.2 A Crazier Situation!

The earlier situation had a 0 payoff point making it an easy approach to a fair game. And it had both the maximin and the minimax values in the same entry. The position of such an entry is called the *saddle point* and is the key to solve such situations, and the position was a *stable one*. This meant that when the project manager or the other members changed their strategy, they were going to have worse outcomes than what the derived outcome was.

However, there are situations where the outcome of such an exercise is not stable. In real-life situations, strategies change during meetings, and the payoffs too could change. The knowledge of the other participant's strategy could provide one with newer avenues for 'optimizing' one's own benefits. We, therefore, give a key mantra here for the project manager:

> **While these are several of the important cases in game theory, the project manager must understand the following two things: (a) Interpreting Strategies and the Utility Functions from Multiple Perspectives and (b) Control of Pay-offs by keeping them Dynamic so as to avoid becoming the 'chicken' in front of his peers and team members. Being tolerant to bad news is a smart and strong strategy to resolve the schedule chicken early on...**

## 13.10 Key Takeaways

In this chapter, we have covered a few tools of risk management. These are traditional tools that have been popularized in Operations Research. The reader is recommended to read elementary OR books to ensure a better understanding of these tools.

The decision tree is a very popular tool used in project management. It is one of the basic tools in decision sciences. Of course, there are many situations where decisions are made using other approaches, but in all the approaches, there are alternatives and there are outcomes. While the alternatives are decisions made by the manager, the outcomes are usually the areas where there is a probabilistic flavor. Hence, this simple tool helps in focusing on them. The strategies that are to be used in making decisions are also an important aspect.

While deriving the outcomes, we also saw that various elements could be clubbed together using utility functions. We will deal with them in greater detail in subsequent chapters.

Another important tool was the application of Game theory. We have seen how this could be used to model project phenomenon like the schedule chicken. Game theory helps in managing and controlling outcomes. Just controlling the payoffs and incisively taking on measures by understanding potential strategies of participants help in developing a good rationale for successful management of projects.

# Chapter 14
# Risk Parameters

**Abstract** As a concluding chapter on Risk Management, this chapter looks at certain risk parameters and how the risk management methodology ties into their definitions and applications. The focus is on parameters, processes, tools, and techniques used for risk control. The fundamental distinction between risk monitoring and risk control is first looked into. This is followed by a new framework called the Consulting Connoisseurs layered view of risk perspectives that is to be applied in the risk management processes. The chapter then goes on to touch upon how risk management ties into scenarios (not strategic scenarios, rather operational case scenarios). In dealing with risk management, we then focus on two interesting and important concepts that distort the practice of risk control viz. the sunk-cost fallacy and the greyhound fallacy. We then move on to another interesting phenomenon called the Last-Mile Phenomenon. The discussion shows how parallel tasks increase the complexity and add to the risks. We also show how controlling risks through meetings also has its own limitations. Following this discussion, we revisit the utility functions for risks and incorporate advanced concepts to define them. We also create a case for the application of systems thinking in risk management. As a concluding concept, we introduce the phenomenon of Micro-planning and discuss on how this needs to be effectively used by the project manager.

**Keywords** Risk management · Risk monitoring · Risk control · Layered view of risk perspectives · Scenario planning · Risk management linked with scenarios · Sunk cost fallacy · Greyhound fallacy · Last-mile phenomenon · Utility functions · Threshold method · Constant stimulus method · Method of limits · Systems thinking · Micro-planning

## 14.1 Introduction

In this chapter, we will try to understand the different perspectives involved in risk control. We will first revisit our four elements with a simple example. We will begin with a sensitive area to make matters clear and visible. While the concepts mentioned here could be numerical, they are not a part of the traditional project management

N. Gurjar, *A Forward Looking Approach to Project Management*,
Lecture Notes in Management and Industrial Engineering,
DOI 10.1007/978-981-10-0782-8_14

literature. However, they will show the reader potential areas of concern when one tries to 'stretch' the limits of the directives in conventional frameworks.

Most project managers today have difficulty in understanding two fundamental concepts:

1. Monitoring: Where a project manager takes information from the reports and compares the data with some values (usually some stored values) resulting in warnings/information to the user of the status by readings or alarms
2. Control: Where a project manager takes information from the reports and compares the data with stored values and then, sends definitive *instructions of action to entities within the project environment* to ensure that the output is affected by subsequent course of action/other input parameters.

In many companies, project managers have difficulty in controlling the outcomes because they do not have *independent jurisdiction* over all the actions that are required to bring the project back on line. Therefore, there is always a grey line between the role of the project manager in actively monitoring and actively controlling a project. This chapter is *about controlling project outcomes and not about monitoring them*, although we will look at the outcomes from a 'monitoring' perspective as well.

> **A Quick Test**
>
> At this stage, the practicing project manager needs to revisit the premises of risk management with fresh information and learnings he has had. In doing so, the following questions are important for the project manager to understand and work on:
>
> **Is the risk management system in your company a qualitative one or a quantitative one?**
> **How do you ascertain the extent of risks?**
> **How do you ascertain the business impact of the risks?**
> **What would be the most significant roadblock to improve the existing system? Why?**

The answers to these questions are more 'elaborate' than a 'quickie'. The reader is requested to answer them with a lot of critical thinking and diligence. While we touched upon this aspect in the previous chapters, we will delve into it, in greater detail, in this chapter.

## 14.2  Revisiting the Perspectives Involved

In our previous chapters, we spoke about four different perspectives. We now introduce the Consulting Connnoisseurs Layered View of Perspectives. For the sake of simplicity, we are building upon a summarization along the lines of conventional logical thinking as shown in Fig. 14.1. In most project situations, it is nearly impossible

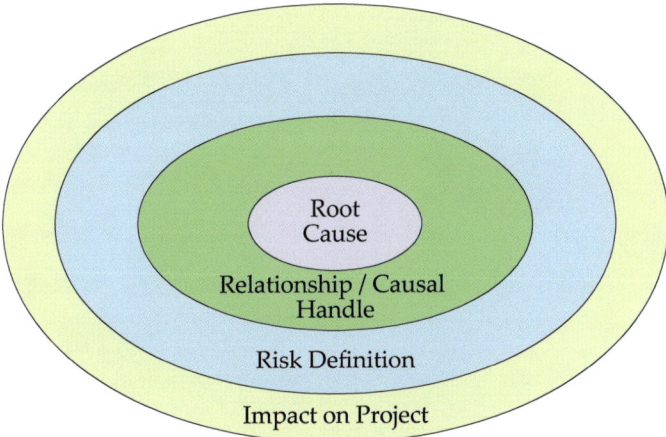

**Fig. 14.1**  Consulting Connoisseurs Layered view of perspectives

to get to the *true root cause* that triggers the risk. Therefore, much of the discussion in the risk management faculty is restricted to 'managing' the risks whereby, it is an accepted *precondition* that the risk will always exist. We will now explain the importance of the perspectives using an example to clarify the fallouts to the reader.

### 14.2.1  A Sample Case

For the sake of simplicity, let us assume a simple 'linear' project having five activities as shown in Fig. 14.2. The first activity takes 2 days, the second takes 1 day, the third takes 2 days again, the fourth takes 1 day, and the fifth activity takes 1 day. Now, the second and fourth activities are the activities involving the approval of output of the first and third activities. In other words, they are certain types of 'check points' or controls used in the project. This could be *internal or external*. In this particular case, let us assume that the review is external.

Let us assume that at the end of the project, the business would have earned USD 10,000. We also assume the cost of the capital is USD 100 per day. Now, given this information, we understand that a delay in the project would cost us a minimum of USD 100 per day. That is, in other words, the impact of the risk of delay on the project.

Since there are two entities involved: the approver and the agency, the next logical sequence is to define the risk. In this case, the risk definition is defined as *Not Obtaining Speedy Approvals*. This is a very common risk that is often contractually defined in most projects. This is also in our risk definition layer. Most organizations are fully aware of this risk and they always ensure that this is explicitly mentioned

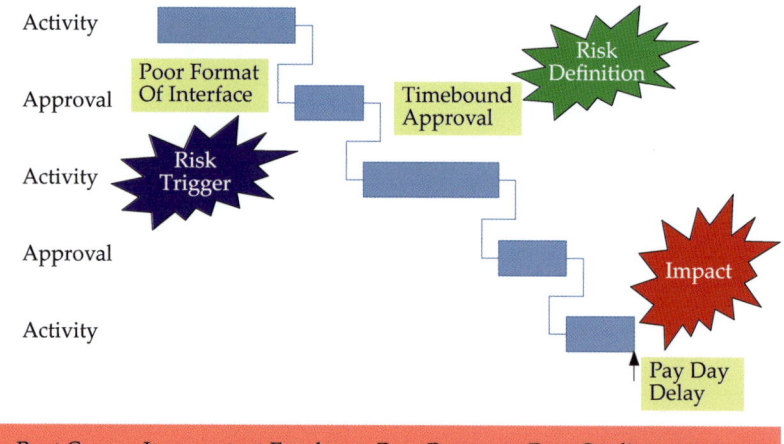

**Fig. 14.2**   Simple 'Linear Case'

and worked around. In fact, several discussions in project reviews revolve around the risk definition layer.

Now imagine that the project does get delayed due to an approval issue. At the review, this is definitely a discussion point, whereby, the approver indicates that the submittal is not *proper* and, therefore, the immediate quick fix is to define a format for the submission. This is nothing but the *causal handle or the Relationship Perspective*. Most project management literature actually addresses issues in this layer rather elaborately. The reader will have seen several books and papers that speak of formats for meetings, follow-up spreadsheets, timesheets, engineering documents, etc. These are nothing but ways and means to address the causal handle perspective in any project situation.

However, the actual root cause could be something else. If one looks at the *root cause of the issue*, one could find several root causes like:

1. Incompetent employee of the contractor
2. Incompetent employee of the approver
3. Poor processes of the contractor
4. Poor processes of the approver organization
5. Poor quality of the document
6. Too much of workload at the approver's end
7. Absence of tools to effectively exchange and process information
8. Differential expectations between the contractor and the approver, etc.

The reader will agree that any one or more of these parameters could actually be causing the situation. This gives a very different perspective on what actually needs to be controlled and what needs to be 'monitored'. However, in most project discussions, the indicator is *mistaken* to be the control variable.

**A Quick Hands-On**

Most risk management literature, however, is silent on these 'root causes.' This brings us to a very difficult situation, viz., what are the kinds of risks that are *normally admissible for management reviews*. Hence, the cardinal questions for the project manager are the following:

> **What are the Root Causes of the Risks?**
> **Which of them are admissible as 'risks' in the project?**
> **How do we get a fair representation of the root causes?**

While the root causes mentioned above might be at a 'lower' level in the hierarchy, the risks need to be classified in different strata to allow appropriate manageability. For instance, an incompetent employee is typically a risk that is extremely low on the hierarchy; it probably could put a huge risk on the project. In other words, the assignment of work to such an employee could be the causal handle to many risks. It is, therefore, essential for the reader to understand that the 'root causes' could be really 'low' in hierarchy, but they could have a profound impact on the business. The key point that is being driven here is that:

> **The risks when 'low' are easily controllable and amenable to quantitative tools and measurement**

This is potentially the biggest road-block to improvising a risk management system. Further, quantitative methods give one the exact nature and an exceptionally good handle of risks and their management. This is pretty much absent in qualitative methods. It is, therefore, important to understand how the control variables are and what the indicators should be.

## 14.3  Scenario Planning in Risks

**A Quick Hands-On**

Consider the upcoming project of yours.

> **What is the *best case scenario*, from a risk perspective, for this project?**
> **What is the *worst case scenario*, from a risk perspective, for this project?**
> **What is the *realistic case scenario*, from a risk perspective, for this project?**
> **What are the *Root Causes and the Triggers of each of these Scenarios*?**

It is very difficult to define scenarios in the business world. While it is relatively easy to ask for 'best', 'realistic,' and 'worst' estimates of a project, the scenario determining these estimates is often more difficult to explain. And rightfully so! Let me clarify what we are driving ourselves up against (out here).

Imagine one is going from Mobile, AL to Atlanta, GA by car. And if one were to ask for estimates, one could come up with an optimistic estimate of 4 h, a realistic one of 5 h, and a pessimistic one of 10 h. And now, if we have to define the scenario for the same…it would be a totally different ball-game: traffic conditions, weather conditions, accident on the way, mechanical faults with the vehicle, driver not driving 'fast enough'…the list could be a really long one. In this whole process, we are only *listing out the risks. Please note that we haven't yet built up the scenario!* And if one has to build a scenario, these risks need to be quantified and defined for each of the scenarios.

When one has the challenge of building up a scenario with such a good handle of the 'potential occurrences of the risks', one needs to *make a judicious choice of the combination.* In other words, a manager is often confronted with the dilemma of defining such a 'balanced and palatable' set of choices. So, the problem in scenario building is now a complex one! The idea of scenario planning is *not to predict the future, but to explore it.* Most project managers and stakeholders have a misconception on this one. They look at the exercise as a 'prediction' and, therefore, fear unwanted accountability due to it.

An excellent summarization on this topic in the context of strategy planning is given by Charles Roxburgh [14] in the McKinsey Insights. He has elucidated a number of points in his work:

1. Always develop at least four scenarios
2. Identify at least 3–5 major variables
3. Scenarios need a 'base' or a 'catchy' case
4. Scenarios must have catchy names.

However, in my humble opinion, building a scenario is an art rather than a science. The important aspect is getting the delicate balance between what the project manager wants to do and what the project context actually perceives. Therefore, it is less about the content of the scenario that is important, rather the way in which a scenario is used. However, in the world of higher mortals, this is always a challenge. Most times, like mentioned earlier, the top management is not very well informed on how the scenario is supposed to work. We will first delve into these issues in a little more detail.

### 14.3.1   A Quick Recap on Fundamentals of Scenarios

The fundamental aspect in scenario planning is that it enables *multiple perspectives* to be built into a case. This is, however, different from that of conventional risk management. The conventional tools believe in understanding the risks and their probabilities and then, trying to control the outcomes. Hence, it is very *specific to the application and the risk at hand.*

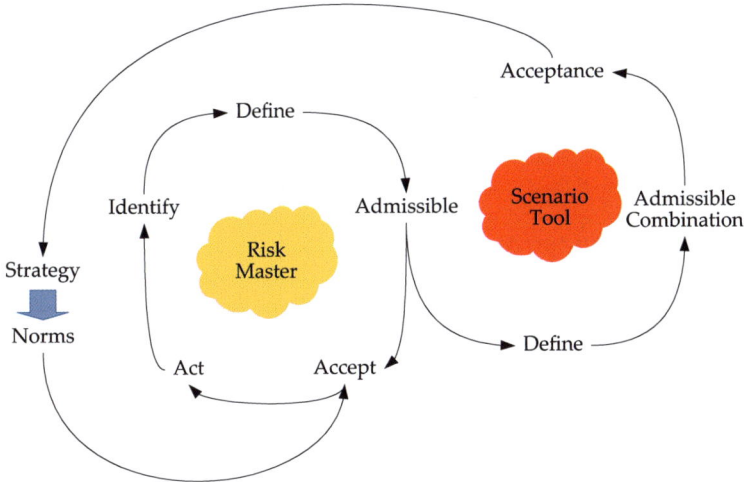

**Fig. 14.3**  How risk management and scenarios are linked!

Scenario planning on the other hand, is a little more generic to give the manager an understanding of a *portfolio or a combination of risks*. Thus, the use of scenario planning is to craft a generic strategy at the business level and identify norms that are to be defined for managing specific cases of strategy. The difference is shown in Fig. 14.3.

Unlike typical risk masters that are monitored and 'controlled' during execution, a scenario planning exercise on project risks is not done continually. Typically, project managers are bogged down with a definite objective and a definite time-frame and a definite budget, among other things! In other words, the outcome is always just one. In scenario planning, one tries to link the scenarios or outcomes to the reality and, this helps the project manager understand *the dynamics of the linkage better.* In other words, they are said to be an *Outside-In Approach.* So, the idea is always in understanding the potential possibilities and seeing how the relationships with the existing project are. This helps in deriving useful strategies and norms to work out ways and means to work around issues.

A scenario is not, however, a tool to be used every other day! It requires the commitment of the management to define the situations and draw meaningful conclusions. An astute project manager must realize that:

**Scenario Planning often helps define new Risk Management Strategies and Norms. Most importantly, they also help identify the 'tipping points' where one strategy has to be discarded and a new one has to be brought in.**
**Since it focuses on how the outcomes link to the current, it directly points to the causal handles and the root causes of risks**

Yet, scenarios are often used in periodic project reviews without much seriousness:-) The result is the TTTH Syndrome (*Talk Through The Hat Syndrome*) that is extremely popular in project circles. What is essential to note is that the scenario, when loosely used, in plain English language, is similar to a guesstimate. Hence, it actually gives a different interpretation altogether.

However, scenarios are extremely powerful as they give a fantastic handle on the way a risk has to be managed and the potential fall-out of a given project situation, especially in combination with a desired or a likely course of action.

### 14.3.2   Two Phenomena to Be Wary of

In building risk management scenarios, two basic phenomena need to be carefully considered and factored in. These two have a tendency to change the perception of the risks. Hence, it is essential to have a scientific method to ensure that one is not overstretching the concept of risk definition and risk evaluation.

**Sunk-Cost Fallacy**

The sunk cost fallacy is an interesting one. One of the more recent works have been done by Sandeep Baliga of Kellog, NW University. A few excerpts of the findings are given below to help people understand.

The term sunk cost fallacy refers to the tendency of humans to stick with something even if they know that it is a bad idea to keep continuing, simply due to the fact that they have already sunk so many resources into the endeavor.

As an example, one could look at a movie ticket that one purchases and sits through despite it being totally boring. Here is another simple example to describe the sunk cost fallacy. Joe owns a business that is losing money. Due to overwhelming competition, there is little hope that Joe's business will ever become successful enough to actually turn a profit. Joe should close down his business, but he refuses due to the fact that he has already sunk so much time and money into the venture. This is the sunk cost fallacy.

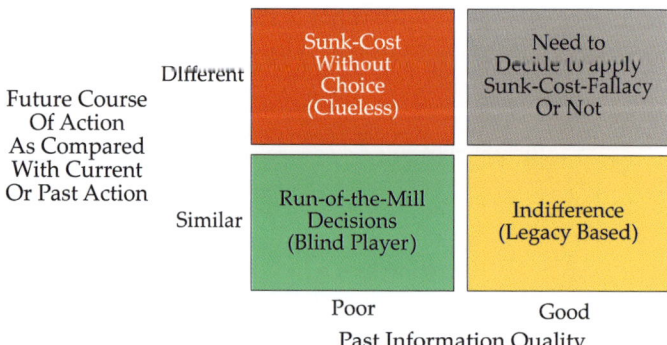

**Fig. 14.4**  Portfolio to understand sunk-cost-application context

Now a project scenario is similar. The first aspect is to understand whether there is a relevance of the sunk cost fallacy or not. This is elaborated in Fig. 14.4. Despite the question of weightage that is often the subject matter of the sunk cost fallacy, one has to look at the following aspects.

1. The derived future course of action under the current circumstances by ignoring the costs
2. The availability of information.

The logical way forward is to understand the future course of action with the available information *ignoring the costs*. If the future course of action is very different from the current course of action and the quality of past information is bad, then the project manager is normally forced to treat the sunk costs as a 'dead-box.' In other words, he would be relatively clueless and make his decisions independent of the past options. Here, therefore, the sunk-cost fallacy does not hold because there is no substantiation except for the claim that *we came so far by this way*. That does not help the manager succeed well. Hence, regardless of his tendency to try and incorporate the sunk costs, the project manager has a serious handicap. He is forced to ignore the past as he has limited insight in any case. The sunk-cost is the only thing making him a *clueless risk taker*. It is like a doctor suffering from amnesia on his own identity! Well, he might be able to treat, but is forced to ignore the past despite what everyone else has to say! Towing the line in the name of sunk costs would only complicate the project situation drastically. Unlike strategy management, projects are typically of shorter durations than the strategic horizons, and the decisions too have a shorter implementation timeframe. Hence, discounting what the project manager does not see, is healthier than doing something without being able to monitor and control. In other words, from a risk definition using scenario planning perspective, it makes sense to completely discount the 'sunk cost' fallacy while identifying the scenarios.

However, at the same time, if the past information is good, one might need to understand whether it makes sense to incorporate the sunk cost or not. This is where the manager needs to focus. However, this decision on the sunk costs becomes relevant only if the *derived course or action is very different from the ongoing course of action*. This is important for the manager to understand. The relevance, therefore, is in defining scenarios for risk management when there is a good handle of the past information. Hence, the project manager needs to know that:

**The project sunk cost could affect the definition of the scenarios in risk management and this is relevant only if there is reasonable insight into the past or the history of the project and if there is a significant change in the course of action anticipated**

The reader must note that our focus here is on the impact of sunk-cost fallacy on the definition of the scenarios rather than on being judgemental about the use of the concept itself.

While the literature also speaks of the merits of this fallacy, the reader needs to understand both the perspectives, while facing a situation, where this is important.

However, at the same time, it is essential to limit the treatment to understanding how this influences the scenarios and ensuring that the representative nature of the scenarios is not significantly compromised. In fact, the research by Baliga speaks precisely of this situation where the sunk-cost fallacy is proved to be a smarter strategy than otherwise in certain cases!

**Greyhound Fallacy**

The greyhound fallacy has its origin from the racing tracks where bettors place their bets. There are three factors that affect the perception of a situation, and therefore, the definition of the scenario for risk planning.

In the racing tracks, these are elucidated as follows:

1. Bettors tend to overestimate the probability of a win by the longshot and underestimate the probability of a win by the favorite.
2. Bettors appear to suffer from the gambler's fallacy (aka *greyhound fallacy*) and underestimate the probability of repeat wins by a dog or horse in the same pole position.
3. Superstition affects the perceptions of bettors.

The second fallacy is what is called the greyhound fallacy.

---

**A Quick Hands-On**

While detailed models can be built on the basis of this fallacy, it is important for the reader to ask himself two questions at the end of a scenario planning exercise:

**Is the scenario as good or as bad as it is?**
**Can the scenario be better or worse than what it is?**

---

There are several probabilistic ways and means to model both these fallacies. However, in most practical situations, the exercise is viewed as academic. We also agree with this view, as the reader must understand that the application of these tools requires

1. a minimum skill level when it comes to the project manager,
2. a management that understands and appreciates the use of such tools, and
3. a pay-off that is able to justify the costs and the efforts.

This is rather difficult in most practical projects and the resulting situation is better evaluated by qualitative judgement rather than any rigorous tool when it comes to detecting the presence of fallacies.

## 14.4  The Last-Mile Phenomenon

We now touch upon an interesting phenomenon in projects called the last-mile phenomenon. In many projects, one often finds that the project is doing 'just fine' despite making reasonable progress. If one takes a closer look at such projects, there are a few things that the project manager needs to know.

A typical project story would read like:

> We made excellent progress in the beginning. We are nearing completion now… Last month the report said we will be operational within 15 days. This month the report says we will be operational within the next 15 days…

From a financial perspective and a management perspective, the initial interpretation is that the project has a lot of buffer to catch-up. However, the 'green' signals always tend to make the management complacent (although most senior managers would claim that they did know that the problem was likely to come sometime later!). However, just like the schedule chicken, the issues are not discussed readily as there are no flags that are raised early on. But once they come to the forefront, 'hell seems to break…'

---

**A Quick Hands-On**

At this stage, the reader is requested to understand the dynamics of this phenomenon in his own organization. In particular, two questions are of interest viz.

**List at least three reasons why the last-mile phenomenon occurs.**
**List three ways in which you can prevent it from occurring?**

---

An intensive exercise can give a lot of insights to the project manager.

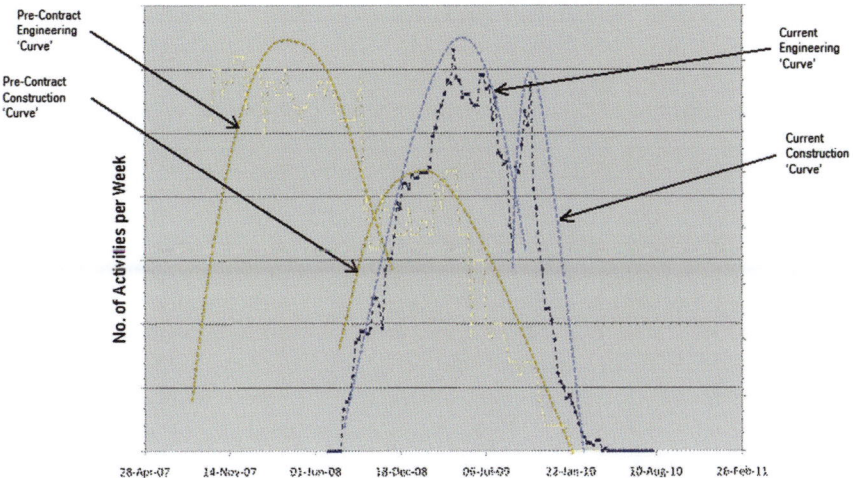

**Fig. 14.5** Typical last-mile phenomenon

**Table 14.1** Managing complexity in projects

| Number of parallel activities per area | Complexity index |
|---|---|
| 1 | 0 |
| 2 | 1 |
| 5 | 10 |
| 10 | 45 |
| 15 | 105 |

A typical last-mile phenomenon is shown in Fig. 14.5. Notice how the original engineering and the original construction curve have transformed over time. The example is taken from a real-case project in the US. As the time progressed, despite having limited progress on the project, no red-flag was being thrown up. However, the intrinsic activity level kept increasing and this started showing steep construction curves. Usually, in the rounds of project reviews, schedules are modified and made 'tight' as they go along. One needs to be aware of this kind of an activity. This is an excellent risk indicator. However, for this indicator to work successfully, there are certain prerequisites on the schedules. We will deal with them in greater detail later in this book.

Ensure that *your Project doesn't get back-end loaded!* This phenomenon starts increasing the operational risk. That is shown in Table 14.1. In this table, there is only one critical activity in a given time window.

The situation starts getting more and more complex as the number of activities that run in parallel keeps increasing. This is graphically shown in Figure 14.6. When more than one activity starts getting critical in a time window, the complexity doubles! Depending on the planning experience, every company can define a cut-off for the complexity. A cut-off can help determine the *project control philosophy and the project management methodology* (daily reviews, additional checks to look at the overall plan feasibility, review of risks, etc.). Hence, this can act as an excellent indicator (like a lead-indicator) while dealing with operational risks.

As we proceed to understand this metric, there are certain pre-conditions that are used in projects that have multiple activities running in parallel. Note that we have not explicitly considered any other parameter like the resources used, the earned value units, or the spectra of durations, etc. This gives us an important point to understand. In all its simplicity, this metric is representative enough when it comes to monitoring risks. Further, the essential aspect is that it need not be treated as an absolute metric. Rather, it could be treated as a relative metric. In other words, after the initial schedule is prepared, the complexity index of the same would be 'acceptable.' Hence, any changes made would need to relate itself to the initial reference or the baseline value of the schedule.

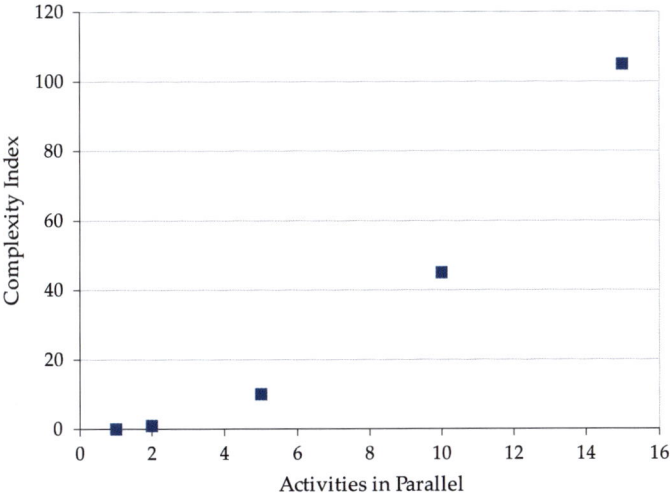

**Fig. 14.6**  Complexity index: Graphical relationship

## 14.5   Control Risk Through Meetings

One of the most popular methods used to manage projects is *the meeting*. The reader might have had a lot of experiences in projects and general management that could reinforce this phenomenon. Most senior managers are 'busy' in meetings the whole day. Meetings might be their process to make decisions, get information, take control, direct, or even directly supervise their colleagues. However, from a subordinate point of view, this could be a fatigue factor as well. In a project scenario, the impact is more predominant. Let us look at the dynamics in this case.

The moment red flags are thrown up in any project, the project manager and the team normally see a surge in the number of meetings. Such meetings are meant to help the owner stakeholders or the senior management understand the situation. There are several issues at play here:

1. Is the project manager competent enough?
2. Is there room to do more in the project (is the manager 'slow')?
3. Are all the alternatives being explored correctly?
4. Are right decisions being made or are all decisions being escalated?, etc.

Central to this line of thinking is that the project manager is not as effective as he could be, for reasons that could be intrinsic or extrinsic.

Let us consider a project that is being executed at a company. Now let us assume that there is a *weekly reporting* performed in this project. If one looks at the time-spent in the project, then there are two broad categories of times. The first category is (a) billable for the end-user and the second one is (b) billable internally. In any project, those billable for the end-user are said to be the 'true productive work' done in the project.

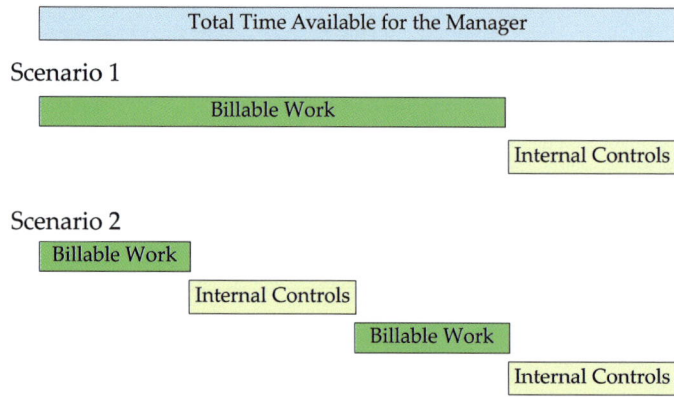

**Fig. 14.7** Productive time schematic for a project manager

However, as the frequency of meetings increases, the time billable for the end-user decreases. This is shown in Fig. 14.7. While the representation may not always be applicable for a project manager, it certainly is applicable to his team members. What is more frightening is the fact that reduction in the billable time can actually have a serious impact on the productivity. For the sake of representing this aspect, let us consider Table 14.2. The table shows the change in values for different average durations for billable work. Very simplistically, if the original billable work was 80 % and the internal controls aka meeting time was 20 %, then the doubling of the meetings in the second scenario actually means the average billable work duration is now

$$\frac{100\,\% - (2 \times 20\,\%)}{2} = 30\,\%$$

So, the productivity then drops from 1 at 80 % to 0.5 at 30 %, which means that 80 min spent in the first scenario would still make 80 min of billable time. However, in the second scenario, 30 min would only make an equivalent of 15 min of time (since the index is just 0.5). This is a dangerous situation. In other words, increased control has actually made the project much slower than what it was. Further, increase would

**Table 14.2** Productivity relationships with durations

| Average duration of billable work blocks (%) | Productivity Index |
|---|---|
| 80 | 1 |
| 70 | 0.8 |
| 60 | 0.6 |
| 50 | 0.55 |
| 40 | 0.5 |
| 30 | 0.5 |

only accelerate the downward spiral and would ultimately result in the dissolution of the team (starting with the project manager).

This trap is called the *Micro-managing Trap* and is again very common in many project organizations. And the best part is that these can be captured. In other words,

> **The Number of Review Meetings and their Durations are often another indicators of project risks. They could also be real time components of the utility functions for risk definition**

It is for the reader to understand how these can be brought into the overall risk framework and understand potential risk parameters that could be of interest. Usually, the meeting duration is a good 'black-box' variable to monitor and track.

## 14.6 Revisiting Utility Functions

We have had several discussions on utility functions so far. However, we have not particularly pinpointed the ways to do so. The reason is delightfully simple: *there are too many possibilities to choose from!* Despite that generic answer, the reader needs to understand some of the ways in which this could be done. In this section, we will touch upon those last few aspects that are relevant.

### 14.6.1 Threshold Methods

These methods actually evolved out of Psychophysics which actually deals with a set of methods and, the results obtained using these methods, relating sensation to the physical characteristics of a stimulus. Gustav Fechner was the scientist who actually developed the basic psychophysical methods that are used today.

In these methods, the concept of threshold is explained simplistically as *the smallest stimulus that can be perceived*. In other words, for each level, the proportion of times that the subject correctly heard the sound is determined in psychophysics. A psychometric function looks like what is shown in Fig. 14.8. The threshold is defined as the level at which the listener achieves some arbitrary proportion of correct detections. The graph of the function is similar to that given by Gescheider. Traffic lights in risk management often work on a similar principle. The important point that every project manager needs to understand in this context is that *the reality of the threshold is the stimulus that produces an arbitrary, but defined level of performance*. In the previous exercises and sections, we have been focusing on the roles of the risk managers and the chief risk officers in defining these thresholds. We have also been insisting on the publishing of these thresholds so that the project manager is able to understand and track his project accordingly.

While there are a number of ways in which one can define the thresholds, the best ones in project management use a combination of the psychophysical concept and

**Fig. 14.8** Psychometric
function

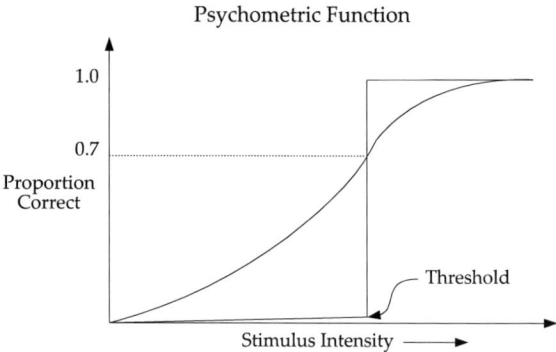

integrate them with statistical correlations. In other words, risk parameters will have
a broad set of normative guidelines that will allow the project manager to 'send the
right signal' meant for controlling the risk.

The moment one refers to statistical correlation, the method of thresholds actually
requires a reasonably good *insight into the history of similar projects at the company*.
Nevertheless, this is precisely the role of the risk officer. In most companies, risk
officers are busy delegating work to the project managers without actually making
an active engagement possible. Using the method of thresholds, one sees a more
concrete approach and helps the overall project management team to understand
how these thresholds *can be integrated in their processes and analyses.*

The advantage with the method of thresholds is that it still retains the *deterministic
flavor* that is often required while communicating with the senior management. The
translation of an analogous function into binary states is an important aspect that
every project manager needs to know. To understand these conditions better, it is
also essential to understand the ways in which these experiments are conducted. The
project manager can get meaningful insights from the procedure.

**The Constant Stimulus Method**

This is the first of the three methods used to determine the thresholds. In this method,
a series of stimuli of several values, some that one thinks people will always be able
to hear, some that one think that they will never be able to hear, and some in between
are presented to listeners and they are asked if they hear each one or not. Then the
number of times they actually hear is proportioned and registered. For instance, "a
delay of 1 day" could be a term that is heard many times. One needs to see if the senior
management is truly 'registering' the information or is it lower than their threshold
level. The proportion of times they truly hear could give us an insight on whether
this needs to be included in the list of thresholds or not.

The reader will agree that this is often seen to be the case. Senior management tend
to ignore initial reports of risks and management issues because they are below their
threshold levels. Of course, the time in the project life cycle is also critical to note
in this exercise. Therefore, a clear identification of thresholds is often times useful

for the project manager. There are situations where the project manager *expects a resolution* from the senior management, yet it never comes by. Hence, there is the genesis of a disconnect about the seriousness of the reporting and the support extended by the senior management.

By understanding the psychophysical concept, project managers can also now realize how the dynamics of the reporting interface are affected. Therefore, an entertaining and sensationalizing report is often times a better way to attract attention? I would leave it to the reader to take this further at his end.

The main advantage of this method is that it provides a complete picture of the sensitivity. It is also relatively easy to administer. The disadvantages of the method are that one needs to have a basic clue of how the threshold is going to be before one starts. And at times, if just the threshold is of significance, without a subsequent desirable course of action, a lot of trials and time are 'wasted'.

**The Method of Limits**

In this method, one starts with a level that one thinks the listener will hear. If they do hear that one, they are presented with a lower level. If they hear that one as well, they are presented with a still lower level and this is continued till they cannot hear the stimulus. One could also start at a very low level and keep going up till the person starts hearing.

This method is efficient, because it tells one, in relatively few observations, the threshold values. One does not need to know where the threshold starts, to begin with. However, this method could also lead to spurious thresholds. That is a risk one needs to be willing to take.

In practice, the project manager might see significantly different reaction from the senior management on project reports that are 'largely similar' in content. This is a practical demonstration of the method of limits. Hence, a risk manager needs to fill in the details on the limits or the values when the risks start becoming significant for the senior management to step in. Similarly, small changes in the reports could also trigger huge changes in the reactions of the senior management. Therefore, the method of limits comes in handy to help define what the organization's interpretation to a particular project is.

The third method is the method of adjustment. This is not discussed here as it doesn't particularly give a good handle in a project management scenario. By understanding the psychophysical experiments and the concepts of threshold, project managers can define a simplistic framework for determining the *risk control triggers* and their values, for their organizations.

The psychophysical experiments also tell us that it is *not always a good idea to continuously monitor a project*. In certain discussions on Linkedin, there was a push by certain project professionals to try and apply 'automated' and 'continuous' type of appraisal systems for projects. Be it appraisal or reviews, the psychophysical experiments give us a clear indication that this may not be a very good idea. It might only increase the manager's fatigue.

## 14.6.2   Combining Parameters

The previous discussion was on identifying parameters and thresholds for them *assuming each of them is acting independently*. However, it might happen that the parameters are used in combination. In other words, this is a concept similar to that of an *indifference curve* used in economics. Combined influences of various parameters are however, more simplistic in the project environment. Simple simulations can help model combinations of parameters.

In the simulations, the potential impact by the parameters can be identified and isolated. This helps the project manager in understanding how the parameters *interact with each other*. In short, it is the application of the *Systems Thinking Methodology*. The methodology is actually based on the tenets of Systems Thinking that are given below:

1. *Interdependence of objects and their attributes*
   —independent elements can never constitute a system
2. *Holism*
   —emergent properties not possible to detect by analysis should be possible to define by a holistic approach
3. *Goal seeking*
   —systemic interaction must result in some goal or final state
4. *Inputs and outputs*
   —in a closed system inputs are determined once and constant; in an open system additional inputs are admitted from the environment
5. *Transformation of inputs into outputs*
   —this is the process by which the goals are obtained
6. *Entropy*
   —the amount of disorder or randomness present in any system
7. *Regulation*
   —a method of feedback is necessary for the system to operate predictably
8. *Hierarchy*
   —complex wholes are made up of smaller subsystems
9. *Differentiation*
   specialized units perform specialized functions
10. *Equifinality*
    —alternative ways of attaining the same objectives (convergence)
11. *Multifinality*
    —attaining alternative objectives from the same inputs (divergence).

Each of these tenets, when applied to any management situation would give an interactive model. Simulations on the model would give the reader a good handle of the impact of the parameters, both individually and collectively, thereby allowing them to make a rigorous analysis.

Needless to add, here are a few words of caution. Not all risks are of the same order. Therefore, while doing a modeling or a simulation exercise, the project manager needs to be aware of this possible problem and allow the dimension to evolve accordingly. In many organizations, such findings tend to trigger the reverse process of redefining the thresholds. This approach, though popular due to its strong logic, is a misplaced one. Repercussions of risks are not always *simplistically measureable*. Hence, the reader must bear this aspect before trying to overhaul the system based on his singular (and often, restricted) experiences.

The final stage is in determining coefficients and functions for the utility. Being from a consulting background, the experience has been that it would certainly not be advisable for the project manager to define a utility function. Our experience has been that it is better to evaluate the overall system, in an integrated manner, without worrying about defining a specific utility function. We will touch upon this again in the schedule analysis.

## 14.7 Micro-planning

We already had a brief discussion on internal controls in the previous sections. This is a tricky subject in itself. There have been several instances where project managers are bogged down with issues related with internal control.

---

**A Quick Hands-On**

**Can you think of the following situations?**

"The boss is back, and he is nosey this time again…"
"Hey, do you want to tell me how to wear my shirt as well?"
"Great, now the font is the top priority!"

**What do these have in common?** The phenomenon is called *Micromanaging*.

The question for the reader is:

> **List two examples to explain the concept of Micro-Managing.**
> **List two reasons when it becomes important to micro-manage.**
> **Is Micro-Managing the same as Micro-planning? Explain your answer.**

---

Despite the general treatment given in the previous sections, micro-managing is often engaged in project environments. And for strong management reasons, micro-managing might be the preferred way to go ahead in a project. This is often deemed to be an *HR Issue* and we will not try to go overboard in looking at this aspect. However, our interest here is in the concept of Micro-planning that is distinctly different from Micro-managing. Although planning is a part of the management definition, the objective of Micro-planning is very different from the objective of Micro-managing.

Just as meetings are used by most managers for monitoring and control, Micro-planning is widely used in risk management and is, perhaps, the most popular tool to assist managers when they have multiple risks.

### 14.7.1   A Formal Definition

Micro-planning is a process in which operational decisions are made, *before* the execution of the job, to the maximum level of detail (lowest possible task). This is different from scheduling that often looks only at the 'time' taken by activities and tasks. It is, therefore, far more elaborate in its approach.

The idea of Micro-planning is to contain risks at the operational level. In many projects, despite the guidelines and the directives given to the managers, there are always gaps between the *strategy-to-operations continuum*. Micro-planning addresses these issues and is extremely powerful in the way it does so.

As a thumbrule, if the reader sees a project with an activity that is less than 1 % of the total duration, you might well be bordering into Micro-planning. Of course, there is a need for a subjective assessment as well. Many times, especially in projects of longer durations, it might be necessary to go for activities that have small durations. It is usually done to check the feasibility of a project and to ensure *smooth coordination of activities (pre-conditions) when there is little buffer.*

### 14.7.2   Criteria to Adopt Micro-planning

While giving Micro-planning its due credit, one must also be cautious about the criteria to apply this tool. The most important criteria are listed below:

1. One has excellent estimates. As one goes deeper into plans, the level of detailing is often constrained by the accuracy of the estimates. Hence, Micro-planning fails if the estimates are not just good enough. While implementing these plans, therefore, a general consensus must be obtained among all the players. Risk-averse team members could end up giving estimates with huge variances. Many times, team members may not be committing to the time estimates. Such a situation is a dangerous one and not particularly useful in Micro-planning.

2. Contractors and all other legal entities involved in the project use one common planning system and the reliability is high. This is again a very important point. One might often find, in meetings, that the contractors pull out a 'different' plan when they are discussing the actual details. This leads only to confusion; because it creates an intrinsic gap between the planning process and the implementation process, thereby, rendering the efforts useless.

3. Issues of rework/quality have not been predominant with participating agencies. This factor directly connects with the reliability of the estimates and the sub-

sequent manageability of the plan. In short, it is evident that with rework and quality issues, the activity tends to go into loops and this would affect the tie-ins that are there in the plan. Often times, this is a critical aspect and could cost the organization a lot as well.

4. Most importantly, one has manpower to update, analyze, and study the plans; and management support for effectively using the tool. This is probably the very first aspect that every project manager must know. Anything linked with a higher degree of detail ultimately results in a greater need for manpower. In many projects, manpower issues mask every other issue, for there is insufficient resource to implement any process or system or even manage the project. The end result in such projects is that they become fire-fighting vehicles each time a risk surfaces…and worse, they jeopardize the entire project rather than 'containing the fires'.

The reader needs to remember that *inappropriate application of Micro-planning can be extremely counterproductive!* Hence, the choice of using the Micro-planning mode to contain risks needs to be a judicious and a well-thought one. The other aspect is that, in practice, it is not possible to shunt from one decision to the other without consequences. In short, once a strategy to micro-plan is adopted, it cannot be reversed in the project unless it flouts the criteria for application.

## 14.8  Key Takeaways

In this chapter, a wide variety of risk parameters and risk management philosophies have been dealt with. Understanding risks from the four perspectives is the first aspect that a project manager needs to understand. The reason for this is multi-fold:

1. Not all the risk management practices use the same perspective,
2. Many risk management practices actually have overlaps in the perspectives, and
3. Often the admissible nature of the risk element determines the subsequent processing from the risk management perspective.

We also looked at scenario planning that is often confused with predictive risk management. Scenarios are developed as part of a strategic exercise and are not a part of everyday risk management. We have amply covered the kind of objectives that these tools have and how they are different from a direct risk management approach. This is important for the reader, who needs to understand that risk management using scenarios is, actually a misnomer and can lead to a confused conceptual framework. Our clarifications were on using the scenario-planning methodology for sound project risk management.

In this chapter, one of the most important focus has been on predictive management. Since this book covers forward-looking scenarios, we have tried to restrict the treatment to help the project manager understand how important phenomena such as the last-mile phenomenon actually transpire in a project environment. The nature of control handles have also been dealt with in great detail. The essential takeaway is that one can *find adequate parameters that help one understand risk from a management perspective*. The last-mile phenomenon is one such example. It is simplistic, yet grossly ignored in most project environments! Our attempt here is to try and focus on how these could be used.

We also had our last brush with the utility functions in risk management. While these are often talked of, in project management literature, our ways indicate that there are also psychological factors that affect risk management. In fact, the method of threshold determination helps elaborate on this process. Needless to add, systems thinking when applied through modeling and simulations is a powerful method to work on risk management as well.

We conclude with a brief overview of Micro-planning, a philosophy in risk management. At the same time, we have also shown that these are different from micro-managing that is dealt with in our discussions on internal controls. Micro-planning is a powerful technique, but is cumbersome. It is highly rewarding when applied correctly, but can grossly falter when applied inappropriately. So the reader needs to make a judicious choice while selecting such tools.

# Part V
# Project Planning and Scheduling

# Chapter 15
# Basics of Scheduling

**Abstract**  As the opening chapter on advanced topics in scheduling, we revisit the basics of the scheduling process. In doing so, the different types of floats are defined and understood. The influence of resources and calendars on schedules is also touched upon. The interesting concept of Resource Leveling is next touched upon. The first treatment of this concept is done using the conventional approach (often found in traditional books and softwares). This is followed by a management reflection of resource leveling using a simplified framework of Consulting Connoisseurs. Apparently, the contrasting results are expressed in detail. The next advanced concept is that of schedule detailing. While most schedule details involve adding newer activities, we cover the tacit requirements of the process. The focus then goes to understanding schedule updation as a management process. While touching this sensitive topic, we also elucidate good scheduling practices that are required to ensure a reliable planning backbone for the project. The chapter concludes with a management reflection of activity relationships in project schedules.

**Keywords**  Project schedule · Scheduling process · Floats · Free float · Total float · Independent float · Interfering float · Resources · Calenders · Resource leveling · Criteria for resource leveling · Activity relationships · Schedule updation · Schedule analysis · Schedule robustness · Relationship analysis

We have covered several aspects of project management in a wide variety of contexts. However, one of the most important areas, the *'brain' of the project management faculty*, is the planning and scheduling area. In PMBoK®, these are covered under multiple knowledge areas. However, for a practical project manager, it is essential to understand these concepts from the perspective of his role. Our frameworks are typically 'integrated' as they cover multiple knowledge areas and multiple perspectives that are important for the project manager. This chapter is exclusively meant to discuss the concept of scheduling and the basic concepts revolving around them. Despite being a book on advanced concepts, we need to revisit certain basics that are commonly misinterpreted and incorrectly used in business today. Schedule management is one such faculty.

© Springer Science+Business Media Singapore 2017
N. Gurjar, *A Forward Looking Approach to Project Management*,
Lecture Notes in Management and Industrial Engineering,
DOI 10.1007/978-981-10-0782-8_15

## 15.1   Reviewing Basics

A schedule is defined as a listing of project activities and milestones with start dates and end dates. It provides us with the information of what happens at what time of the project. The schedule is one of the few tools which connect the future to the present and to the past in a single view. We discussed about the EVA which was an accounting innovation along the lines of project schedules. However, by tradition, the schedule has far more information and *robustness* in determining the way a project is mapped.

---

**A Quick Hands-On**

The reader is requested to find out the different representations of the schedule that are used by the different user groups in his company. Specifically, the reader is requested to focus on

**The information content in each of the representation**
**The ease of understanding the representation (at a personal level) and**
**Two modifications to make the format more useful than the one being used.**

---

There are various ways in which schedules are represented. One of the most popular methods is the tabular representation of the schedule shown in Table 15.1. The table simplistically lists out the various activities and their dates. This is a very popular format because of its simplicity. It has a quick summarization of the dates and the activities, and helps the project manager have a quick glance at the overall project.

The second most common way of representing a schedule is the Gantt Chart representation. Here the activities are shown arranged across the time scale. The advantage with the Gantt Chart is that it provides one with additional information on sequencing of the activities. Thus, a project manager can quickly understand how one activity ties with the other. This is an advantage in most smaller schedules. However,

**Table 15.1** Tabular format of a project schedule for a software project

| Activity number | Name of the activity | Start date | End date |
| --- | --- | --- | --- |
| 1 | Engineering (architectural and DB design) | 01 Sep 2014 | 15 Sep 2014 |
| 2 | Development | 16 Sep 2014 | 15 Dec 2014 |
| 3 | Testing | 16 Dec 2014 | 10 Jan 2015 |
| 4 | User Handover | 11 Jan 2015 | 15 Jan 2015 |

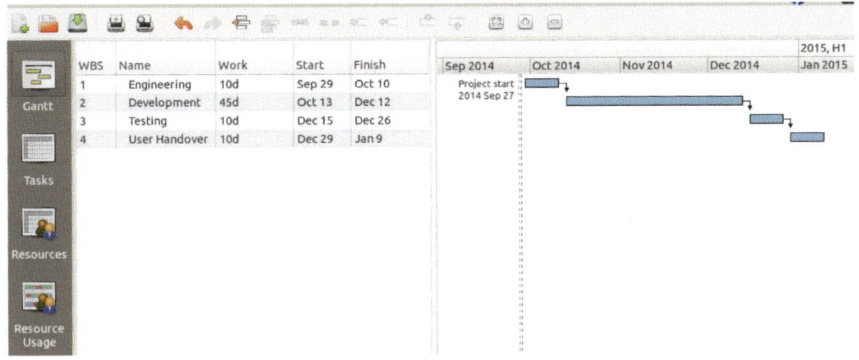

## Another Software Project

**Fig. 15.1** Gantt Chart representation

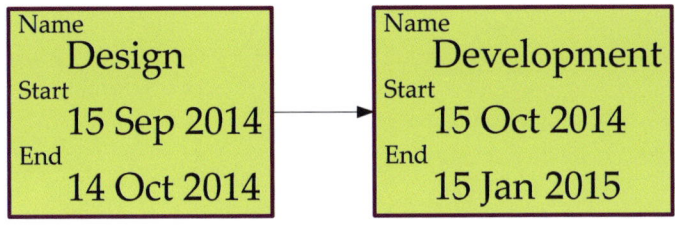

## Node Format

**Fig. 15.2** Activities as Nodes

as the schedule size increases, the utility of this feature (showing the relationships) actually decreases. In any case, this is an extremely popular way of representing the schedule as it shows the *layers of activities on a timeframe* (Fig. 15.1).

If one looks at basic Operations Research theory, the sequencing information is represented by AON and AOA models. AON is the Activity on Node approach and AOA is the Activity on Arrow Model. Both these have been popularized in the literature. However, I have found limited popularity in project management practice all these years. Some softwares do use the AON approach, though in a limited way to describe PERT charts. Though this is a representation that also shows the relationships, it does not visually help one differentiate a 'large activity' from a 'smaller one.' A snapshot of two activities from such a schedule is shown in Fig. 15.2. So a massive schedule will have several such blocks. The distance between the blocks has no relationship with the time lines. Moreover, the spread of the entire project too has no link with the timelines as such. This is, therefore, a more difficult way of modeling a project. However, the advantage of this representation is in the software development industry. In the software development industry, the entity listing and the database schema views are similar, making this view 'less alien' than the other industries. That is, subtly hinting toward the orientation of different users in the project context.

In this book, however, we have been using the Gantt Chart representation of the projects as it relates well with the default views of most scheduling software.

## 15.2  Derive the Schedule

The next step in scheduling is to actually derive the schedule. The two fundamental requirements of a schedule are

1. Defining Activities
2. Defining Relationships.

Activity definition involves the determination of the set of actions that would be carried out in the project. In other words, some project experts actually say that these are phrases containing 'action verbs.' They use the support of English Grammatic to differentiate the Work Breakdown Structure from the activity itself. The difference in the characterizing question is indicated as the WHATs—indicating the WBS and the HOWs—indicating the activities.

A whole lot of engineering methods are applied to activities. These are meant to provide one with several details:

1. Process: How it would be done?
2. Resources: Who all would be required?
3. A Time Estimate: Typically based on some standard Industrial Engineering Tools and Techniques.

The next aspect is that of relationships. There are essentially three types of relationships that are recognized in the literature. They are shown in Fig. 15.3. To make

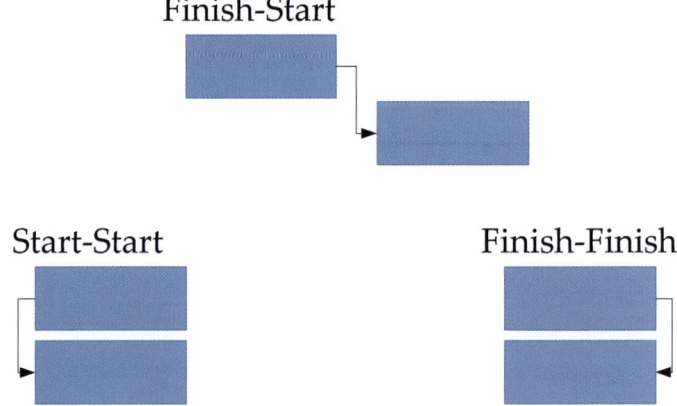

**Fig. 15.3** Relationship types in project management

modeling possible for the vast variety of conditions, a new term in the relationship definition is introduced. This term is called the *lag*. The lag is nothing but the difference in the dates of the two activities:

1. Finish to Start: Start Date of the Successor—Finish Date of the Predecessor
2. Start to Start: Start Date of the Second Activity—Start Date of the First Activity
3. Finish to Finish: Finish Date of the Second Activity—Finish Date of the First Activity.

While this is very elementary for any project manager, one of the important concepts is that of the passes and floats. Most project managers have a difficulty in understanding the passes and the floats. However, before we go into the details of the floats, it is essential to get a feel for the passes in the project.

**A Project Schedule is a Derived Outcome of Inputs given by the Project Stakeholders. The only possible things to define are the activity names, durations, and relationships.**

While deriving the schedule, there are two reference points that one can use:

1. The Start Point of the Project
2. The End Point of the Project.

Each of these would give different outcomes. Some try to use both; but that is bad scheduling practice.

Let us take the case of a sample project as given in Table 15.2. Note that this project has two paths that run in parallel. A is the first activity as it does not have a predecessor and D is the last activity as no other activity has it as the predecessor. While doing the forward pass, the purpose is to evaluate *how all the activities are scheduled with reference to the start date*. This is simple to identify. If one has to find out completion dates with respect to A, then B will be completed on Day 17, C will be completed on Day 27. Similarly, E will be completed on Day 16 and F will be completed on Day 22. However, D needs both C and F to be complete before it can start. In other words, the earliest start date for D is governed by C. Hence, D has an end date of 35 Days. What is essential to note here is that F completes on Day 22, and then there is 'no activity' along that path for sometime. This period of 'buffer' is called the *float* of the activity.

**Table 15.2** Activities and relationships in the project

| Activity number | Name of the activity | Duration (days) | Predecessor |
|---|---|---|---|
| 1 | A | 7 | – |
| 2 | B | 10 | A |
| 3 | C | 10 | B |
| 4 | D | 8 | C, F |
| 5 | E | 9 | A |
| 6 | F | 6 | E |

We will now perform the backward pass. In the backward pass, we know that the reference point for any calculation is going to be the end of the project. Hence, now C and F need to end latest by Day 27. So, taking it further, B has to end latest by Day 17 and E has to end latest by Day 21. Similarly, A can end latest by Day 7. Now knowing the durations of each, one can identify the float or the slack in the project.

One sees that activities A, B, C, and D have dates that are unchanged, regardless of whether one goes by the forward pass or the backward pass. Hence, it is on the critical path and does not have any buffer or float or slack while executing the project. The other activities do have floats. In other words, the project manager can 'relax a little' as far as these activities are concerned.

The dates calculated on the forward pass are called the *early dates* of the activity in the project schedule. The dates calculated on the backward pass are called the *late dates* of the activity in the project schedule. Note that *neither of these dates indicates an advancement or a delay in the overall project schedule*. The resulting schedule is represented in tabular form as shown in Table 15.3. Note that the *Start Date is the beginning of the Day while the End Date is the close of business of the Day*. Hence, a direct subtraction of the dates could confuse the reader. One needs to factor in the actual dates in terms of working days that are contained. So an activity that starts and ends on the same day requires 1 working day.

The floats are simple and easy to calculate. They are the difference between the Early Finish and the Late Finish of any given activity. This is the most simplistic definition that needs to be used. Of course, it is unlikely to see a discussion in any project review that explicitly makes a mention of terms like *Early Start* or *Late Finish*. However, one does hear the term floats or slacks often in meetings.

Any project manager needs to know the basic mechanism in which these are calculated. The first and foremost aspect is to understand that floats occur only when there are multiple paths in a project. So, if one were to have only one path throughout the project (like the waterfall model), it would never have a float for any activities. That is, all the activities would be along the critical path. Hence, the concept of float is relevant only when the *branching is high*.

**Table 15.3** Schedule in tabular form

| Sr. No | Activity name | Duration (days) | Early start | Early finish | Late start | Late finish |
|--------|---------------|-----------------|-------------|--------------|------------|-------------|
| 1 | A | 7 | 0 | 7 | 0 | 7 |
| 2 | B | 10 | 8 | 17 | 8 | 17 |
| 3 | C | 10 | 18 | 27 | 18 | 27 |
| 4 | D | 8 | 28 | 35 | 28 | 35 |
| 5 | E | 9 | 8 | 16 | 13 | 21 |
| 6 | F | 6 | 17 | 22 | 22 | 27 |

## 15.3  Slacks or Floats

If one looks at network theory, there are lots of views and definitions of Floats. While they are seldom used in practice, it is important for the reader to understand these terms and where they need to be used. The terms are important from a typical classification perspective and for the project manager who wants to understand how to interpret and leverage from his schedule control philosophy. This is easier said than done, though there is a lot of potential to consider here.

**A Quick Hands-On**

**Given a network for a project, identify how delays are to be interpreted. Which delay needs to be a red flag and which need not be one? And among the ones identified, which would actually be a serious red flag?**

In other words, the relevance of floats is dependent on the way they are perceived. Thus, the next step would be to understand them in greater detail and have a sneak peak of the background of how it would be used. We will deal with the application in detail in later chapters.

In literature, there are five basic types of float. They are defined as

1. Free Float Early
2. Total Float
3. Independent Float
4. Interfering Float
5. Free Float Late.

All these terms are explained in Fig. 15.4. The project manager now gets a clearer picture of how the project is to be structured.

**A Quick Hands-On**

While we are trying to clarify the situation, a number of questions might be occurring at the reader's end:

**Is the independent float a gift? Can free floats be used in claim management? Can consequential damage be adjudged based on the use of interfering floats? etc.**
**Hence, which float is applicable to which situation or decision framework the most?**

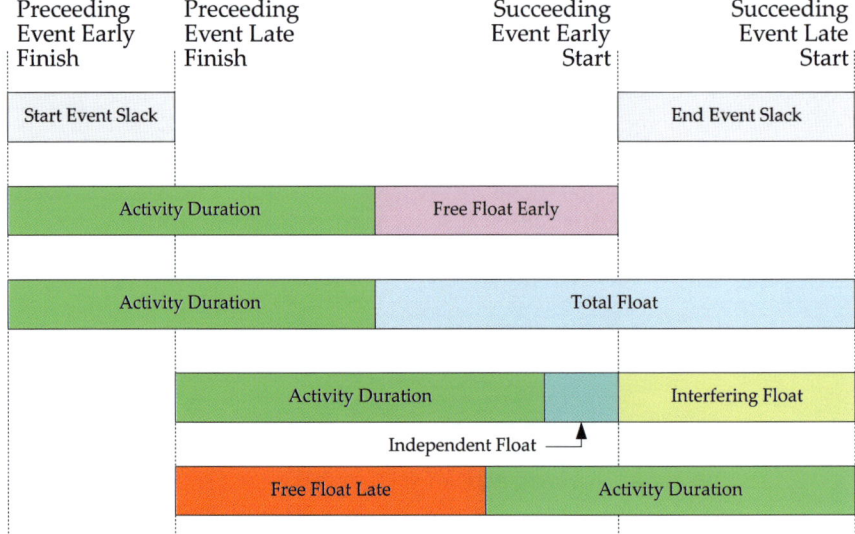

**Fig. 15.4** Demystifying floats

These are natural questions and the reader needs to understand how to strategize and use these terms. A lot of the aspects on admissibility need to be checked and verified.

## 15.4   Resources and Calendars

The basic schedule build was done using the activity definition and the relationship definition. However, in real life, each activity also requires the factors of production to be successful, that is to say, man and machine in addition to capital (land is neglected). Hence, it is essential for a project manager to understand how well a schedule can be built when one incorporates this information.

Resources, in most project environments, have their own schedules of working. For instance, an operator might be available only for 8 h in a week. Similarly, the engineering department might work only for 5 days each week although the site might work for 6 days. With such constraints, the project plan needs to be built. This makes the entire exercise more complex than it seems, because it alters the calendar time required for the project.

Let us consider two activities A and B. Now if both the activities are requiring 7 days each, they could be scheduled one after the other as shown in Fig. 15.5. In doing so, they are completed in 14 days. Now, imagine that there is a resource required for all the 7 days and the resource works only for 5 days a week. So, if one starts the project on a Monday, it would be completed on Day 18. In other words, it takes a good *4 additional days to complete the project*. Such calculations are quick and pretty easy to model in any modern project management software. However, there is also a catch while one uses such software.

Two Activities of 7 Days Each.

Both Need a Resource that only works on 5 Days each Week

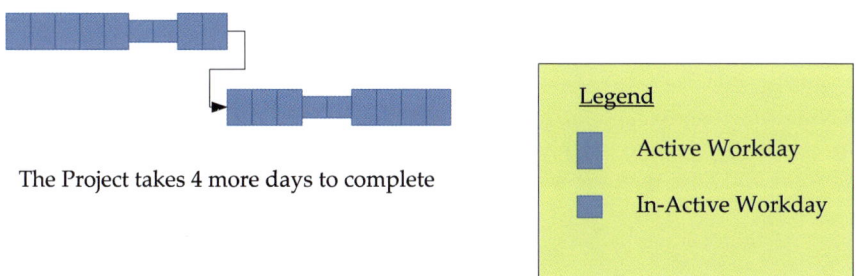

The Project takes 4 more days to complete

Legend

Active Workday

In-Active Workday

**Fig. 15.5** Resource calendars and schedules

**Continuing with our previous discussion, does the use of a resource change the float available in an activity? If yes, then how is the project manager supposed to treat this?**

The problem with resources and calendars is that they start making the schedule more complex to develop, comprehend, monitor, and control. And in this entire exercise, it becomes more and more important to understand how to develop the schedule.

For instance, if the resource we are speaking of, has a planned vacation from the 15th day to the 25th day, then a delay of 1 day in the project start will actually mean a delay of over 10 calendar days in the completion of the second activity. This leads to just one phenomenon: *chaos*. Therefore, although it is a great thing to be able to model resources into projects for scheduling purposes, the difference in the calendars could actually be misunderstood to be the 'invisible hand' that is 'skewing' the schedule. However, this is one of the important aspects when it comes to actual modeling of schedules.

## 15.5   Resource Leveling

Another interesting situation, in schedules, occurs when the same resource is required in tasks that occur in parallel. In other words, the resource is nothing short of being (technically stated) overloaded. In such a situation, there are multiple options for the project manager. Most times these are more complicated than one can imagine. Let us understand some possible ways and means of doing this.

**A Quick Hands-On**

Imagine there are two activities that have the same requirement of resources. In the initial schedule, where the activities were described, they ran in parallel, as the resource information was not clubbed. However, now that they both need the same resource, the project manager needs to refine the schedule.

**Explain how one could do the resource leveling the best**

Among the many strategies available, the following are some of the 'simpler' ways in which one can perform resource leveling. It should be clear to the reader that these algorithms are to be applied after doing a first round without resources. This way, one will have multiple ways of comparing the schedules.

### 15.5.1   Strategies for Leveling

1. *Earlier Start First*:
   This is a relatively easy algorithm. Simplistically, it says that the activity which is assigned a resource already should get the first priority. Hence, it is relatively easy to plan.
2. *Earlier Finish First*:
   Like the earlier start first, in this algorithm the activity that finishes earlier is assigned the resource at first. This helps the project manager show increased progress.
3. *Shorter Duration First*:
   Similar to the earlier finish first, the shorter duration activity is taken up early on, so that the resource can focus on the larger one later. This helps in a kind of warm-up and ensures that the resource is available in an uninterrupted manner to the later activity.
4  *Sharing Activities*:
   This is a common method used in scheduling. However, it has limited success as the resource keeps getting pulled in all the directions.
5. *Longer Duration First*:
   At times the longer duration activity requires more skill and technique and, is often, more critical in the broad spectrum of things. Hence, project managers might want to allocate resources to the activity with the longer duration and try to expedite the schedule there.
6. *Higher Bill Rates First*:
   This is a popular algorithm followed by project managers who are focused on their billing cycles. It helps reduce the exposure of the company by easing the cash flows.

7. *Longest Subsequent Path First*:

   In this methodology, the longest subsequent path is taken into consideration. The longer the path or more the activities remaining, the greater the risk of failure in the later parts of the project. Hence, this is another alternative.

8. *Critical Path First*:

   This is a no-brainer for most project managers. However, the definitions of the critical path could change. For some, it is the path with no or little float (TF < 5 days, for example). For others, it is the length of the path that is critical.

9. *Importance of the Activity Rank First*:

   Sometimes, a project manager might have different criteria than just looking at the schedule network. For instance, a crane sequence has to be such that the crane is accessible and can be brought to a specific location. Hence, the area might determine the importance and not other parameters.

10. *Using Free Floats*:

    Similar to the critical paths, one can also determine ranks based on free floats.

These are some of the simpler ways in which a project can be leveled for resources. However, it is important for the project manager to understand that the *choice of selecting an algorithm should be a conscious one*. In other words, the manager needs to apply the algorithm carefully.

If the scheduling is automatic, there is no possibility to normally allocate different algorithms for different resources. Thus, the resulting situation should be clearly understood from multiple perspectives. In the current schools of thought, a lot of research has been done on resource leveling and the impact of the same on the project. However, there are several project managers who practically try to use these software results for their own projects. In doing so, it is important for the reader to know that most commercial softwares have solutions that *give results after evaluating many situations*. Therefore, they are expected to give a better solution than can be worked out by the manager.

In typical projects that have more than 800 activities, it is often extremely difficult to study and evaluate the schedules from a leveling perspective. Hence, a software that can intrinsically understand and model the needs of the project manager would give fantastic results. Interestingly, many companies also try to track their costs using the scheduling software. The principle of costing is a simple ABC or Activity-Based Costing Principle.

## 15.6 Despite All Said...

In my own experience, however, the resource loading and the calendar setting of the resources on a global schedule are not very interesting, and there are very strong reasons for the same. Given a situation where there is indeed resource loading that needs to be evaluated and accommodated, one needs to understand the following dimensions from a management perspective.

1. *Duration of the Overload*:
   While resource load could peak in projects, the duration of the overload is a critical parameter. If the peak is for a short time, one could try and manage it by actually working long hours or looking at possible ways in which one could make the employee work lesser without compromising on the work the employee needs to deliver.
2. *Availability of the Resource in the Market*:
   Again, a trivial factor! If the resource is readily available in the market, it might make sense to ramp-up the resource for the duration.
3. *Replacement or Augmentation Time*:
   Another important factor. If the augmentation time is too high (for instance something happens during the production release of the software and there is no one else to take it up), then there would be limited choice but to work overtime and get it through despite the best intentions.
4. *Productivity Constraints*:
   Productivity constraints are significant. If the resource being overloaded is a person of very high productivity levels (like a team leader on the project), it may not be possible to get a second person with comparable productivity.

These are some of the practical problems that also affect the choice of whether to actually go for a leveling algorithm or not.

Most importantly, however, one needs to understand the overall scenario on the project. The resource and the time value of money are two dimensions that determine the strategic continuum. For instance, in a USD 100 million project, the time value of the project assuming a cost of capital of 3 % is close to

$$\text{Time Value of the Project} = \text{USD 100 Million} \times \frac{3}{100} \times \frac{1}{52}$$

which comes out to be

$$\text{USD } 60000/\text{Week}$$

That is a pretty high value if one is considering the ramp-up of some manpower. Thus, the strategies for such project scenarios can be given by the portfolio shown in Fig. 15.6. It is clear that this strategy, discounting the legalities involved, is in use in several companies. Software developers and project personnel in developing countries, especially, are asked to work for longer hours without actually 'booking them' on the project.

Regardless of the region-specific practices, it is commonly observed that project personnel are overloaded in most projects. Many times, there are issues like those described above (availability, etc.), but it has been observed that there are instances when the overloading is a conscious decision of the senior management. In such situations, the project manager does not seem to have a *budget* for the right level of manpower. Such a situation actually implies that the resource leveling exercise, if any intended, is not going to yield any outcome. Moreover, in complex projects,

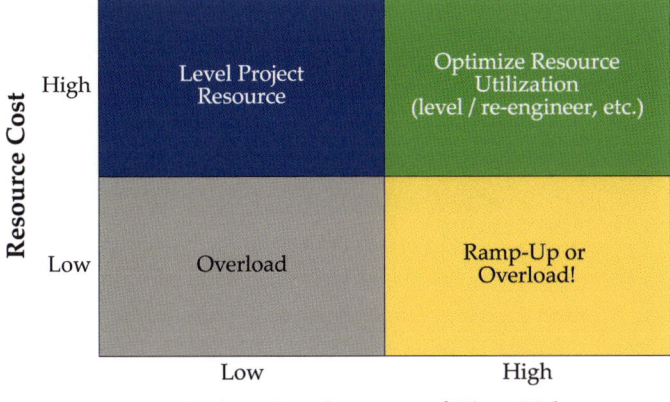

**Fig. 15.6**  Leveling resources: making the call!

with the complexity of calendars, one might often find a huge number of calendars that make the schedule an overly complex animal to deal with.

Summarizing this point, therefore, the reader would now probably agree that

**It is better to schedule a project without resource loading and use one common base calendar of activities when one is using a scheduling software. Resource ramp ups can be dealt with using activity codes that provide a more reasonable platform to work and evaluate the schedules.**

Subsequently, based on the circumstances in the project, the resource in question, and the project portfolio map, the strategy can be chalked out. This gives the project manager a better handle of his own project working rather than going in for some strategy that is likely to be shot-down by the owner stakeholders.

The other aspect is that of adjusting productivities when one actually ramps up on a project. This is a very tricky situation. One needs to look into this in greater detail. We will be covering this topic again in later chapters.

## 15.7  A Quick Understanding of Relationships

We have seen that project schedules are defined by two parameters: the activities and the relationships. While looking at relationships, it is a common practice to confront the basis of schedules. The fundamental business pressure is to *complete the project within the shortest possible time*. Such a view actually questions the basics of project sequencing.

## Typical Start-to-Start Relationship with a Lag

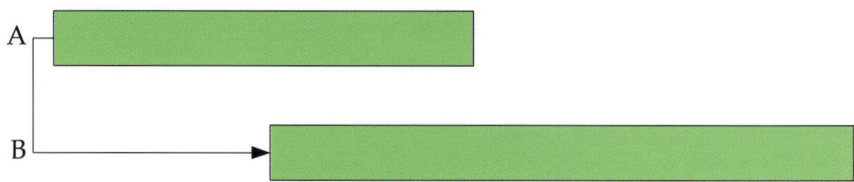

## What is the Reality of the Relationships

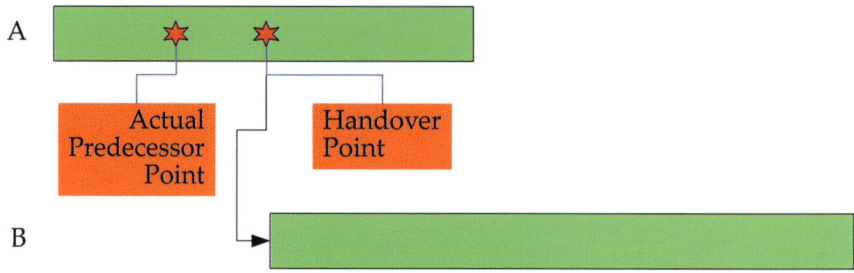

**Fig. 15.7**  Identifying true relationships

For instance in a project, let us assume two activities A and B are predecessor–successors. They are connected by a Start-to-Start relationship with a lag. Now this relationship is a difficult one to comprehend. For instance, let one assume the SS lag is 7 days. Now, let us say that activity A has started. However, does it actually mean that B will start on the 8th day? In reality, the start of B is dependent on the progress of A. This is, therefore, in the true spirit of scheduling, a bad practice.

To explain the dynamics better, one needs to refer to Fig. 15.7. The activity A actually has two points within its duration, that are of interest. The first is the point where the actual work required for B to start as the successor is complete. The second is the point where this information or intermediate product is actually handed-over to B (in short, a take-over point for B). Hence, going by this logic, activity A has too much of 'hidden information.' If one looks at the take-over point, the relationship has now changed to a Finish-to-Start relationship. It is common to find many project schedules having SS relationships of FF relationships. However, these are only gross-level relationships. In reality, *every project relationship is an FS relationship.*

Hence, the project manager needs to ensure that his high-level schedule is replaced early on with schedules that have definite Finish-to-Start relationships. This is, in many project environments, a difficult task to achieve, especially due to the fact that it burdens the project manager and the planner with the humongous task of detailing out the schedule. However, once detailed, it is easy to use the schedule for the remainder of the project.

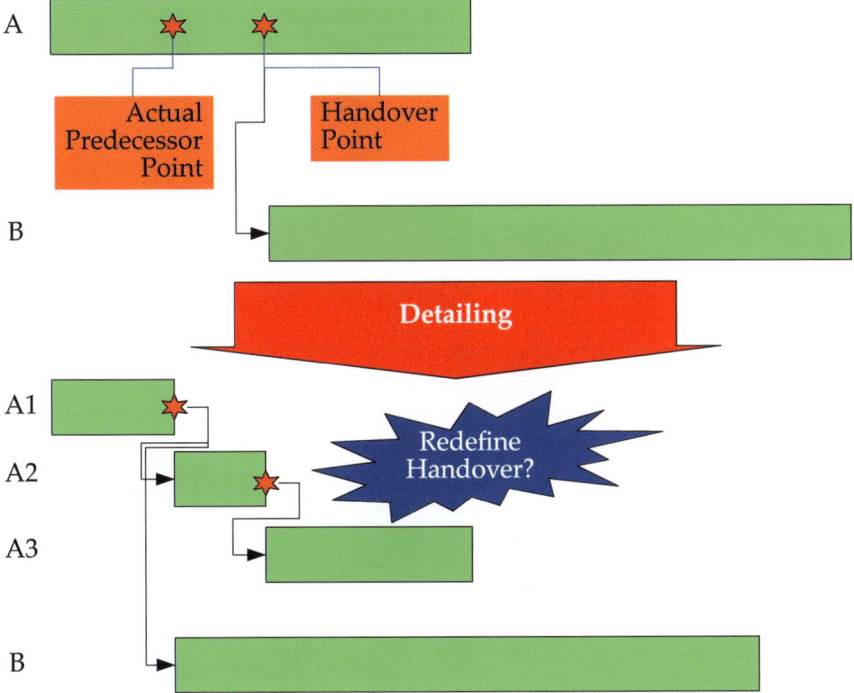

**Fig. 15.8** Detailing of the schedule

Thus, the schedule now transforms into a more detailed version as shown in Fig. 15.8. This schedule has a possibility of optimizing the time frames further, by moving the activity B further up to the actual point in A when the predecessor condition is fulfilled.

One also finds schedules, in practice, where the lags are 'negative.' This is another absurd situation especially if one is looking at 'forward looking scenarios.' In short, there are a lot of challenges to build good and robust schedules. Further, in many projects, there are multiple *constraints that are imposed on activities* in order to fulfill certain conditions. These are typically tie-in points. However, ever-so-often, the project manager might find these points to be vaguely defined and at the insistence of certain team members. Forcing constraints need to be clearly driven exclusively by *external conditions in the project* (that need to be incorporated). For instance, award of contract is an external condition that can be 'pegged' using a constraint. However, when constraints are imposed where they are not actually required, it involves a foul system of scheduling and an extremely unhealthy planning philosophy. When such contracts end up in litigation suits, they could cost the organization heavily.

To cut the long-story short, therefore, the reader is advised to ensure that

**Project Schedules are to be developed based on the Finish-to-Start relation-ships as far as possible (if there is a will, it could well be 100 %)**
**Further, constraints are to be used only for the start of the paths and exclu-sively for extraneous conditions.**

These are certain important guidelines for the reader to follow.

## 15.8   Updating Schedules

In many organizations, one would find that the schedule development process itself is not a satisfactory one. This reflects badly on the schedule quality and does not actually allow a project manager to perform any rigorous schedule analysis. This element together with the fact that most organizations have few people who understand the actual dynamics of the scheduling process, pose as huge road blocks for appropriate application of the schedule information in managerial decisions.

There are several organizations that *do not update their schedules*. This is not a very wrong strategy. In order to update a schedule, one needs to ensure quite a few preconditions:

1. *Robustness*:
   The schedule has to be robust. In other words, it should have good estimates and a sound network that are predominantly FS relationships.
2. *Accuracy*:
   The schedule should reflect, fairly accurately, the actual work process and sequence.
3. *Predictability of the Scheduling Engine*:
   One of the primary drivers in updating schedules is the understanding of the behavior of the scheduling engine. We have seen in our previous treatment on resource leveling that the algorithm has a strong bearing on the outcome of the schedule. Most commercial softwares like Primavera, MS Project, etc. have lev-eling algorithms that combine two or more criteria. This makes the schedule more optimal. However, the flip side is that the resultant schedule is less intuitive and 'predictable' for the project manager. Thus, an update could significantly change the picture of the schedule.
4. *Availability of Information*:
   This is another critical aspect for the project manager to consider. In most projects, this is a challenge. It is quite possible that the project manager gets field infor-mation on just a limited part of the project on a specific day. Such a cycle of information flow would actually compromise the accuracy of the schedule. In fact, it is not possible to update a schedule effectively if the information is not accurate and is not coming at the same time.
5. *Check on the Accuracy of Information*:
   Many times, information for the update tends to carry with it the 'momentum' set by the previous weeks. For instance, there are projects that are 99 % complete,

but the last 1 % often takes an enormously long time. This was also dealt with in the Last Mile Concept that we discussed earlier. However, the critical aspect here is to understand that a higher progress reported at any time in the project is not normally negated. Hence, the trend continues till the end of the project. So, updating a schedule makes sense only when the project manager has a reasonable mechanism to verify the accuracy of the update information.

6. *Schedule Analysis*:

Finally, schedule analysis during the update process is a must. Most organizations just punch in data through time sheets and through some manual reports. They hardly evaluate the schedule. Even when subcontractor schedules are submitted, they are initially just tied-in with the master without carefully deciding the options. Such a situation gives rise to a reactive management strategy. This again implies that the updates, though done regularly, are not used to leverage management decisions effectively.

These are some of the key aspects to understand while updating schedules. We have already touched upon the preference that the reader needs to give to update the schedule (using the *remaining duration* approach rather than *physical progress*).

## 15.9  Basic Schedule Analysis

Earlier on, the topic of schedule analysis has been touched upon. The fundamental question for a project manager is to understand how his project plans are being laid out. While the ground rules have been clarified, a lot of 'tweaking' occurs in the world of projects. All said, every project plan needs to undergo a basic schedule analysis. We will qualify a few of these issues here.

### 15.9.1  *Robustness*

We now look at the dimension of robustness.

**A Quick Hands-On**

We have already described this from a network perspective. However, from a management perspective, the question is fairly straightforward:

**Is it possible to lose or delay a project by 10 days in 7 days?**

It often happens in the project scenario that the project slips beyond the reporting cycle. For instance, one might find a specific activity to have delayed beyond 3 months in just 1 week. This is common, especially when one is dealing with the *schedule*

## Poor Schedule having Multiple End Points

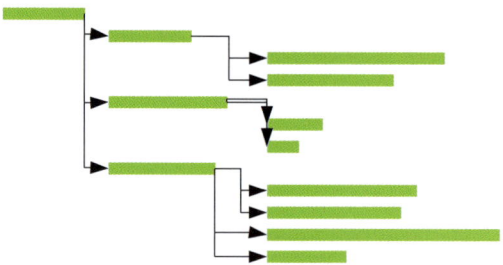

## Poor Schedule having Multiple Start Points

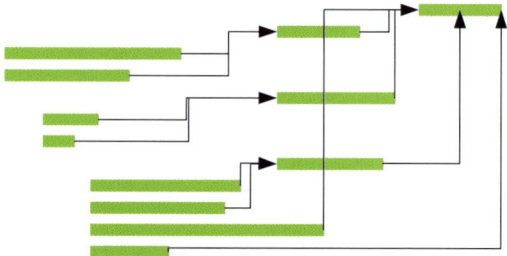

**Fig. 15.9**  Poor scheduling practices

*chicken.* However, it is *also seen to occur in situations where there is no sign of the schedule chicken phenomenon.* Determining this is tricky for any project manager, but if one looks at the detailed plans, one will find a 'thick pad of assumptions' that actually drives the entire schedule topsy-turvy.

Therefore, any schedule has to be clearly understood from the perspective of its management robustness. The question, therefore, is how can one check the robustness of a schedule? It is relatively simple. Since most schedules are developed using software these days, the robustness needs to be checked with respect to the actual capabilities of the software.

The first aspect in the project plan is that every project must have a definitive start and a definitive finish date. This is mandatory. There are several project plans where I have seen multiple starts and multiple finishes. This is, simplistically put, a poor scheduling practice as shown in Fig. 15.9. A good schedule must have one single start activity and one single end activity. An important corollary to this finding is that

**If a Project Manager sees too many external start points, then the model of the project plan is incomplete. In other words, a broader scope is to be used to model the project with reasonable efficacy.**

## 15.9.2   Variances in the Plan

The second aspect to consider is the *variance of durations in a project plan.* The general principle of a project management system is that one details the project as the timeline progresses. In other words, the degree of detailing is initially kept low and this is later increased to include more information on the project.

To a certain extent, this is a good strategy. However, it is always more relevant, from a management perspective, to have a set of *place holders* in the project plan. Regardless of the strategy, therefore, the place holder approach will help a person understand and tailor the schedules in a balanced way. This is seen in Figs. 15.10 and 15.11. The place holders help in firming up the work method in the overall project plan. They also ensure that the user is aware of the detailing or the *quantum of activities* that are likely to show up in the subsequent phases of the projects. Hence, the project manager too gets a clear visual aid to manage the schedule better. Although the place holders are for making it a visual aid, the project manager must understand that *it is not complete or accurate, yet it is close to what is being targetted.* Most project managers fail to do the latter in their basic schedule analysis. The result is that they quickly disown the schedules and are straying in the waters…pretty much rudderless! This phenomenon is a dangerous one. Hence, it is necessary for the reader to understand and use it to prevent project failure.

The second test is a simpler one. It has to do with the variance in durations. In order to ensure that the project plan is usable, the variance in the durations of individual activities must be within certain limits. As a thumb rule, typical project activities must have a duration that lies between two or three times the update cycle as shown in Fig. 15.12. This same variance model is also applicable to the costs involved.

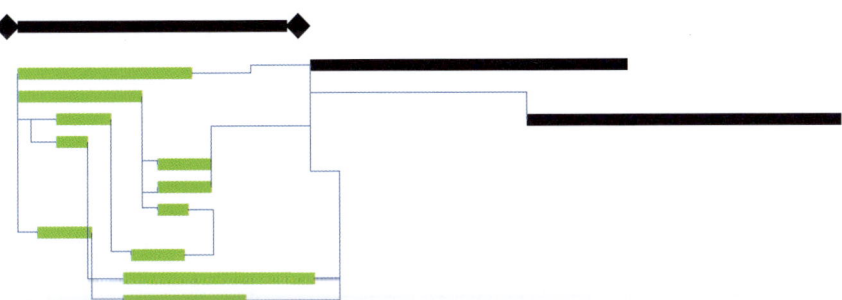

## High Variance in Detailing along the Timelines

**Fig. 15.10**  High variances in detailing levels

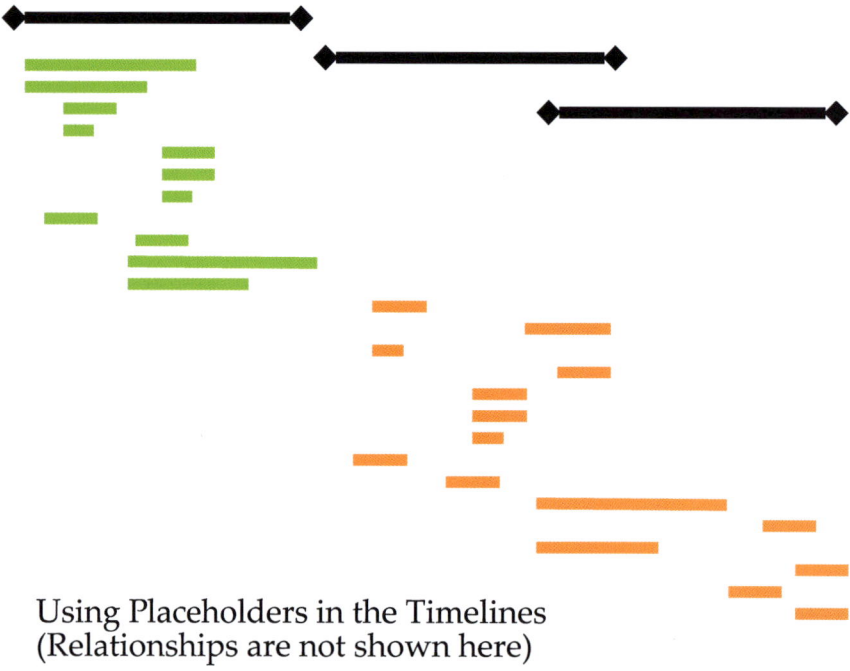

Using Placeholders in the Timelines
(Relationships are not shown here)

**Fig. 15.11** Using place holders to have better detailing

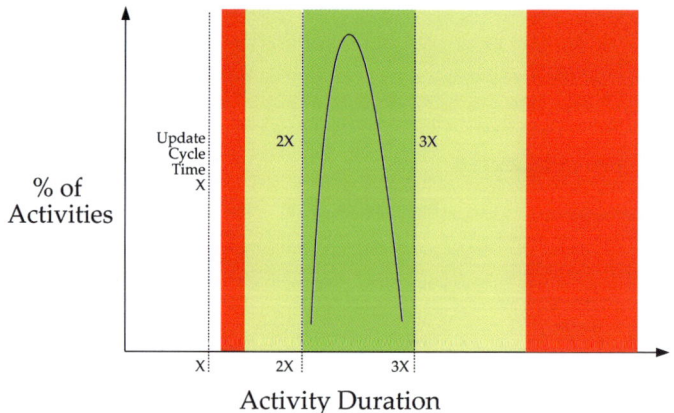

**Fig. 15.12** Variances in durations

## 15.9.3 Activity Assignments

In any project plan, the activity assignments should be unique. In other words, each activity must have a clear and distinctive set of entities:

1. Doer
2. Supervisor
3. Reviewer
4. Manager.

In smaller projects, the supervisor may be absent. Further, an independent reviewer too may be absent. The activity needs to have only *one of each entity* assigned to it. This is a simple way to define a clean schedule.

### 15.9.4 *Clear Activity Definition*

We have already touched upon this in our general treatment on the schedule derivation. However, one of the points that lacks clarity is that of take-over points. In most project plans, the take-over points or the handover points are not clearly defined or specified. This actually allows the gray area to set in. For instance, mentioning that the area should be clean at handover is an interesting one. The arguments could go both ways. While the contractor might insist that the area is clean, the recipient might claim that it is not clean enough!

Clear take-over points are required to ensure that one does not have expensive claim suits that plague the project team for years after the project is complete. While most of these are contractually managed, they are, unfortunately, not managed using project plans, the end result being that the project plan actually reflects a different version of the *reality* when compared with the contract. This needs to be carefully addressed by every project manager.

The other aspect is in defining the driving relationships in the project. An activity might have multiple predecessors; however, only one of them truly satisfies and drives the FS relationship. This is an important concept for the project manager to know. Hence, in any project, the relationships need to be clearly focused on and *cleaned* to ensure that they are accurate and complete and follow the work logic. This is shown in Fig. 15.13. The software, unfortunately, cannot distinguish between the *must-have, the nice-to-have, and the extras included to pressurize the client.* Hence, in the schedule depicted, the simple activity that is included to 'pressurize the client' gets on the critical path. In other words, this serves no purpose for both the stakeholders in the project. We will touch upon this again in subsequent sections.

So, clear activity definitions involve clear take-over points, clear relationships, unconstrained tasks where unnecessary, work and schedule following the same plan, and, the identification of driving relationships that are important and often, *different from the relationships indicated in the plan.*

The reader needs to understand that

**A good Project Plan is like a detailed Map that takes the Project Plan from its Current 'Place' to Success. Therefore, the more the detail in the plan, the more likely it is to be successful in its objective.**

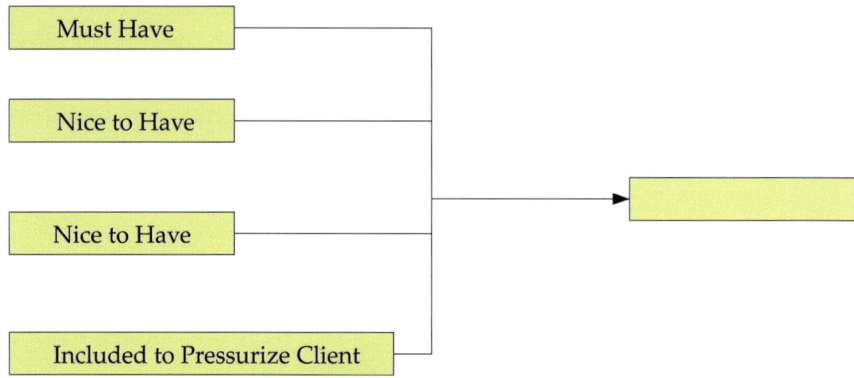

## Evaluation of Predecessor Activities

**Fig. 15.13** Predecessors and relationships

Given this logic, it makes sense to have the entire map ready so as to ensure how one has to reach from one location to the other. This perspective is in sync with our claim that the detailing in a project through actual development or even place holders would give the reader a better chance to achieve project success.

## 15.10   Key Takeaways

In this chapter, the overall schedule concept has been revisited. Although this is a fairly elementary treatment of the project planning process, it covers some of the critical areas involving the schedule development.

It is recommended that schedules need to be independent of the resources and one need not complicate the scheduling process by including several resources and calendars. The simpler a schedule, the easier it is to track, evaluate, and manage. As the complexity increases, the demands on the detailing as well as the resources to develop, monitor, and maintain the schedules increase. Hence, we subscribe to the view that these can be handled *more effectively* outside the schedule software.

We have also touched upon the basic schedule analysis that every project manager needs to perform during and after the project plan development. This actually helps identify a few areas that could snowball into significant concerns during the course of the project.

# Chapter 16
# Advanced Planning and Schedule Analysis

**Abstract** The chapter begins with a detailed treatment of the WBS and the common approaches used to define the WBS. In doing so, we have tried to explore an interesting concept called the 3D WBS that combines the faculties of Geography, Product, and Activity. We then describe a new concept called the WBS Maturity Model that helps the manager understand the potential of his WBS. We then move on to an independent topic that covers the phenomenon of Multi-tasking in projects. We provide a numerical analysis and an alternate perspective to this concept. Following a better understanding of this subject, we move on to the challenging area involving the establishment of successful baselines in projects. In doing so, we touch upon a new framework that looks at the Uncertainty Portfolio in Baselining. We delve deeper and refine the Portfolio to one of its critical components (Time to Closure) and define an additional Uncertainty Portfolio for the Time to Closure. We then conclude the chapter by helping the reader understand how Baselines are linked to the Business Case.

**Keywords** Work breakdown structure · Geographic breakdown structure · Product breakdown structure · Activity breakdown structure · 3D work breakdown structure · WBS maturity model · Multitasking in projects · Project efficiency · Level of detailing · Planning time · Myths about baselining · Dimensions in successful baselining · Uncertainty portfolio in baselining: effects perspective · Uncertainty portfolio in baselining: strategies on time of closure · Baseline linkage with business case

Having touched upon the bare basics, we will now look at a few advanced tools in project management.

## 16.1 WBS and Its Link with the Project Plan

Simplistically speaking, a WBS is the Work Breakdown Structure. However, due to the way many managers get bogged down with this concept, it is also called a 'work-

© Springer Science+Business Media Singapore 2017    245
N. Gurjar, *A Forward Looking Approach to Project Management*,
Lecture Notes in Management and Industrial Engineering,
DOI 10.1007/978-981-10-0782-8_16

till-the-project-manager-breaks-down' structure! A procedure to define the WBS is given in the blog of Praveen Malik [16], as mentioned below:

1. Think in terms of outcomes (nouns) rather than tasks (verbs).
2. Think in terms of deliverable (For instance, a Project Manager can start with phases and sub-projects then go to deliverable).
3. Think about achievements (small/big) that can be shared with customer/management and which are likely to be appreciated.
4. The whole work (at top level) is unstructured. The WBS (by dividing into smaller components) gives it a structure.
5. Divide the work to a level where the Project Manager can assign responsibility to an individual/department/organization.
6. Divide the work to a level where the Project Manager can easily monitor and control—not too low level where it becomes headache and not too high when definition is not complete (work cannot be satisfactorily measured).

This is a simplistic process and the reader might find it rather intuitive and logical to implement. He goes on to give the 7 reasons as follows:

1. WBS is a pictorial representation of scope; it represents total scope of work. A picture is worth more than a thousand words and hence, easy to understand and follow.
2. WBS represents the total scope and, hence, it can act as a checklist for the project. A project manager can easily tell what all has been accomplished and what is remaining in the project.
3. Each successive level of the WBS provides a basis for more accurate estimation of effort, duration, resources, and cost.
4. The deliverable, as represented by the WBS, can be easily monitored and tracked.
5. The Historical WBS can act as a templates for future work, specifically for some repetitive processes. It results in easier and faster project planning.
6. Smaller work can be easily assigned to individual team members and they can be made accountable.
7. Most importantly, it reduces the project risk.

That summarizes the conventional perspective.

### A Quick Hands-On

The question for the reader is not just to understand how to define a WBS. The cardinal question is, in fact, the following:

**Is it important to have a WBS?**

The plain and simple answer to this is *no*. It is important to know that you have identified all the work of the project. The WBS is just one approach to help the project manager do that. In addition,

1. On most construction projects, Master Formats serve the same purpose.
2. On smaller projects, you may be able to skip the WBS and go straight to a network logic diagram (as long as you start from the back and work towards the front).
3. On Agile software development projects, a WBS is unlikely to provide value to an iteration.

And well, if you do prepare a WBS, it's important to do it right. There is a lot of Mis-information about WBSs in the literature, and there is also a lot of conflicting terminology. For example, not everyone uses 'work packages' or 'cost accounts'.

### 16.1.1  3D Project WBS

Jean-Yves Moine, on a forum on Linkedin.com, gave a new concept called the 3D Project WBS. This concept is similar to the overlay of multiple perspectives that we had mentioned in our earlier mantras, the only difference being that he has tried to elaborate on the perspectives to help create a tool that could be a useful control mechanism for any project manager.

Our perspectives defined are like the axes along the project cube, according to him. He has identified three key perspectives as:

1. Areas (GBS, Geographical Breakdown Structure), sections in the space, localizations on the site—meaning WHERE?,
2. Products (PBS, Product Breakdown Structure), civil works components, equipments, materials—meaning WHAT?,
3. Activities (ABS, Activity Breakdown Structure), actions, processes—meaning HOW?

Of course, we also had a WHO dimension in our project perspective.

In Equation: The only significant innovation is his equation called

$$WBS = GBS \times PBS \times ABS$$

According to him, the WBS is a crossing between these three breakdown structures. The time schedule task is at the last level of the WBS, included in it. However, we don't subscribe to this view. It is *more useful to define the WBS using one perspective and refine it using the others*. In refining, one is actually qualifying the information available in one perspective by adding fresh information to the activities and NOT the WBS. He further includes the OBS in the WBS equation.

Despite the problem in using the model, the control point that he has addressed is an interesting one:

The control points are at the interfaces of the WBS. So, while developing the WBS, any changes along any parameter viz., the organization (the OBS), the area (the GBS), the product (the PBS) or the activity (the ABS) implies an interface. In his proposition, he has said that a cub constructed on caliberated axes could help the project manager quickly identify the interface points. And if effectively constructed, the graphical nature of the tool could be leveraged to make it a powerful one.

This is the key takeaway from his theory. The Work Package has multiple perspectives to consider. And although one goes by a single perspective, the awareness of the nature of interfaces can help the reader *choose his project perspective more wisely.*

A preliminary cause–effect relationship is provided in the Auditor's Notes on the GAO Schedule Assessment Guide [18]. They have listed out 10 best practices and have defined the effects if these aren't followed.

### 16.1.2  So, Where Is the Problem?

Despite the wonderful definitions and their underlying concepts, projects still get late and have issues! So, if one is anyway going to be bogged down with issues after defining good WBSs, there is something in the *line WBS that isn't being captured effectively.* One needs to delve a little deeper into this. Fugar [17], in his study, has given an excellent background to this subject.

Some of the key reasons of project delay, in construction projects, in his work, are the following:

1. Material

    (a) Shortages of materials on site or market
    (b) Late delivery of material

2. Manpower

    (a) Shortage of unskilled labor
    (b) Shortage of skilled labor

3. Equipment

    (a) Equipment failure or breakdown
    (b) Unskilled equipment operators

4. Financing

    (a) Delay in honoring payment certificates
    (b) Difficulties in assessing credit
    (c) Fluctuation of prices

5. Environmental

    (a) Bad weather conditions
    (b) Unfavorable site conditions

6. Changes

   (a) Client initiated variations

   (b) Necessary variations

   (c) Mistakes in soil investigation

   (d) Poor design

   (e) Foundation conditions encountered on site

7. Government action

   (a) Delays in obtaining permit from municipality

   (b) Public holidays

   (c) Discrepancy between design specification and building code

8. Contractual relations

   (a) Legal disputes

   (b) Insufficient communication between parties

   (c) Poor professional management

   (d) Delay in instructions from consultants

   (e) Delay by subcontractors

9. Scheduling and controlling techniques

   (a) Poor site management

   (b) Poor supervision

   (c) Lack of program of works

   (d) Accidents during construction

   (e) Construction methods

   (f) Underestimation of costs of projects

   (g) Underestimation of complexity of projects

   (h) Underestimation of time of completion

Hence, any planning and scheduling analysis must explicity address the linkage between these issues from a project context. This doesn't happen very *clearly and explicity in most real life projects*.

## 16.2  WBS Maturity Model

In continuing with the previous section, there are a couple of dimensions that are important for the reader to understand:

1. WBS definitions are often very 'personal' to the likes and tastes of the project manager.
2. The WBS definition has not had any specific conceptual framework to evaluate the quality of the WBS.
3. The quality of the WBS often reflects in the quality of the subsequent plans.

**Fig. 16.1** Progression of
WBS maturity

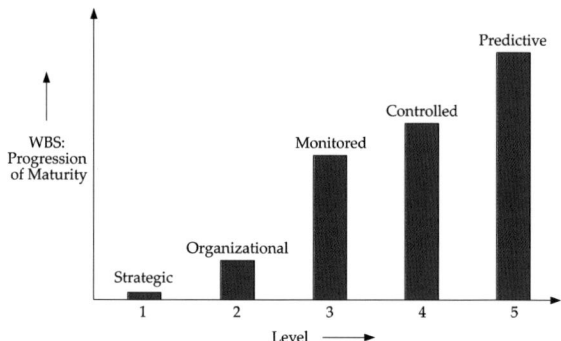

Keeping these in mind, we have identified a progression of the WBS as shown in
Fig. 16.1. The WBS progression extends into the subsequent activities as well. So,
although there is a generic progression, the repercussions are many and the reader
needs to appraise his own plans with the model.

The definition of the WBS progression model is given below.

1. *Strategic*

   The lowest WBS level is the strategic level. In this WBS, the project is very
   rudimentarily planned to evaluate the basic status from a *Strategic Level*. The
   final plan is used for very limited objectives like

   (a) Project Status review
   (b) ROI/NPV Assessment
   (c) Portfolio Decision Making
   (d) Issues with Major Partners
   (e) Strategic Interventions

   Such a WBS is usually based on *what the owner stakeholder is looking for from a
   project*. Needless to say, it is devoid of details. The number of levels is restricted
   to 2.

2. *Organizational*

   This is the Level 2 maturity WBS. This is usually prepared at the beginning of the
   project, for the executive, to have a better handle of the project. The WBS here is
   a little more detailed, but still retains its 'generic' character. In other words, most
   activities will have durations that are at least 4–6 times the update cycle time. The
   purpose of this WBS is basically to

   (a) Create a *general scope document and basic plan*
   (b) Tell the reader how the project is organized
   (c) Have details of a few Departments/Areas/Packages, etc.
   (d) Outline a general responsibility
   (e) Define Work Elements
   (f) Be a basis for Contracts.

In other words, this WBS is more *contract and function-centric*. Many project managers prefer these kinds of details as it helps them track their projects from the contract perspective. This WBS is typically of 3 levels and it often changes as the project progresses and fresh information comes in. The characterizing question in this level is: *who is doing which part of the project?*

3. *Monitored*

Unlike the previous two levels, this level is significantly different. The main difference is in the purpose. This WBS and plan is meant for organizational review. Unlike the previous two that were more oriented toward the management (project/senior/owner stakeholder), this WBS and plan is oriented toward a *measurement-based philosophy*. The main feature of this WBS and plan is

(a) Measurement oriented
(b) Task aggregated as opposed to Work Package.

So, a package here is further refined to allow fairly independent sub-packages that are measured. Note that the measurement philosophy plays a central role in detailing out the WBS as it goes along. While this is different from the convention, it is true that at project level, a purely 'theoretical' method of organizing a project scope does not work well as not everyone has the same take on the concepts and the belief in the *gains* from a specific method of developing and organizing a project plan differs significantly from person-to-person.

To clarify the measurement perspective, this WBS is more resource- and allocation-centric. Its detailing is very high as it directly connects with activities and entities at the operational level. The mode of development is, however, a reactive posture. In other words, despite the detailing, there are likely to be surprises that the project manager is bound to see. The characterizing question in this level is: *What is the definition of the problem?* The WBS must satisfactorily answer this question

4. *Controlled*

This is the fourth level WBS and project plan. It is very similar to the monitored level. The fundamental objectives here are, however, a little more elaborate than the Level 3 covering issues like

(a) Package refinements that need to be used as control measures
(b) Cause–effect capture in the plan.

Needless to say, this level of maturity is geared toward *resolution of issues* rather than just reporting the issues. So, this is a proactive approach toward the WBS and the plans. The characterizing question of this level is: *how should the current problem be resolved?* Again, the WBS must satisfactorily answer this question.

5. *Predictive*

This is Level 5 of the Maturity Model. In this method, a lot of focus is laid out on the predictive causes and potential effects that could occur in any project. It allows the project manager to have total project control at all times and fairly contains surprises born out of chronic symptoms. The key or characterizing question this level addresses is: *is there a problem likely to come in the future course of the project?* Like earlier, the WBS must satisfactorily answer this question.

Many project managers might believe that the latter 2 levels could be kept outside the maturity model, as they believe these are core areas that constitute the competence arising through experience. Moreover, the level of detailing is often a political decision (context issue). However, if the project plan is done correctly, it benefits the project manager the most and ensures a better chance of overall success.

## 16.3   Multitasking in Projects

While discussing resource leveling and allocation, one of the aspects that invariably comes to the forefront is the multitasking of members in the project.

**A Quick Hands-On**

At this time, the reader needs to answer certain fundamental questions like:

**Should one multitask in a project? If yes, how?**

While multitasking itself is a complex issue, the overload condition described earlier forces most managers to actually multitask in every project. Regardless of the environment, there is always going to be an element of multitasking for the project manager.

Handling of multitask depends on personal traits and developing of different skills such as

1. Planning: which helps the manager to prioritize his/her tasks
2. Education and knowledge: which helps the manager to solve the task in short time
3. Training: which increase the manager's skills to perform the task in good quality in minimum time
4. Focusing: which helps the manager to know the parameters of the issue and run around it quickly
5. Careful listening: allowing the manager to quickly understand, and make decisions.
6. Experience: giving the manager the required vision and quick decision-making
7. Passion: which allows the manager to love what he/she is doing, and move forward with high ability
8. Delegating the tasks, then quick monitoring, discussing, short meeting, etc. That will increase the ability to handle more than one task at one time.

Personal traits, when it comes to handling of multitask in one time, is different from one person to other. So, it is the duty of the team leader to chose the suitable persons, who can handle the tasks, based on their personal traits.

Additional tips, in a more prescriptive manner, are referenced from Linkedin.com as the following:

1. If you want to make it better, treat it as a problem. Spend a few minutes at the end of the day reflecting. Can you identify any Lessons Learned at the end of the day?? If yes, make a note of them. Make a note of tomorrow's high priority tasks.
2. Do not operate in an 'interactive mode'—responding to messages and email as they come in. Schedule times to check and respond.
3. Think about priorities. Our human nature is to switch to tasks that we are more comfortable doing, not necessarily the most important.
   As something comes up, classify it as 'must do now', 'must do today', or 'must schedule'.
4. *The real problem is not usually the number of tasks, it is the number of times that we change tasks [and the associated 'context switching' overhead]. Checking your email 27 times a day is an example of a potential inefficiency.*
5. If you know that you will be making phone calls, cluster them together to minimize task switching.
6. Just step back and think about your particular work patterns from the point of 'how could I do this differently'. I know that I am my own worst enemy when it comes to too much task swapping going on.

The idea of context switching is further elaborated in the Critical Chain Project Management. However, all said, one has to understand this concept from the perspective of improving productivity to the fullest.

From a modeling perspective, there are two components that are of priority when it comes to evaluating multitasking. Let us try to understand this by means of an example. Let us take a typical project manager who performs a task. Now, if the project manager has one thing on his focus, it is going to be his *effective productivity.*

So, in our model now, there are two factors to consider:

1. The planning effort
2. The tasks to be covered.

Mathematically, speaking, we have the following conditions.

1. The Planning Time is going to be proportional to the duration of the task. Hence, longer the task, more the time to plan the same.
2. The Productivity is proportional to the planning time.
3. The productivity is proportional to the level of detailing in plans. So, a higher level of detailing would ensure better productivity.

This is shown in Fig. 16.2. We define high productivity thresholds as the high value assumed ideal for the project.

The Level of Detailing is going to drive the Planning time. This is shown in Fig. 16.3. However, note the general trend in the curve. This increases sharply in the beginning and then slowly levels of. The planning time also increases with the duration of time that needs to be planned. This relationship is shown in Fig. 16.4.

Now the total time is actually the sum of the planning time and the execution time. This is a fixed number. The progress of work is actually going to be the product of the

**Fig. 16.2** Level of detailing
and maximum productivity

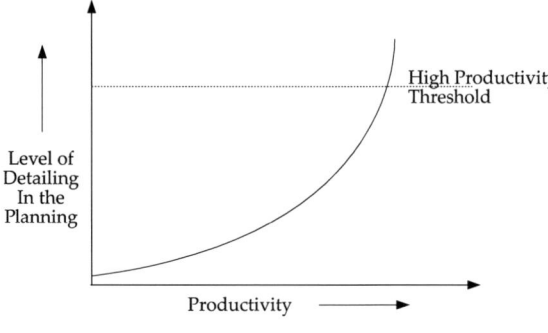

**Fig. 16.3** Planning time as a
function of detailing

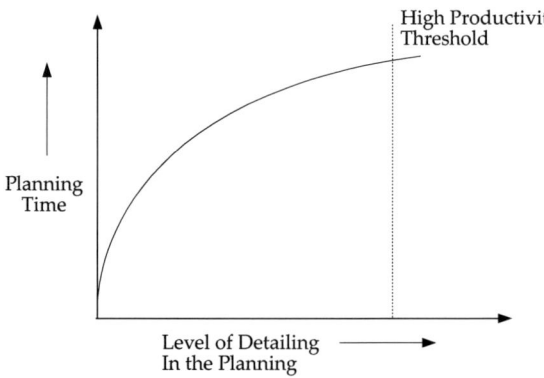

**Fig. 16.4** Planning time as a
function of duration of the
task

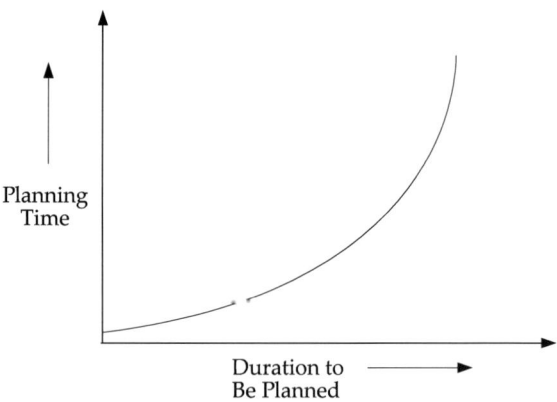

execution time and the maximum productivity. Hence, this is a problem amenable to
be solved using OR approaches.

$$\text{Processing Time for the Time Bucket} = P_T$$

$$\text{Execution Time for the Time Bucket} = E_T$$

For any management situation, we have the following relationship

$$\text{Total Time for the Time Bucket } T = P_T + E_T$$

Going further with the formulation, we have the following parameters, in addition, to consider

$$\text{Level of Detailing} = L_D$$

Therefore, the constraints are arising from the equations given by

$$\text{Productivity } \eta \propto L_D$$

$$P_T \propto L_D$$

$$P_T \propto [T - P_T]$$

And the objective function is given by,
*Maximize*

$$Z = [T - P_T] \times \eta$$

This formulation gives us a good handle of how much multitasking a project manager can use. We are using simple proportionality in this case. The real world could be more intricate and complex than what is shown in this case. Nevertheless, it is easy to understand how the progression runs. Since we know that there are $n$ activities to multitask, we actually can have a penalty function to adjust the main objective function. This will help evaluate the maximum number of tasks that can be done. Alternatively, one can also set a minimum value for the Execution Time.

Thus, going by the current scenario, we get that

$$Z \propto P_T \times [T - P_T]^2$$

**Fig. 16.5** Variation of objective function with planning time

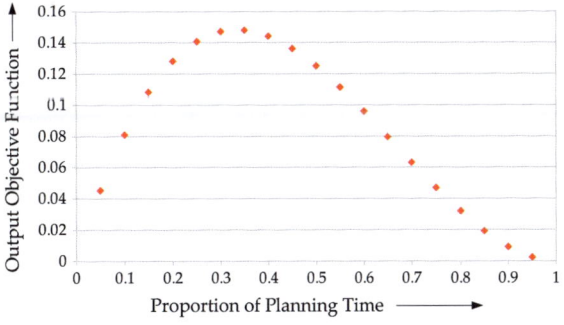

Thus, the peak of the function is around 35 % of the Total Time as shown in Fig. 16.5. Ideally, in a multitasking environment, the minimum duration should be around 2 h on each task. Therefore, one can *effectively multitask with three or at the most, four tasks*. This can be numerically ascertained in most environments. However, it needs a little bit of modeling and number juggling to come to the right conclusions.

## 16.4   Establishing Successful Baselines

One of the most important exercises in project management is the establishment of the project baseline. Most project managers often have difficulty in *successfully defining* the project baseline. In this section, we will delve into this aspect to understand certain key aspects.

### 16.4.1   Some Myths About Baselines

In the real world, there are two orientations of project managers. There is a huge class of project managers who call themselves the field guys: Rolling up their sleeves and getting ready to do anything required. The other class of project managers call themselves the corporate managers: More focused on the corporate management perspective. Both have their advantages and disadvantages. However, depending on the school of thought, their approach and perceptions are many times different.

**Baseline Is Paperwork**

One of the common misconceptions about baselines is that many think this is just paperwork meant to 'satisfy' the needs of some 'ISO guy' or some 'stakeholder' at the company. The manager believes that it is just a waste of precious time on paper as it is of little relevance to the overall project! Nothing can be further from the truth, yet this misconception prevails and we will touch upon the aspects again as we move forward.

**Baseline Can Change**

This is another interesting observation most managers often have. They believe that the future is uncertain and the knowledge at the current time is limited (and inadequate). So, a baseline is an additional burden for a plan that will anyway change! While it is true that many projects do have to change baselines, this misconception is like putting the cart before the horse! Obviously, the approach is a 'plan for chaos' rather than being a 'plan to manage' one. People shy away from committing because they view the situation as one with limited gains.

As a forward-looking project manager, one needs to, however, understand the use of the baseline. To do that one needs to first look at the basic definitions of baselines.

## *16.4.2  Define the Baseline*

In Project Management, a Baseline refers to the accepted and approved plans for executing a project. Project baselines are, generally, approved by project management team and those are used to measure and control the project activities.

Baselines give the project manager an excellent way to understand project progress (by analyzing baseline vs. actual) and forecast the project outcome. Baselines are important inputs to a number of project processes and, the outputs of many processes raise change requests to these baselines.

The important aspect to understand is the basic components of the Baseline:

1. The Business Case
2. The prime contract and the documents leading to it (RFPs, Negotiation Documents, etc.)
3. Project Execution Plan (PEP)
4. Project scope document
5. Project schedules
6. Project estimate
7. Project staffing plans
8. Project Work Breakdown Structure
9. Project process maps (some people prefer including these as well)
10. Internal Management of Uncertainties and Risks (again some people like to include these in the baseline definition).

In most companies, however, baseline is treated as a project schedule just because there is *a button in some scheduling software that is called the baseline!*. Hence, the overall understanding gets skewed when one has an incomplete picture to start with.

This step is typically completed in the early days of the project life cycle. As the project progresses, the team must revisit the original baseline documents and evaluate the potential need for new documents. A baseline can and does comprise of any number of different documents, each establishing a well-defined metric by which the execution team will track, measure and deviate against (via the change management process).

If one is to single out three features of the baseline, they would be:

1. Enforceable
2. Published
3. And should be treated with discipline so as to not change it at random/at will!

However, we will now look at additional areas that determine the 'true success' of the baseline process. This is different from 'just defining a baseline' and being done with. As the reader will go through this detail, the dynamics will become clearer and there will be a better insight into the elements of success associated with the baseline process.

### *16.4.3  Dimensions to Ensure a Successful Baseline*

The following dimensions are important to understand how the success of the baseline establishment procedure is affected in any project environment. The reader needs to understand these areas and their repercussions.

**Information Content**

While defining a baseline, the content that is included in the baseline needs to be checked for its quality. Ideally, the information source is a good indicator of the quality of the information content. There are three possible sources from which information is incorporated into the baseline:

1. Estimates or guesstimates of the team
2. Preliminary market feedback
3. Specific case-wise feedback from potential project partners of the current project.

It is customary in most project situations to give the last source the highest weightage. However, the reflection on the experience of any project manager comes from studying the deviations between the sources. A truly seasoned project manager will be able to ensure *minimum deviation* between his own estimate and the actual values that come from the market.

   This also does not mean that one directly goes for case-specific feedbacks. Many project managers plead ignorance, especially when they are asked for estimates, stating that they need to talk to potential project partners/contractors, etc. This does not solve the problem at hand, rather only creates a huge risk for the project. Owner stakeholders need to understand this fact while using the baselines. For, in short, this strategy will make the process unsuccessful.

**Entity Involvement**

A successful baseline needs the *active involvement* of various entities. Again, in most projects, there is an involvement, but it is often very limited. Unlike a business case, a baseline is extremely detailed and needs to be *respected for its detail!* Often times, however, the detailing is a part of the *delegation process* which really takes away any active involvement, from the seniors, that is desired. In the ideal situation, therefore, the involvement has to be clearly identified and defined:

1. Project Manager: who needs to be the owner and the key driver of the process
2. Project Planner
3. Other Team Members
4. Colleagues from End-User Departments (post project completion)
5. Colleagues from Controlling and Finance Department
6. Lastly, an owner stakeholder who needs to be comfortable discussing the 'devil in the detail'.

If one doesn't have the involvement of these entities, it is likely that the baseline application would get compromised during an early phase in the project.

**Baseline Process**

Once the entities are involved, the methodology becomes important. There are three common ways in which a baseline is established:

1. An iterative process that involves several successive iterations of the baseline documents until there is a common agreement between all the entities involved.
2. A two-stage process whereby a draft is tossed out in the open and the baseline is democratically decided based on inputs from the plan. This is like a quick meeting approach where all the involved persons will come together and provide their inputs. They would then complete the discussion and simultaneously incorporate it in the baseline.
3. A forced baseline where the project manager or some delegated person (like a planner) shares a baseline, and tries to 'push' it through to the other members. The differences remain and are, at times, the beginning of issues in management alignment.

Normally, the iterative agreement approach would be the preferred one as everyone is able to involve themselves better and bring out a baseline that can be owned by all. However, this cannot be generalized and a lot of parameters come into play. Nevertheless, the reader needs to be aware of this dimension from the perspective of evaluating its impact on the success of the baseline process.

**Distribution with Respect to Decision-Making**

Baslines are developed and used in companies using different strategies:

1. In most companies, the baselines, although developed, are only partially published. These companies believe in a kind of a *Need-to-Know policy*.
2. In others, the baselines are published, but are not easily amenable for decision-making. For instance, a cost baseline is at such a 'high level' that the project manager is often unable to get any meaningful information for his decision-making.
3. Lastly, there are companies where the baselines are published in detail and are in usable levels of detail and formats to ensure that the manager can make the right decisions.

Its a no-brainer that the last type of baselines are the ones that mark a successful process.

**Treatment of Uncertainty**

We have cursorily touched upon the treatment of uncertainties in a project plan earlier. However, when it comes to establishing a successful baseline, we need to look at

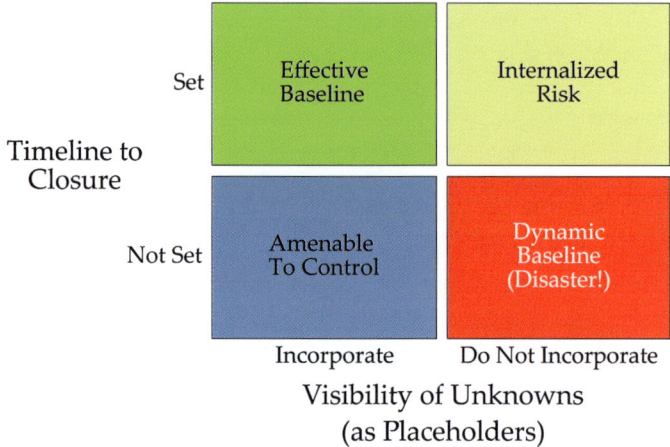

**Fig. 16.6** Consulting Connoisseurs Uncertainty Portfolio: Effects perspective Model

two basic dimensions: (a) Visibility of uncertainties as placeholders and (b) Time to Close the 'uncertain' elements to ensure a good baseline.

The model is called the Consulting Connoisseurs Uncertainty Portfolio Model for the Effects Perspective and is used for evaluating the success of the baseline process. These are given in Fig. 16.6.

There are four conditions that are a direct fallout.

1. If there are placeholders in the baseline with a regime that ensures timely closure of these placeholders, the baseline is the most effective one from usability and long-term perspective.
2. If there are placeholders in the baseline and there is no 'active mechanism' to minimize the uncertainty, the baseline isn't adequately controlled. However, most times, as the project progresses, there is a possibility of this visibility to become a trigger and force the team to set a deadline to close all uncertainties. Usually this is done at the first instance of a milestone slip!
3. If the placeholders are not shown in the baseline, yet there is a timeline for closure, there is a possibility that the entire baseline is completed on time. However, this could, many times, potentially result in scope, budget, time, or even manpower 'adjustments' when they surface. Moreover, the lack of an explicit list only internalizes the risks of the process in the organization. This is unnecessary, though it could happen in many instances.
4. If one doesn't use placeholders and one doesn't have any timeline to close the 'open uncertainties', the baseline would be an extremely dynamic creature and such a project is definitely on the road to disaster!

Thus, the reader must take cues on how to handle these situations by taking adequate precautions to ensure one isn't bogged down with the pitfalls.

There are several instances where large projects go 'hay-way' because there is an improper baseline defined to start with. The project manager, in such projects, often

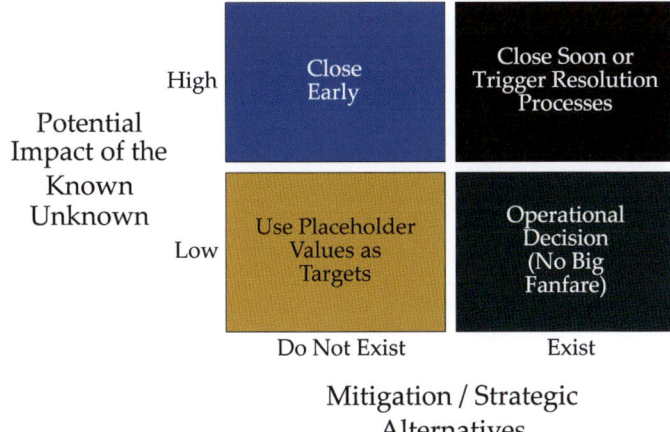

Fig. 16.7   Consulting Connoisseurs Uncertainty Portfolio: Strategies to view 'Time to Closure'

ends up in a fix. One of the key triggers to the chaos/mess is (often) found to be the absence of placeholders to define the *known unknowns*.

If one looks at the intrinsic nature of the 'known unknowns', then one has a slightly different perspective. We call this the second part of the Consulting Connoisseurs Uncertainty Portfolio model. This helps in putting the parameter 'time to closure' in the right slot. The key drivers are: (a) the mitigation perspective or the strategic alternatives perspective and (b) the potential impact of the risks. Often times, a project plan is revisited late in the cycle. In fact, so late that the marginal benefits of the costs incurred are often too low to justify the same. This is, therefore, summarized in the strategic map. The directives are summarized in Fig. 16.7. As one can see in the figure, the time to close is directly governed by the presence of both strategic alternatives as well as the impact of the uncertainty.

If the impact is high and there are no strategic alternatives (for instance, a prime contractor selection in the baseline), then, it is imperative to ensure quick resolution. If alternatives exist and are identified, one needs to quickly look at triggering the resolution among the alternatives as well. If the impact is low and there are no alternatives, then the project manager should try to use the place holder guesstimate as his 'official target'. And if the impact is low and there are alternatives, then the decision could be operationally handled in the baseline rather than making it complex and stalling the baseline process.

## Management Commitment

Any management initiative needs to be reinforced. And the baseline is a very important management activity. Yet it is very often not adequately reinforced. A lack of this reinforcement actually renders the entire exercise useless.

While it is agreed that a detailed baseline is a 'technical document', most project stakeholders need to be trained in using the same. In fact, in many organizations, *the rule of 5 Whys* is used by the senior management. In other words, any deviation is confronted with 5 simple whys: for the reasons and the reasons of the reasons. For instance,

1. Why are we late? Due to the Contractor.
2. Why is the contractor late? Due to payment issues.
3. Why are there payment issues? … and so on.

By the time the senior management goes to the fifth why, they understand where the problem stems from.

However, to ensure that they are indeed able to trigger such a rigorous process, they need to use the baselines correctly. Hence, there are two strategic postures that the management uses in such cases:

1. Use baselines in a binary reporting format
2. Don't use baselines or a flexible reporting format

whereby, a binary reporting format only allows for two states: 'baseline green' or 'baseline red'. In such a situation, the management's commitment to the baseline process is extremely strong and this reinforces the success of the baseline process. On the other hand, when the management doesn't use the baseline or allows a flexible reporting format, the sanctity of the baseline is lost! A baseline is a primary tool in project management. When the management decides against using it, they only compromise their own interests in the project, and this starts giving confused signals across the board with differential priorities that are mostly ad hoc in nature.

**Linkage with the Business Case**

Though the business case is an integral part of the baseline, its relationship with the other entities is not established in many circumstances. The other component documents of the baseline are actually derived and linked to the basic business case of any project. However, *in the use of the other component documents, an active linkage with the business case may or may not be explicitly defined.* This implies that three different cases arise out of this situation/parameter:

1. The linkage is not established. Therefore, each document is a stand-alone, with limited influence on the other. In short, every functional area becomes like a 'free bird'. Decisions are mostly operationally verified and the business case is protected only when the owner–stakeholders decide to review the project. This is a typical scenario where the information is given on a 'Need-to-Know' basis or when the structure is extremely hierarchical in nature.
2. The second case is when a linkage is established, but is not fully respected. In such situations, the project manager doesn't truly have the competence to incorporate the relationships to his decision-making. In other words, the 'uncertainty' is the first scapegoat in such cases. However, it should be made clear that this is more of a competence issue than anything else. The use of the baseline is only to highlight the deviations and is less proactive in nature.

3. The third case is when the linkage is established and made explicit and is also respected. In this kind of a situation, the project manager actually is fully aligned with the requirements of the organization and the baseline process is indeed successful in meeting its objective.

After looking at the dimensions, the reader can assess his own project scenario and try to improve the baseline process. The baseline is the document that connects the *project dream* to the *project reality*. Hence, if it isn't successful in doing that, the project success gets significantly compromised.

## 16.5 Key Takeaways

In this chapter, certain advanced topics were covered. We began with the understanding required in the development of the WBS. The WBS needs to be linked with the subsequent plan. This is an important connect. Yet, there are very few tools that help in the *verification and control* of the subsequent plan with the WBS. Some of the upcoming areas of development and tools had been discussed the 3D Cube model for the WBS being the most promising development in recent times. However, given its preliminary state, it will take a few iterations before this concept would actually be a part of any standardized method or process.

In continuing the discussion, we also saw how the WBS and the subsequent plan could be rated in terms of a Maturity Model. Taking cues from the CMMi framework, an effective classification has been defined in this work. The maturity level is like the *length of the lever* that the project manager has. The more the maturity level, the more the leverage that the project manager could get from using it.

Following this, we touched upon the area of multitasking and demonstrated that *although there is a lot of advice and literature on this subject, one can actually derive the efficacy using a rigorous method as shown*. A quick application of modeling and simulation demonstrates that a thumb rule of around 35 % of time must be devoted to planning in order to ensure effective execution. This model can be refined to the specifics of any project manager's work context and can give more accurate and predictable handles on multitasking.

Lastly, we covered one of the most ignored areas in project management viz. *successfully establishing a baseline*. A lot of literature speak of baselines, however, there is no literature that explicitly addresses the method of successfully establishing one. Most treat it as a 'run-of-the-mill' exercise that is seldom used in the right perspective. One of the key reasons could be that baselines are never treated as an 'exit gate' and hence, it is always treated to be an operationally relevant exercise and not a strategic one. A lot lies in the perspective. However, this is far from the truth as we have seen in the section that first defines the components of a baseline. Hence, the definition needs to be clarified at the outset. The section covers these aspects, the myths, the actual understanding of the process, and gives valuable insights and pointers to the reader on defining and establishing a baseline that is high on the index of management effectiveness.

# Chapter 17
# More Advanced Planning

**Abstract** This relatively large chapter begins by touching upon a case study that deals with changes in WBS structures during the project execution. It touches on the nuances of such changes and the sensitive management dimensions associated with such changes. We then move on to Integration in Planning. Today, there are technological products that could model integration (particularly 4D models). However, we have also focused on lighter versions of 4D models. In doing so, we have elucidated the complexity of perspectives involved in 4D planning. We have also touched upon modeling critical issues and areas through appropriate (and innovative) schedule analyses. In the subsequent sections, we have introduced the Integration of Schedules with Logistics viz. Deliveries, Parallel Activities, and Infrastructure. This section demonstrates how the schedules are useful in their integrated forms. We then focus on integrating schedules as schedules (incorporating one schedule into another). In doing so, we touch upon the challenges in updation along with the influence of the phenomenon of Retained Logic and Progress Override. Another important concept that is dealt with is that of selective updation of schedules. We then debate on the philosophies of Activity-based Scheduling versus Milestone-based Scheduling. We touch upon some useful and critical aspects of Activity Detailing from an Integration Perspective. And finally, we touch upon the concept and the application of the Critical Chain Project Management.

**Keywords** Work breakdown structure · Changing WBS · Perspectives in WBS · 4D integration planning models · Lighter versions of 4D models · Complexity in 4D planning · Analyses of critical areas using 4D model methods · Integrating schedule with logistics · Integrating schedules as schedules · Retained logic · Progress override · Selective updation of schedules · Activity-based scheduling · Milestone-based scheduling · Critical chain project management

In the previous chapter, a few interesting aspects about planning and scheduling have been dealt with. Our discussion at the end was on how one needs to establish a successful baseline. We would now take that as a starting point and discuss a few issues around it.

© Springer Science+Business Media Singapore 2017                                265
N. Gurjar, *A Forward Looking Approach to Project Management*,
Lecture Notes in Management and Industrial Engineering,
DOI 10.1007/978-981-10-0782-8_17

## 17.1   Change in WBS and Baselines of a Project

We have already dealt with the WBS document earlier. However, everything is subject to change in a project environment. One needs to understand how this change transpires in a real project scenario.

**A Quick Hands-On**

Reflect on your project experience. In particular, try to understand:

**Why should a WBS change in the middle of a project? How was the change process? Who were communicated about this change?**

Alternatively, the other question can be formulated as

**Why are there so many changes in Projects mainly in the execution phase?**

One of the most common reasons given to the change in the WBS is poor requirement documentation. This is an exceptionally common reason in the software industry. However, it is not just poor requirements documentation, it is poor budget, poor schedule, etc. When budget and schedules are forced on a PM who did not develop them then the execution to plan will be off. In addition, a good PM knows that there is no schedule without risk and they plan for the risk; a person who just develops schedules does not know how to plan for the unknown. This is another popular reason.

The uncertainties that were addressed in the previous section too have a bearing on the format and could drive the change. However, I would like to touch upon a case study that would be interesting to the reader to understand the dynamics of schedule changes. These are shown in Fig. 17.1. This interim report only clarifies that the project was actually *baselined and re-baselined 5 times!* The value of the project was around USD 100 million and it dealt with the development of a utilities facility that was attached to a large manufacturing facility.

The figure is fairly straightforward, but the amount of changes in the baseline is interesting to note. What is more interesting is that the perspective of building the WBS also changed over the course of the project. This is shown in Fig. 17.2. Notice that the overall definition and the detailing for the entire project changed over the phases. From 17 elements in the original plan, it was brought down to 12, then, with negotiations, the number changed to 30 and later it went up to 50. What is important for the reader to know is that there was *very little that changed at the scope level*. However, everything changed in the way the project was being modeled and the way the plans were being laid out. In short, the handover and the internal control points were being continually reworked. This was a part of the overall *catch-up plan* that was being forced through by the management.

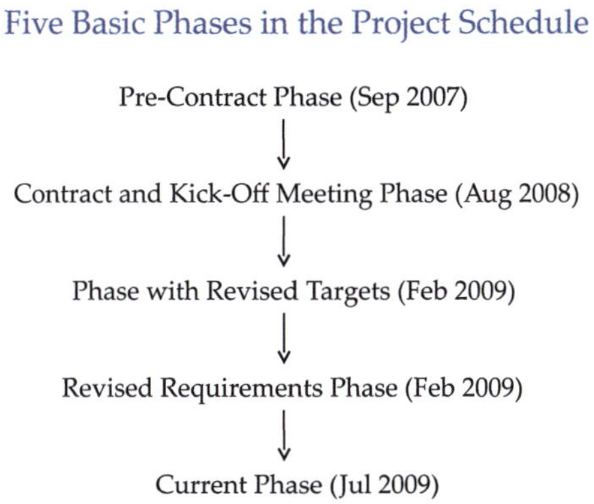

## Five Basic Phases in the Project Schedule

Pre-Contract Phase (Sep 2007)

↓

Contract and Kick-Off Meeting Phase (Aug 2008)

↓

Phase with Revised Targets (Feb 2009)

↓

Revised Requirements Phase (Feb 2009)

↓

Current Phase (Jul 2009)

**Fig. 17.1** Major triggers for WBS changes as defined in phases

## Different Perspectives to WBS Definition posed A huge Problem for Direct Analysis

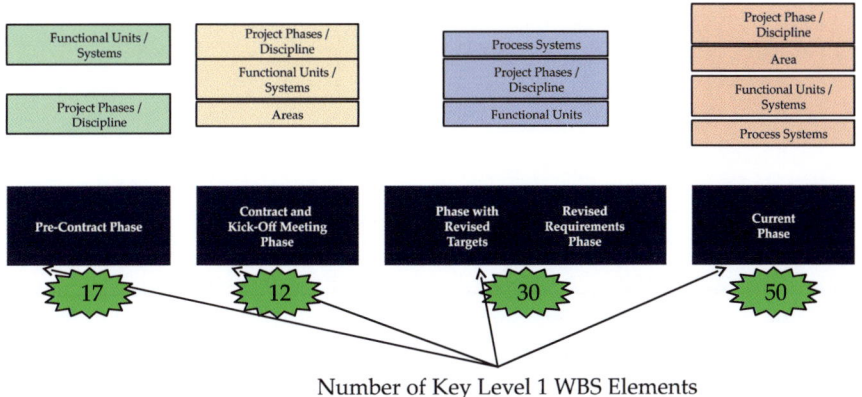

Number of Key Level 1 WBS Elements

**Fig. 17.2** Perspectives in the WBS

The project saw significant delays and slips. The system readiness dates moved significantly during the course of the project. Murphy's Law in this project seemed to say that the *deliverable that one needed right at the beginning was the most complex and difficult one!* In short, since this was a utility, its tie-in points were critical. However, the course of engineering development was such that it came to the tie-in points fairly late (from the design perspective). So, the contractor responsible for the turnkey solution was on a delay mode from an overall perspective.

The breather for the project manager was provided by the system requirement dates. The points that needed the tie-in were also getting delayed by the other contractors! Hence, his delay was being 'masked' by the delays in the system requirement dates. Nevertheless, the net effect was still a delay of at least 18 days for the overall project. So, a utility costing USD 100 million was going to delay a program that costed several billions! So, the saying in projects… *Losing the battle for the want of a nail!*

This situation was tricky for several reasons:

1. The budgets were running short
2. The designated manager was appointed a little too late (in fact, after the prime contract was awarded)
3. The uncertainties were so high that it gave the entire part-project the benefit of doubt. Hence, none of the 'red flags' were actually considered serious and there were bigger fires to put out elsewhere…

However, at the end of the day, *the top brass started getting interested in the status of the project as it was affecting the overall project ROI.* In doing so, the development of the schedules was such that they were not amenable to a good analysis. Hence, the project required additional resources to establish the links and put it in perspective. However, futuristic planning also had little to do with the past problems.

Hence, the overall project was being reassessed to see what future interventions would hold good. Five different modes of analysis were employed to establish the potential to 'optimize' the schedule. They included

1. Detailed Requirement Analysis
2. Package Definitions
3. Technological Variants
4. Schedule Rationalization
5. Temporary Solutions.

These were elaborately explored in the project.

The first step was to revisit the requirements. This was the starting point or the feeder to the subsequent rationalization process. After long and exhausting meetings, the actual requirements, environmental regulations, site conditions, and technological alternatives were analyzed and a lot of brainstorming sessions were held to actually come up with the right solutions.

The point of interest for us, in this chapter, is the change in the package definitions. This is shown in Fig. 17.3. The figure shows how the work packages defined in the contract were *totally reworked* and they did not quite have any semblance with the original schedule. Such a drastic U-Turn in the WBS philosophy is rare to find, making this an interesting live case. When the change is so drastic, it simply reflects on the management skill and the acumen of the lead manager!

Schedule rationalization was done with the help of the procurement team as well as using an expert group. It was observed that certain packages could be merged so as to get a better leverage in terms of contract management and discounts based on the procurement volumes. The expert group was called in to give a more objective and detailed estimate as well as execution plan for the project. Further, the site management PMC and experts from other offices at the company were also brought in to evaluate the situation and recommend the best possible way forward. Some budgetary adjustments helped the designated manager to expedite a few of the contracts with a price escalation within acceptable limits.

Technological variants were also evaluated in detail. Some components, particularly the settlers, were postponed to give a good buffer by creating a resource pool

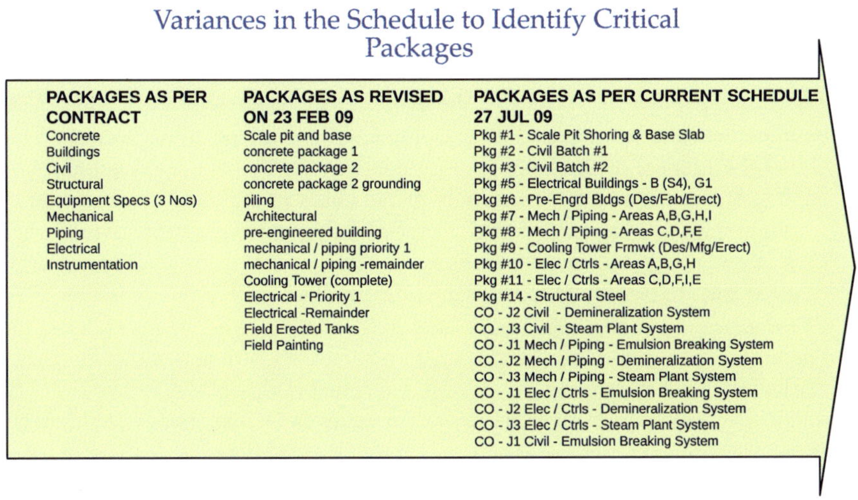

## Variances in the Schedule to Identify Critical Packages

| PACKAGES AS PER CONTRACT | PACKAGES AS REVISED ON 23 FEB 09 | PACKAGES AS PER CURRENT SCHEDULE 27 JUL 09 |
|---|---|---|
| Concrete | Scale pit and base | Pkg #1 - Scale Pit Shoring & Base Slab |
| Buildings | concrete package 1 | Pkg #2 - Civil Batch #1 |
| Civil | concrete package 2 | Pkg #3 - Civil Batch #2 |
| Structural | concrete package 2 grounding | Pkg #5 - Electrical Buildings - B (S4), G1 |
| Equipment Specs (3 Nos) | piling | Pkg #6 - Pre-Engrd Bldgs (Des/Fab/Erect) |
| Mechanical | Architectural | Pkg #7 - Mech / Piping - Areas A,B,G,H,I |
| Piping | pre-engineered building | Pkg #8 - Mech / Piping - Areas C,D,F,E |
| Electrical | mechanical / piping priority 1 | Pkg #9 - Cooling Tower Frmwk (Des/Mfg/Erect) |
| Instrumentation | mechanical / piping -remainder | Pkg #10 - Elec / Ctrls - Areas A,B,G,H |
| | Cooling Tower (complete) | Pkg #11 - Elec / Ctrls - Areas C,D,F,I,E |
| | Electrical - Priority 1 | Pkg #14 - Structural Steel |
| | Electrical -Remainder | CO - J2 Civil  - Demineralization System |
| | Field Erected Tanks | CO - J3 Civil  - Steam Plant System |
| | Field Painting | CO - J1 Mech / Piping - Emulsion Breaking System |
| | | CO - J2 Mech / Piping - Demineralization System |
| | | CO - J3 Mech / Piping - Steam Plant System |
| | | CO - J1 Elec / Ctrls - Emulsion Breaking System |
| | | CO - J2 Elec / Ctrls - Demineralization System |
| | | CO - J3 Elec / Ctrls - Steam Plant System |
| | | CO - J1 Civil - Emulsion Breaking System |

**Fig. 17.3** Redefined WBS packages

that could be available for the remainder activity at the site. Certain technological considerations and performance guarantee issues helped the postponement of another expensive set of technological equipment. Together, these bought both resources as well as saved on the cash flows for the project.

And finally, where the technological problems related to the tie-ins with the existing systems were concerned, temporary solutions were pressed into place to ensure that there was no 'undue pressure' on the remainder of the project. The last mile phenomenon was definitely 'upbeat' in this project. We have already discussed this in previous chapters.

This project case shows how the baseline process and the re-baseline of the project took place. It is just to provide the reader with a perspective of how involved a re-baseline process could be *when the owner stakeholders get involved and are interested in the details*. Needless to say, the time of their involvement also could have had a significant bearing on the project progress. However, this is a separate debate that could be taken up and the reader is free to reflect on additional dimensions that could affect this case.

The one aspect the reader needs to understand is *how often should one be allowed to change a baseline?* This is probably a question that keeps coming to everyones' mind when one sees baselines. Many books actually recommend a minimum stability period for a baseline. However, there is another school of thinking that has the Counter-question *Should one go for an unrealistic plan knowing only too well that it wouldn't work?*

Many companies, therefore, believe that it is ok to change baselines as long as one needs to. The downside of this strategy is that the management keeps 'changing baselines' at will. The owner stakeholders soon start seeing an unexplained impact on their business case! This is, therefore, a delicate balance to understand. The following factors then come into play:

1. *The Planning Horizon:* Typically, the distance between the baseline and the re-baseline needs to be at least 1.5 times the length of the planning horizon. This allows the first set of activities (those underway) to be completed without any hassles. Recurring baselines will only create chaos in the teams and, instead of working, they will be anticipating changes leading to lower productivity levels. This is essentially true with baselines that do not have any significant scope changes and are largely driven by schedules, quality, costs, or other issues.
2. *The Replanning Time:* Even when one goes for a baseline, in most cases, the implementation cannot be immediate. In other words, if one were to say that a particular activity has to be ramped up, the actual ramp-up would take time. At times this is fairly immediate. However, in many cases, the change could take a significant time to ramp-up. Hence, any re-baseline activity needs to ensure that a change in the current plan is truly feasible and manageable within *the current realities of the project.*
3. *The Rework Cost:* If an activity is already underway, one should not normally expect a rework in the activity. However, if there is a rework expected (like a foundation being cast…that needs to be recast to new specs), then the re-baseline should be immediate.

In most projects, the scope objectives are fairly established and are unlikely to change *drastically* over the course of the project. This is like a journey from New York to Washington, where one might have to finally correct course to go to say Annapolis or Gaithersburg. So, much of the journey is going to follow the same route, in which case a baseline too early is not going to really affect the course of action. In other words, any correction till one is at say Elkridge or so. This is, therefore, the planning horizon condition. However, once past Elkridge, the replanning time and the rework cost become more dominant and affect the course of the project.

In any case, the decision to use a number of baselines is more than a simple management decision. Sometimes, slow 'delay creeps' are introduced through the new baselines. Similarly, budgets too are 'adjusted' slowly and steadily. This is similar to the concepts in behavioral economics that cloud decision making at several levels. We have also seen this aspect in our discussions using psychophysics.

## 17.2   Integration in Planning

In the previous chapter, the reader was introduced to the concept of integration. Simplistically put, *integration is the process of bringing in inputs from various faculties for the purpose of decision making.* Since the focus is management information for decision making, any integration tool is supposed to collate and provide information in a format that allows the reader to understand the message early on and, at the same time, is providing the relevant information for decision making. One of the simple integration tools is the 4D modeling software.

TILOS is an example where the plan is represented along the timelines as well. Although there are several others, we are using this example for illustrative purposes only. We wish to state on record that *we aren't beneficiaries of such an unintended promotion of the product*. Yet, we are using it to give the reader a better perspective.

If one looks at their website, they describe 10 benefits of using their system. These include

1. Plan a project naturally by directly drawing it on the screen: Fully graphical interface for planning time-distance-diagram with full support of the critical path technology, enhanced by location and production constraints. Provides maximum flexibility in organizing your plan.
2. Flexible view system allows maximum customization to your specific needs: Place as many sub plans from the same or other projects into the same view to show the overall project or more detailed plans. Allows integration of data.
3. Import/Export functions: TILOS allows the import of graphics and synchronization with distance axis as well as activity, links, resource and other site data such as elevation and productivity profiles. Eliminates planning mistakes.
4. Clash detection and clash avoidance: Specify the distance between crews to avoid collisions, but if it happens TILOS will tell you where. Realistic planning using construction specifications.
5. Distance-based calculation: TILOS calculates the work parameters based on the distance values. For example, in rail construction a sleeper is installed every 0.667 m. Based on the length of the task, TILOS calculates the number of sleepers needed and the number of railcars needed to transport them. Full control over costs and resources.
6. Powerful resource calculation: TILOS supports flexible resource calculations based on the task quantity and time parameters to model the construction process and driving parameters precisely. Saves time planning by using company templates.
7. Library: TILOS supports template activities in a library with their display and resource attributes. Drawing a task as a simple line means: setting the speed of the process, calculating the work parameters, the resource assignments, and their cost in the background with little extra effort. Highlights changes in order to prepare claims.
8. Baselining allows the comparison of the planned versus actual schedule to identify and highlight any differences. High quality output of your professional plans.
9. High quality output: TILOS produces high quality drawings in any size, similar to CAD systems. Quick return on investment and consistent savings on new projects.
10. Saves time and money. A client study has shown, that even on the first project, the planning time was reduced by 50% from 40 to 20 hours.

A schematic sample case study is also included in Fig. 17.4. This is a simplified version of what TILOS provides the user. On the horizontal, is the line that is the road from A to D (divided into segments with end/intermediate points as A, B, C and D).

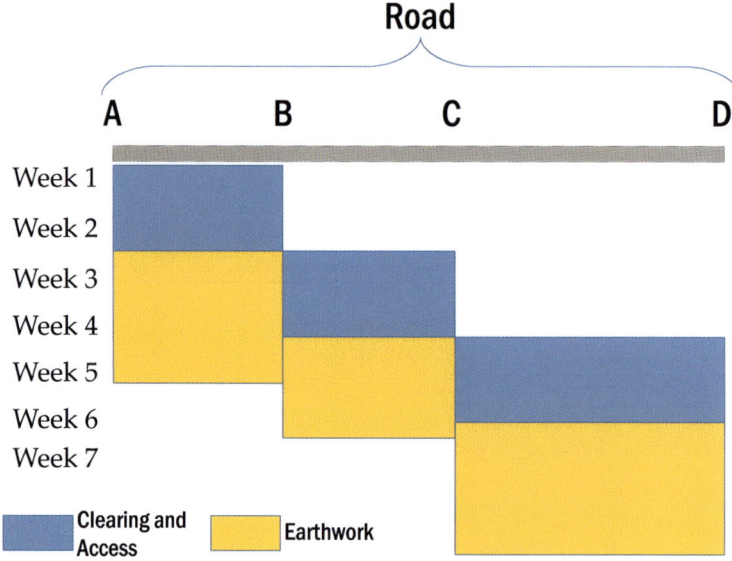

**Fig. 17.4**   TILOS Case Study: A Simplified Schematic

This is typically a drawing. Along the vertical is the timeline. So, for instance, it says that clearing and access provisions from A to B would be done in the first 2 weeks. Thereafter, the earthwork activities would be carried out. So, in one shot, the manager understands how his project is moving. The software allows for dependencies as well.

TILOS provides the user with a new perspective.

## 17.3   Light Version of 4D Models

An engineering team, being predominantly comprised of engineering professionals, often makes the use of 4D models based on engineering drawings. They are at ease and find it far simpler than most others. Though the engineer is trained to have good spatial, analytical, and general management abilities, like we said earlier, too much of information is often difficult to put into one single tool. TILOS is an example of a linear planning model. That means, it actually incorporates one single dimension of planning. However, most sites actually require at least two dimensions to be incorporated to define a meaningful plan.

One of the simpler variants is shown in Fig. 17.5. So, in the figure, the black triangles indicate the location where work starts. Since red is the primary thread color, work begins along the red arrows from the black triangles. In this drawing, work begins at the left extreme end as well as the central portion simultaneously. The grading, excavation, and subsequent concrete work follows this sequence. Since the figure only clarifies the methodology, it is closer to a model of the project execution plan than being a schedule document. The yellow threads start only after the red

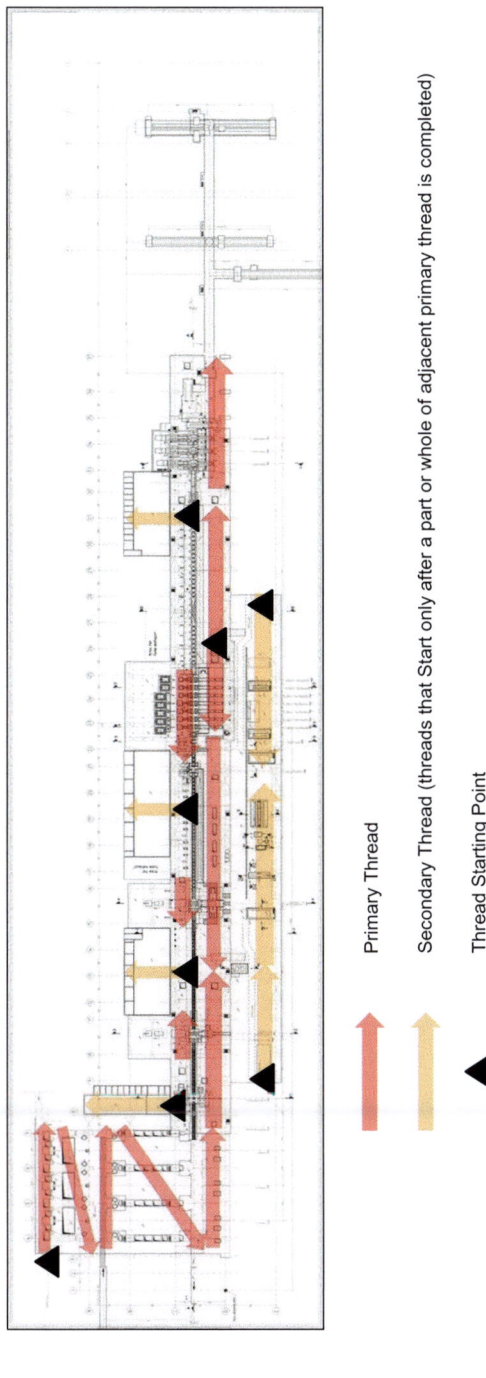

Primary Thread

Secondary Thread (threads that Start only after a part or whole of adjacent primary thread is completed)

Thread Starting Point

**Fig. 17.5**  Infrastructure sequence of release by area

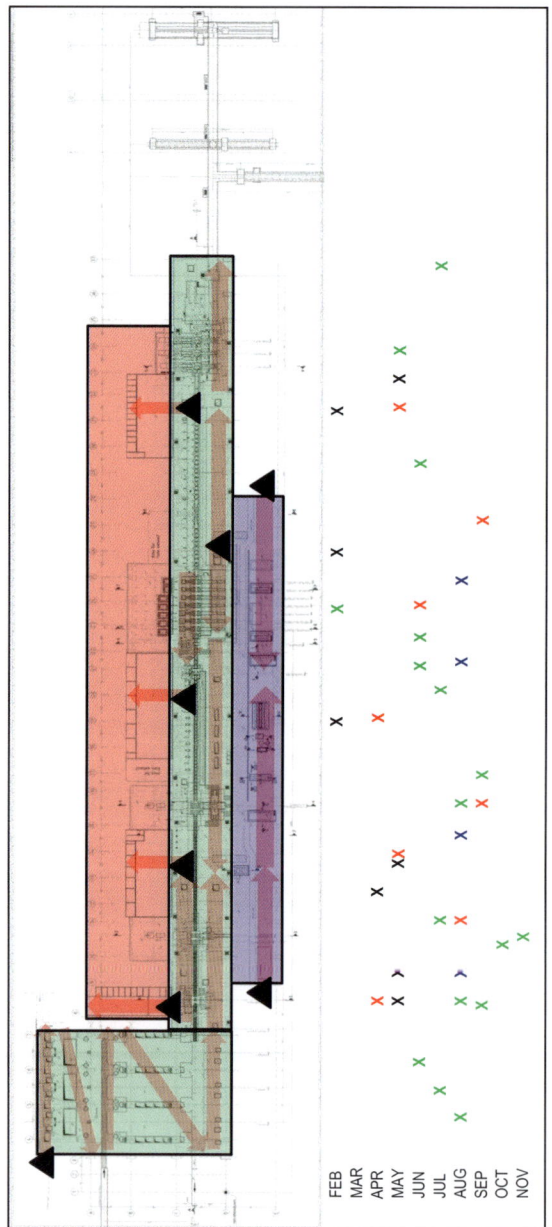

**Fig. 17.6** Integration of execution plan and the schedule

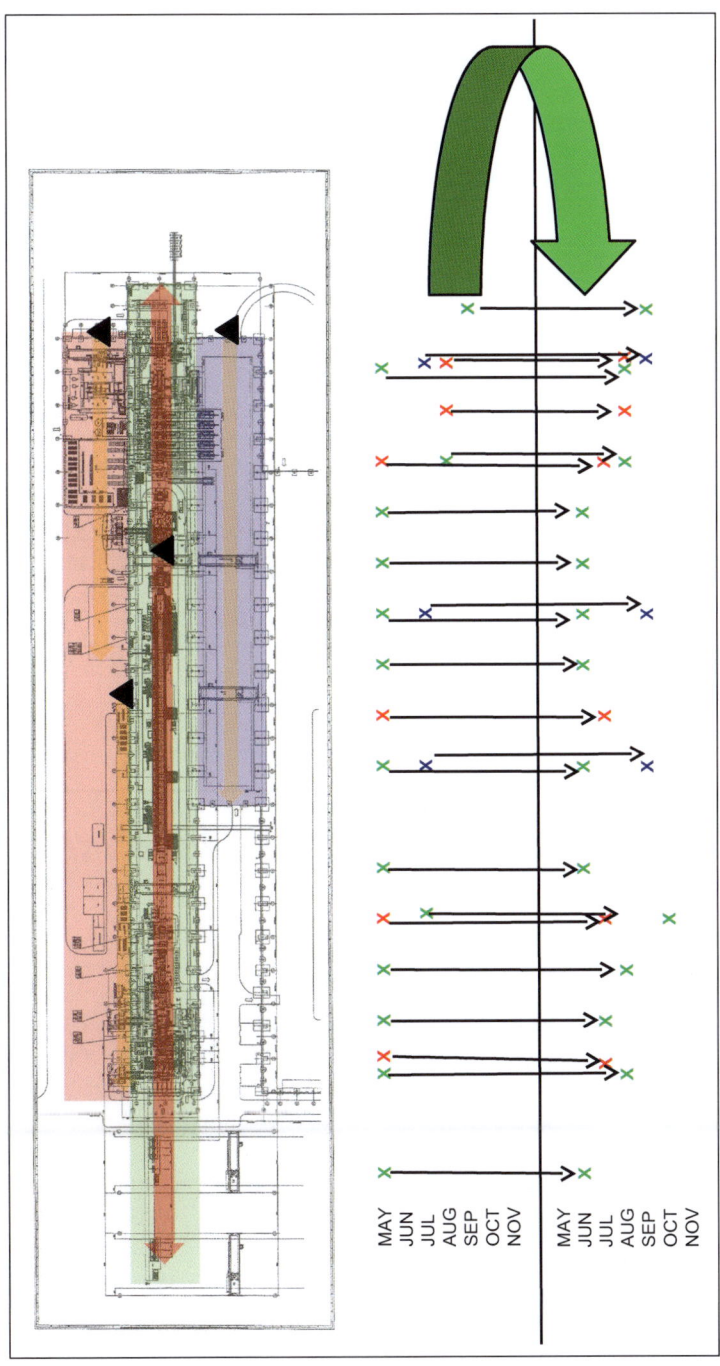

**Fig. 17.7** Re-negotiated schedule of release for a contractor

threads that are parallel to them are completed. Thus, one gets a sound idea of the execution methodology. This is important in defining the relationships in the project. The sequence will define, fairly rigidly, the sequence of execution.

The next step is to integrate this plan with the schedule. This is shown in Fig. 17.6. In this figure, the handover dates are shown in the table. The schedule had over 4000 activities. However, the key handover dates for equipment erection were quick to grasp from the representation. In short, in one simple view, the reasons for the sequences, the derivation of the dates and, most importantly, the tentative dates of handover are all spelt out. Like the TILOS software claims, a picture is definitely worth a thousand words! We are talking about at least 40,000 in this single representation!

Again, in a multi-contractor scenario, the release dates of one function become the start dates of the other. And the requirement dates need to be clearly spelt out as we go along. This is further described in Fig. 17.7. The result of a negotiation over the handover dates is described with respect to the execution methodology.

What we are essentially indicating here is that with a strong tool like TILOS, this could get complicated (and expensive) to develop. However, such simple descriptions are often the best starting places for most integration tools. A tool like TILOS would give formal and advanced functionalities for better project management.

So, one has a good way of analyzing integrated information and making informed decisions with such simplistic 4D modeling techniques.

## 17.4   Complexity of Perspectives

Any plan needs to clearly define the following details:

1. What needs to be done: in the terminology of the user
2. How it needs to be done
3. When is it required
4. Who would work on it
5. How it ties-in with the other stakeholders of the project.

Most project managers focus on the first few items, but they are not too keen on overloading the team members with additional information. The essential reason is the complexity in defining and explaining the various perspectives. However, if one does it correctly, this seemingly complex issue can be resolved by clarifying the scenario using a simple graphical perspective.

To develop this perspective, let us look at Fig. 17.8. In this figure, three different perspectives are indicated. Notice that one area of civil corresponds to several areas of the equipment. This makes the overall analysis and communication extremely complex. This results in huge coordination needs, and potentially, issues. However, if the tie-ins are clear, most managers will not have any difficulty in defining their priorities and, more importantly, *communicating it with the teams*. When communication is not free within the program, there always comes the question of the haves and the have-nots of information! In other words, it starts breeding political influences. Coordination and interface management become key in such areas.

While looking at these tools, the reader must understand its repercussions. Some of them are given below:

1. Integration always requires people who understand all the relevant perspectives that are required to incorporate the right information into the decision-making process
2. Appropriate information availability is a must to begin using any integration tools
3. The tool itself needs to be simple enough to be usable
4. The design of the integration tool is dependent on the issues that need to be addressed in the project
5. The same integration tool may not have the same utility along the life cycle of the project. Therefore, the project manager must understand when to 'pull-away' from investing too much in tools.

Integration tools require a very strong understanding of modeling and simulations. This is because, when improperly constructed, it causes more confusion than it would assist in decision making.

The traditional approach to project modeling has always been the engineering perspective. In all the projects, since the engineering team takes up the project activity at the very beginning, they are the ones that bring in the conventions in terms of terminology and the perspectives. However, this has serious limitations. The engineering perspective is often different from the execution and, still more different from contractual and the operational perspectives. Thus, an overlap in perspectives is often the better way to take this forward. In fact, the engineering perspective favors a component breakdown approach for the WBS because it helps the engineering team track their activities better.

Analysis of integration also yields valuable insights into the way a schedule is constructed. For instance, in a schedule where there are different functions working in sequence, the handover points are often critical to understand. For instance, the schedule was showing that several areas were critical from a schedule perspective as laid out in Fig. 17.9. This situation was based on a specific handover condition that required a fully sealed and complete building for installing the equipment. However, it was later observed that these handover points needed to be correlated with the actual assembly activity that was proposed in each of the areas. The fundamental problem was that the equipment supplier was given a set of preconditions in the contract that were different from the erection agency's contract. Ultimately, the delays in the project meant that both the set of preconditions were not truly being met. This is an extremely complex case and we will not delve into the details here. More so, because it requires a lot of site knowledge and background information before one could truly understand the figures involved here.

Before going into the responsibility discussion, it is necessary to understand the actual nature of the activities. Again here, a method similar to a *virtual walkthrough* can be adopted. Again, there are software tools available for the same, or one could go for a customized schedule analysis. The results can help understand such critical

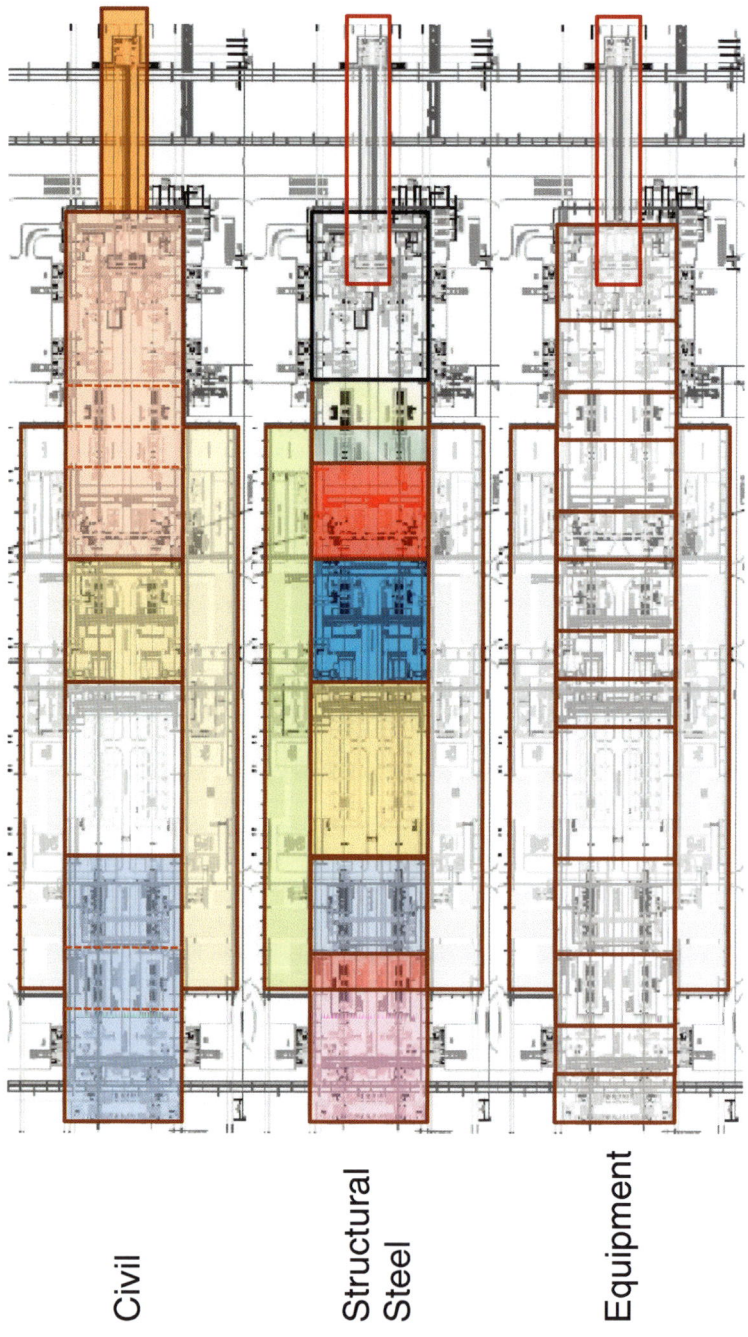

**Fig. 17.8** Clarifying multiple perspectives is often complex

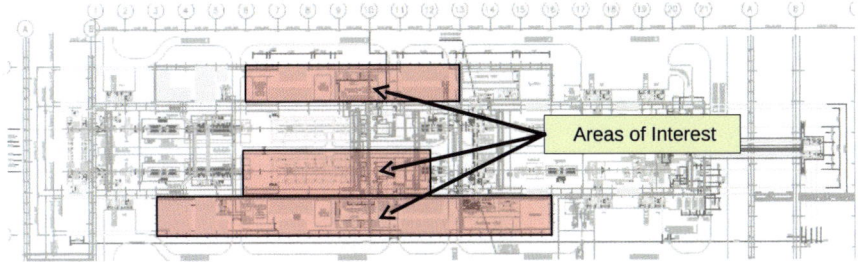

Areas of Interest:
1. Auxiliary Bays (Both Sides)
2. Furnace Area
3. Zinc Pot / Coating Area

**Fig. 17.9** Critical areas in the schedule

issues like preconditions. It also helps align various contractors to the goals of the project. In effect, such tools are extremely 'disarming' for they take away the ammo from any manager (of any of the stakeholder agencies) who is trying to cause hurdles for the project completion.

To take the case further, the issues were basically on the availability of the cranes and the FAA Warning lights at the site. This is shown in Fig. 17.10. At the same time, without seeing this in isolation, an actual walk through the three areas was done from past experiences. The first activity in the zinc pot area, one of the critical sections, was with heavy lifts of 160 tons as shown in Fig. 17.11. Ideally, an open bay would

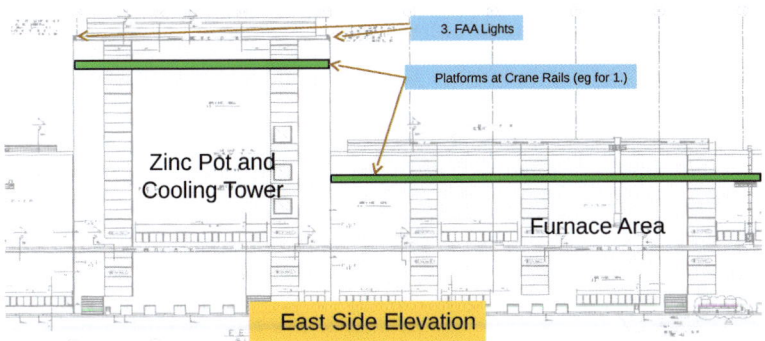

1. Completion of Stairs, Ladders, Hand rails, Platforms, etc. (Misc. Metals) needs to be expedited
Alignment of Crane Rails is easier (of shorter duration) with miscellaneous metals in place.
2. Crane for the tower is assumed to use the crane from the Steel Erector.
Alternate methods will require additional time for the crane erection.
3. Procurement of Warning Lights as per FAA Regulatory Requirement needs to be expedited

**Fig. 17.10** Key issues driving handover delay

Massive Heavy Lifts in an Open Bay could be an Advantage?
Pre-Conditions may not be logical???

Activities in the first
45 days

Erection Estimated
Tonnage:
~ 160 Tons

**Fig. 17.11**   Heavy lifts in the zinc pot area

be deemed as an advantage for heavy lifts. This contradicted the preconditions. Again, the same happened in the furnace area. The first activity was to erect the structural steel which again had little to do with an enclosed building. In fact, an open building was more beneficial in this case. Both these examples also show the reader that *generalized terms and conditions for handover or take-over are not always applicable to specific project cases.* Hence, the reader must be clear about how these need to be treated in detail (Fig. 17.12).

Advanced planning tools provide for both the options to the project manager, especially in large projects:

1. Specialized conditions of work that can easily be evaluated, monitored, and controlled
2. Increasing productivity at the site as they outline plans better even in general conditions.

The general conditions and potential conflicts are also easily identifiable using these tools.

Like we have shown here, the perspectives involved are often too complex and it requires a deeper evaluation of the overall planning methodology. This is seldom the case, where the project manager is too bogged down with operational decisions and finally, it boils down to countless discussions and emails without actually giving both the parties the true picture of how their plans 'interact' and 'overlap' with each other.

The Steel Structure is
independent of the Bay
Pre-Condition

Estimated Tonnage of
the entire Steel:
~ 640 Tons
Support Structure:
~ 191 Tons

Estimated Duration:
35 Days

**Fig. 17.12** Structural steel as first activity in the furnace area

## 17.5  Integrating Schedules with Logistics

We will now look at another perspective: that of *site logistics*. Schedule analysis is often done to check delays, monitor progress, etc. However, if it is done well, the schedule can also give one a good insight into the site logistics. Logistics is often a crucial component, yet is many times handled outside the scheduling domain. The reason is that most physical planning methods are considerably easier when they are done separately.

This is basically for three reasons:

1. Man, material, and equipment are numerous. So, it is difficult to track using a scheduling tool that would be clouded with unnecessary detail
2. Legal issues in receipt are dominant in determining the right kind of logistics management system
3. More importantly, site logistics change with developments on the site. So, it is not a 'static' activity in the true sense. Hence, a baseline is difficult to design and arrive at!

Yet, it is fairly easy and many times required to define the logistics from a predictive perspective. In other words, what would 'hit the project site' and 'when.' The reason being that *greenfield sites often require some preparatory work at the site (that needs to be taken care of) before the actual material starts coming in*. For instance, a simple delivery schedule is shown in Fig. 17.13. This shows the itemized deliveries of a particular equipment line as it is coming in. This helps to understand the nature of the commissioning strategy as well. If one sees the schedule, it is evident that there is some equipment that is coming in after the contractual start date for cold commissioning. Needless to say, the erection times and the way the plan is integrated ensure that hot commissioning is on time. In an integrated schedule having several 100 lines of code, such a representation gives a better understanding to the project manager. Now, the current schedule was reporting a 'green traffic light' but on closer introspection, there was a need for better control over the deliveries. Most managers

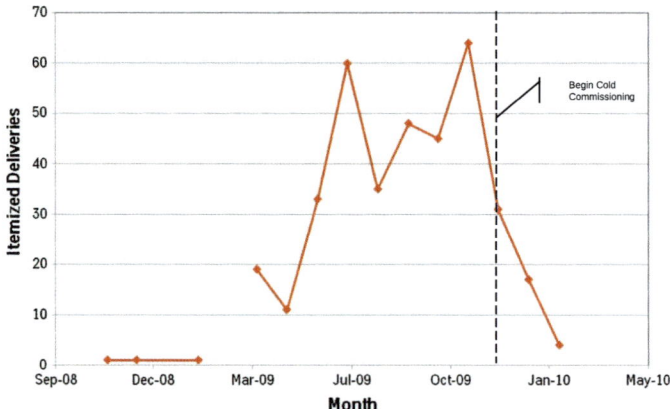

**Fig. 17.13** Itemized deliveries in a project (major items) over time

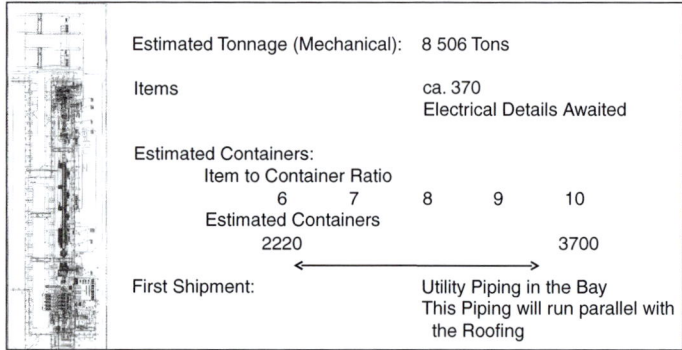

**Fig. 17.14** Converting schedule information into tentative container numbers

following the traditional approach insist on 100 % deliveries, but that is an extreme stand that doesn't help anybody. In particular, it doesn't expressly make the project a 'better managed' one.

Now taking the 'predictive faculty' a little further, the team needs to understand the number of containers that are likely to be shipped. This is described in Fig. 17.14. The figure clearly shows how the team needs to expect between 2200 and 3700 containers of sea-worthy packing for the project. Such a large number of containers require special logistics planning. For the receipt, the inspection, the handling, the storage, the transfer to the bay, the disposal of container packaging material, the tagging, the planning, and the issue are all critical. In short, in large programs, the schedule information gives an estimate of the work-load volumes and the way ahead.

The logistics can further be evaluated in conjunction with the subsequent erection activity. This is shown in Figs. 17.15 and 17.16. One of the figures shows the deliveries and the other one shows the erection. Clearly, the correlation between the two is a logistics requirement. The huge amount of erection activities needs closer scrutiny.

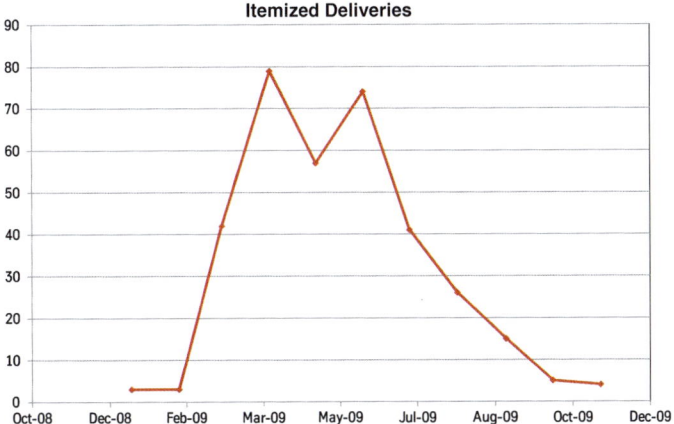

**Fig. 17.15**  Delivery of major items

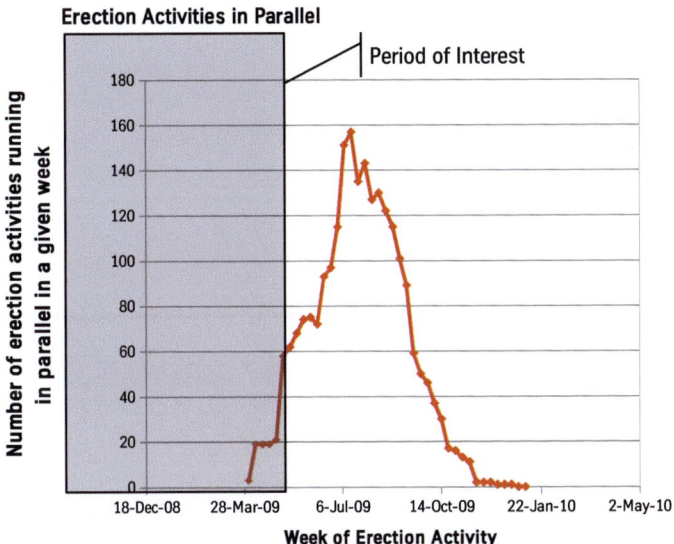

**Fig. 17.16**  Erection activities for the deliveries in Fig. 17.15

It can be seen in Fig. 17.17 that the main areas that are contributing to the peak activity can be identified. This further helps assess the loads on the erection cranes as shown in Fig. 17.18. Note that this is a prediction that is over 6 months prior to the actual set of events. In short, the use of the right kind of integration tools would help assess several parameters like logistics and handling resources. It also tells us more about how the entire project is going to move and the degree of control is drastically improved.

**The main contributors are areas 2,4,8,9 and 12 each accounting for at least 50% of the overall activity**

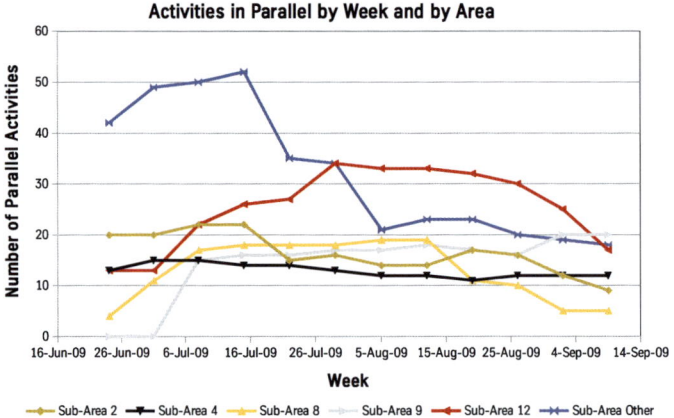

**Fig. 17.17**   Area-wise split of erection activities

· **Overall crane requirements are critical; especially if one is to incorporate the erection of both lines in the bay!**

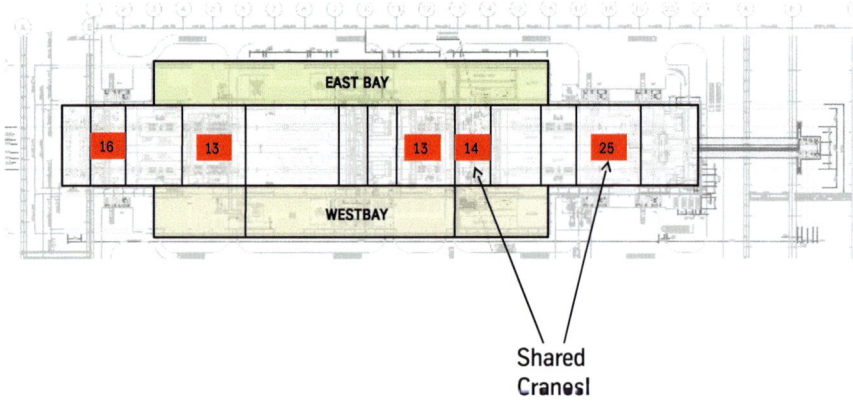

**Fig. 17.18**   Crane requirements in various areas

Such analysis and representation go way beyond the traditional Gantt Charts and the typical milestone evaluation that is seen in most projects. Merging areas needs a good handle of the actual site conditions and the actual performance requirements/impediments. In most projects, resource deployment is a critical aspect. For cost issues, the resources are kept 'lean.' However, the project manager starts seeing repercussions as they go along the execution. This is a complex issue in itself.

Most project managers come from a school that it is *impossible to model project chaos!* Unfortunately, nothing can be further from the truth. It is possible to clearly identify and develop strategies by going in for predictive analysis. For instance,

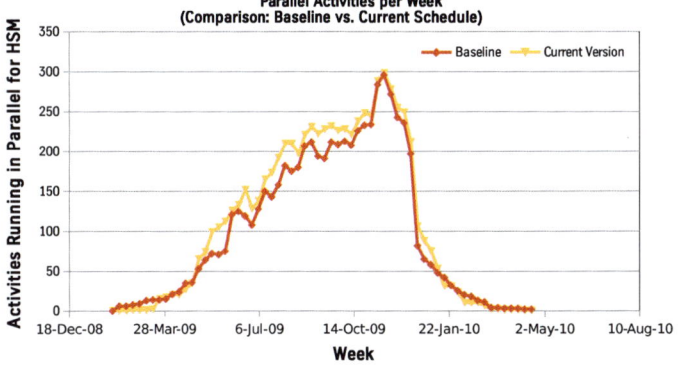

**Revised schedule increases activity overlaps and project risks; however, keeping the maximum the same**

**Fig. 17.19**   Re-baselining analysis

in the above case, the space requirements for the logistics storage/warehouse, the regulation of material entry at site, and the transport and resources for erection all could be easily determined. Since a project environment develops over time, most *big problems* are not exactly 'visible' to the project manager. Integration tools like the ones described here can help the manager and other stakeholders understand the magnitude of the problem and evaluate options *before* it strikes them. In other words, advanced planning tools are nothing but predictive and analytical aids to ensure good risk management.

An overlap analysis of re-baselining with activity and logistics is also possible. Figure 17.19 describes one such situation. The evaluation is again a quick overview of how the re-baselining has been done. More activities have been made parallel in the beginning of the project. The peak of the activity has not been changed. However, the overall increase in the average activity levels is definitely a cause of concern for the project manager. This is beautifully described in the tool. Depending on the project challenges, such a representation becomes a good starting point for discussions and often helps the stakeholders understand the subsequent implementation better.

Unlike what most literature speaks, a lot of project success is on the cohesion one develops within the project groups. Advanced planning methodologies provide one precisely with those better perspectives that assist managers ensure that their teams have a very cohesive view of how the project is moving forward.

## 17.6   Integrating Schedules as Schedules!

In the previous sections, our focus was on the integration tools and how they interface with other areas in the projects. In this section, we will look at specifics on how schedules can be integrated with other schedules.

While one works in projects with multiple agencies, it is often required to integrate information across multiple platforms. However, while doing so, there are a few

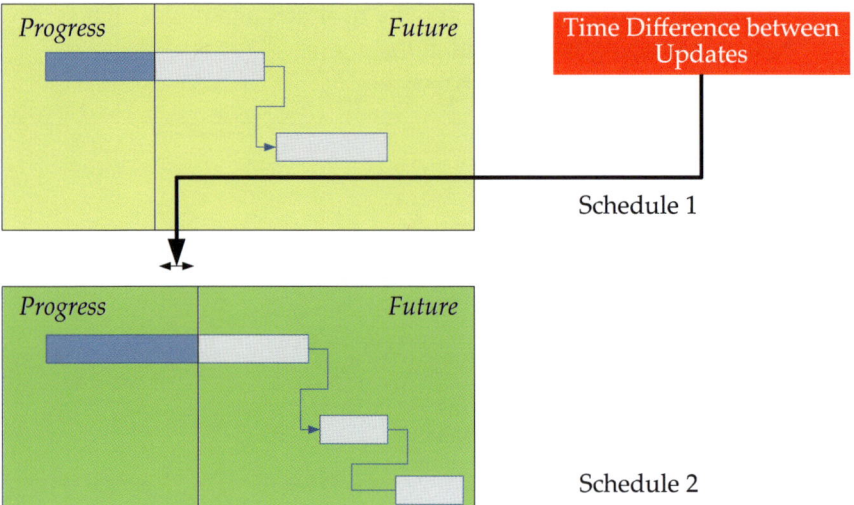

**Fig. 17.20** Different update schedules

considerations that are important. We have already seen EVA methodologies and their bearing on the out-of-sequence activities. In this section, the details of how this logic works are given.

It does not require any great expertise to understand the update methodology for project plans! The logic is often fairly simple. If there are two different schedules, they need to have the same update date. This is a mandatory requirement most planning software that are used today. The situation is shown in Fig. 17.20. Here, both the schedules have different update lines. If the difference in the update lines is not too large, the schedule with the earlier date can be taken as the reference schedule and the difference between the dates can be added to every *ongoing activity* of the subsequent schedule. So, the first activity of Schedule 2 will have a remaining duration given by

$$\text{Remaining Duration} = \text{Remaining Duration} + \text{Time Difference}$$

This has two fallouts:

1. The current activities get distorted as their remaining durations are increased. However, this is quickly restored with the next update of Schedule 1.
2. Items that would be work-in-process as per the update line of Schedule 1 in Schedule 2 would be shown as complete by following this process. This would distort the schedule information in the vicinity of the timeline
3. The nature of the project progress of past activities is not visible or amenable for further analysis.

Yet, with this methodology, one is able to move forward with a better understanding of all the subsequent activities. If the activity durations are reasonably small and if there are a long sequence of activities still to go for completion, this method would work out. However, the drawbacks could be significant in most cases, which is the main reason why this algorithm is not found in any standard scheduling software. The software usually just asks for the date of reference and then, blindly starts assigning values. Hence, while integrating schedules and getting a reasonable analysis of events, one might have to use manual intervention in many cases.

### 17.6.1   Retained Logic and Schedule Override

While working on schedules, the out-of-sequence activities pose a significant problem. One has already seen some of these effects in the EVA methodology; however, we will briefly explain how this change looks like.

In any schedule, there is a sequence of activities that are planned out and laid. However, in practice, there might be several workfronts that are simultaneously available. When a contractor changes the priorities of his workfront, he actually begins *a successor activity before ending a predecessor activity*. In other words, the software witnesses an 'out-of-sequence' error. The software has multiple options now.

1. To treat both the current activity as well as the updated out-of-sequence activity as parallel activities, thus, breaking the relationships/predecessors of the out-of-sequence activity. This is probably closest to the current situation.
2. To treat both the current activity and the out-of-sequence activity as parallel activities. However, the remaining tasks in the out-of-sequence activity will still follow the relationship logic that the original activity had.
3. To short-close all activities till the out-of-sequence activity as completed on that date. Thus, the subsequent activities are only those that follow the out-of-sequence activity.

There could be several different options in how a software could treat this. Oracle Primavera® has two commonly used methodologies for the treatment of such sequences. These are called (a) Retained Logic Method and (b) Schedule Override Method. These are options in the software that need to be understood by any project management. The first option mentioned is nothing but that of the retained logic option, while the second option is the schedule override option. These are shown in Fig. 17.21. In the progress override scenario, activity A suddenly becomes noncritical. This actually distorts the project information. For instance, if work starts *before the approval is given for the activity*, despite being a Finish–Start relationship, the schedule can record progress. However, in the process of doing so, if the activity A is taken off the path AB, then it could hide the delay in the completion of activity A. For instance, if the approval is delayed, the path may not show it as a critical activity. This again is a dangerous conclusion. To avoid such issues, some softwares have the functionality to make open-ended activities critical! Though from a network perspective, this may be absurd to the manager, from an accountability perspective, it does make sense.

**Fig. 17.21** Retained logic
versus progress schedule
override

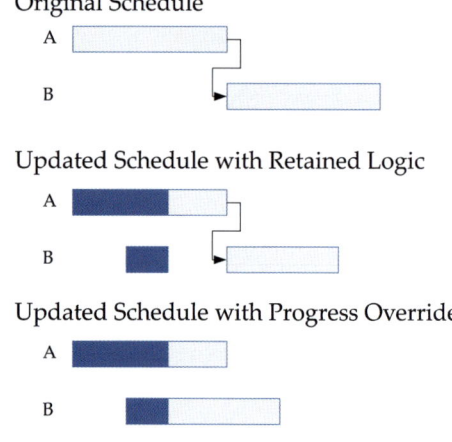

A detail-oriented activity like project planning and scheduling needs to have sufficient checks and balances to ensure that the information and reports do not ultimately boil down to a 'GIGO' (Garbage-In-Garbage-Out) model. Hence, a basic understanding of what actually comes out of a software system is very essential for effective management of the project. In particular, the manager must be able to identify if the situation is caused due to

1. Error in input data of progress
2. Error in the logic or sequence of the project plan
3. Improper build of the schedule; not synchronizing with the logic
4. Improper output due to the wrong programming logic options selected
5. Improper output due to a limitation of functionalities in the planning software.

Each of these is supposed to be identifiable as different sources of errors while dealing with multiple schedules. However, there are few software systems available that actually cover this aspect in detail. Our objective here is to sensitize the reader on how the planning programs work on popular functionalities available in the market. When the projects become large (essentially programs), the probability of using the tools increases. However, with these tools, the need to ensure that the plans and schedules are reflecting current, correct, reliable, and analyzable information is of prime importance. For instance, in a project where I worked in the US, the schedule had over 100,000 activities. This was way above a normal benchmark, but justifiable for a multibillion USD project. So, in effect, each line was tracking around USD 40,000 or more, that, in all rational contexts, was a reasonable cost head to track.

Given such a situation, it is necessary to ensure proper controls. Fortunately, out-of-sequence is one way to check and identify errors in the schedule. However, for adequate controls, the effects need to be understood.

## 17.6.2   Selective Update of Contractor Schedules

Modern software has brought several changes to the ways in which projects can be managed. One spoke of the errors in the 'progress data' input. Further, changes in the logic, etc. are often easier when *one compares the schedule with its own previous version* rather than on an integrated master schedule. There are four popular ways in which this is achieved in practice.

One of the functionalities that are offered by certain software is such that it defines a virtual scenario of the imported schedule *before* it actually integrates with the other schedules. This functionality is called 'Reflections' in Oracle Primavera®. It shows multiple 'what-if' scenarios on the schedules and includes cases where one has selectively accepted a schedule. For instance, a contractor might send a schedule with a change in the duration of an activity. The functionality will immediately flag the activity. Now the master planner can decide if it makes sense to accept the change or leave it out of the import. Such a functionality has multiple uses. It helps do a quick check on the changes. It also shows the effects of selected changes as output scenarios so that the project manager can get an exact handle of how the project plan is likely to change.

Another commonly used method is to have a well-defined (restrictive) policy when it comes to the rights of contractors. Rights management is popular because it can easily be setup and the control layer is independent of the activity layer. In other words, this is like the use of software. One could check how the product pans out or could use the operating system to limit the use of the software. The rights management system does precisely that. WBS based, activity based, project based, resource based, and product based (timesheets, master scheduling engine, etc.) are various types by which rights administration can be done. In certain scenarios, one would prefer to 'disable' the schedule button from the software for the contractors. This ensures that contractors who are 'scheduling' do not skew the overall project plan. In an environment where rights are strictly monitored and controlled, there is always a possibility that one set of users are *unable to benefit from the system* as much as they need to. This becomes a very difficult situation and most agencies with 'lesser rights' tend to use alternate platforms to model, monitor, and control their activities. The end result is that the project will have two different realities, and most times the one in the integration environment is a compromised one.

The third approach is to control the Linkages between the contractor schedules and the master schedule. This is a must-use method and works very well in most environments. The movement of milestones helps the master planner understand the dynamics of the schedule during each update. It is commonly recommended by many agencies that there needs to be a hierarchy of schedules. The recommendation is to have an overview plan, that is an integrated master schedule, which is subsequently broken down into each part schedule. If the correlations between the progress between the activities in the master and those in the detailed schedules are clear, this method is a clean way of providing the project manager with an excellent snapshot of the entire project. However, maintaining the schedule in its form is a complex activity by itself and, oftentimes, requires significant manual intervention (and work). A direct multi-schedule tie-in works in environments where there are cohesive groups across organizations and there is a mechanism to address the concerns at the interfaces. Usually, the integrated master schedules insulate the project manager from such issues that start surfacing at the ground level during the course of the project. Therefore, the derivation of the integrated master schedule is a critical step.

When one talks about collaborative environments, in addition to controls, there is often a need to make provisions to restore the information to a previous 'normal' state. For instance, when one often finds unexplained changes to a project plan, it is *cheaper at times to repeat the steps than to try to figure out what went wrong*. Thus, a 'factory reset' kind of an option is often adopted. In other words, backups are used to control the integration process. There are multiple provisions here. Most project scheduling softwares automatically have backups inbuilt for each of their update processes. However, in addition to those backups, the project manager might want to exercise explicit backups at his end to ensure that the information is representative and can be worked out in subsequent analyses. This also helps one compare two plans: a backed up schedule with a current update. Doing so, one can find the deviations quicker than otherwise. Hence, this too is a popular approach to integrate schedules and evaluate them.

## 17.7  Activity Versus Milestone-based Scheduling

We have been looking at the interface between planning and other functions. However, the concept of planning and scheduling varies in the project management community. The advent of integration tools has seen two fallouts:

1. More transparency
2. Better Management Control.

However, the project manager can have some fundamental philosophical questions while taking on the reigns of a project.

---

**A Quick Hands-On**

Given the several possibilities, it is often essential for the project manager to choose between two polar positions:

**Is it better to manage a project using Activity-based Tracking or is it better to go for Milestone-based Tracking?**

Personal preferences of the manager definitely come into play here. But the logic behind these two philosophies is important to understand and explore.

### 17.7.1  A Few Situations to Consider

A lot of projects do have certain basic problems to start with. We take the following to demonstrate:

1. In many projects, the strategic objectives are clear, but the scope remains unclear for a long time. For instance, it might be known that a bank wants to go for an IT upgrade, but the exact scope is not clearly spelt out in the beginning. In such situations, one understands stakeholder expectations, but isn't clear WHAT needs to be done. Defining activities for the project is less likely to be achieved. Since the WHATs are themselves not known, the activities may not be defined well to start with.

   While many project managers do state this as a reason, it is, in my humble opinion, a *massive systemic coverup!* There is less sanity in such a project environment where the stakeholders or the project managers are not sponsoring a full-fledged requirement gathering phase. The tendency to 'not commit' to a scope in the beginning is probably the worst for any project. The end result is that most projects need to be sufficiently 'padded' for costs that would come in before the actual requirements are known. In any case, the project manager *may not himself be fully responsible* for this situation.

2. In the case of large projects extending over long durations, there is always a situation of large uncertainties in the project. Most activity-based models are largely focused on deterministic timelines which are discomforting when the uncertainties are not actively managed or controlled. Hence, in such a situation, seeing a detailed approach to achieve the result borders between the 'feasible', yet 'unrealistic', elements of project plans.

   The simple principle that the project manager follows out here is *detail out only when one is entering that period of the planning horizon.* As discussed earlier, such results often skew the overall plan. The advantage of having a deterministic plan is to *simply, stick to the plan!* If the project manager decides to keep changing the plan, he will obviously be more comfortable with a milestone-based system.

3. One of the common reasons for not detailing any project schedule is that the prime contracts are not awarded. While some might be awarded, the remainder contracts that are also on the critical path are yet to be awarded. The logic is that the agency that is awarded the contract would plan on how they should achieve the milestones. This is another example of procrastination that simply keeps delaying the detailing. While it is good to have a trusted partner on board, such a strategy will often postpone the bad news and give the managers a breather. In the end, the information about slips will be out.

   However, this strategy does have merits at times. In certain types of projects, durations could be *very strongly correlated with the resource availability.* For instance, an activity might take a fraction of the duration depending on the resources that are utilized. However, this simple inverse relationship does not always hold. In other words, as a thumb rule *the duration of an activity could be brought down to around 65–70 % of its nominal duration with changes in the resource patterns.*

Of course, this varies from project-to- project and activity-to-activity. Hence, in such a situation, the duration is dependent on the resources the agency would deploy. Therefore, a schedule could be a little too speculative and unrealistic to start with.

That being said, it is essential to have a nominal duration for every project. This is because the nominal plan will be a 'typical' resource loading condition and there could be several agencies that could be qualifying such typical resource conditions. The moment one starts speaking about a very specific pattern of resource loading, one is actually going to see an increase in the contract price. Second, as mentioned earlier, the infinite resource situation does not truly reduce the duration to 0. The typical spectrum of compression is around 20–25 % of the typical duration. Hence, expecting too much of change and, therefore, postponing the activity detailing in a project is definitely not a very meaningful strategy. Moreover, a plan actually provides the project manager with a basis to start the discussions and negotiations with the agency. Therefore, it is always a better strategy to go for activities.

4. In a project where multiple paths and sites are possible, there is always a potential change in execution plans due to the then-site conditions. In such a scenario, building a logic and a sequence for the overall schedules is not the best way to go forward. This is primarily because the schedule of activities, once made, does act as a baseline. However, the schedule itself may not be a critical aspect. To clarify it further, it is the total quantum of work that is more important than the way the work would be sequenced. So, essentially, a critical path does exist, but the allocated duration is longer than the nominal critical path. In other words, we can clarify this with an example.

   In a project, if three parallel fronts exist and these three paths have say 30 days, 50 days, and 40 days, respectively, then if one tries to sequence these activities with a successor, predecessor relationship, the durations could vary. However, if the completion milestone is say 60 days away, then there is some 'flexibility' that is already built into the schedule. In such a situation, it does not make sense to try and figure out the activity sequencing. Hence, once the path lengths are found out, it is best to leave them out and allow the agency to pick and place their priorities by incorporating their preferences into the schedule. So, in this case, it might be wiser to have a schedule management philosophy based purely on milestones. The agency would work out the details.

5. One of the main tipping points for project controls is the availability of adequate manpower. Unlike other functions where one can start small and then grow, project control area needs to have a minimum staffing level that can ensure building-in the details in the project plan. In many environments, however, insufficient manpower is available to detail. In such a scenario, only piece meal plans would be made. Hence, the argument is in favor of a Milestone-based methodology.

6. In a multiparty environment, there is every likelihood of the project plan being revised and re-revised until the plan is baselined. This period is treated as a 'blind' plan period in the project. In such a period, it is always good to have an alternative that is easier to evaluate the changes. Milestone-based methodology helps in such

cases. Naturally, the project manager must have a very strong planning acumen to quickly evaluate agency plans.

7. One of the most complex environments to model is the one where iterative work is expected. With each iteration, the scope changes, based on the evaluatory findings of the last activity. Sometimes, the scope changes are marginal, but many times, they are extremely different from the original model. In such environments, a large amount of probabilistic information is used. Creating a deterministic project plan is often extremely difficult in such cases. We have already addressed the complexity of the model in our earlier treatment on the life cycle. However, in such situations, it is many times easier to work with milestones since the variances due to the lack of information could be significant. A combinational approach where one plans the immediate activities but uses milestones for the remainder of the project also helps.

8. Some projects are driven by chaos! In short, one experiences changing priorities throughout the life cycle. Such projects need short planning horizons. Such environments usually are better managed by milestone-based methodologies. Here, the variances are so high that there is less return in detailing and bring in the control over the chaos. Interestingly, chaos is often generated by the leadership team (owner stakeholders) that gives confusing directives to the project manager. The end result is often that the replanning effort exceeds the replanning benefit. In such a situation, it is better to go by a milestone-based method.

## 17.7.2  Milestone-Based Management

While we do see there are possible candidates for milestone-based management of project plans, the reader needs to understand the essential fallouts of such a decision. We will touch upon a few of these here.

### Blind Spots Could Increase

In a milestone-based management philosophy, the only visible tracking is that of a milestone. A project plan could be overly ambitious to cover up the slippage that it is incurring. Hence, despite having information, the monitoring could be deemed *blind* at times to the realities at the ground level in the project. Blind spots are where the information is not particularly visible to one due to his own perspective.

### Surprises in Schedules High

The result of the blind spots is that one often sees surprises in the project. In other words, things start 'hitting the project' from unknown directions. The trick in overcoming these two aspects is by defining milestones from multiple perspectives and monitoring not just critical events in the project, but also the drivers of the events. For instance, approval of the software architecture is critical to quick development of the code. The approval might be given by another agency. Hence, it is important to track the same as well.

**Resource Fluctuations Not Easy to Control and Monitor**

Milestone-based management many times requires significant resource fluctuations to ensure that the milestone does not slip. Thus, the focus on the milestone could mean fluctuations need to be planned for and made possible. In typical project environments, infinite resource is a mythical condition. Even getting a reasonable level of resources is tough at times. Hence, the loss of control on the project could come in too soon. In other words, in such a situation, resource pool constraints are not incorporated and the visibility might be important at times for the project manager.

**'Too Trusting' on the Contractor**

Probably, the most important aspect while going into a milestone-based management methodology is that the project manager must have a good level of trust on the agency that is performing the task. While this can be contractually established, the reality also needs to be checked at times.

In short, a milestone-based management system requires very strong management skills and is, generically speaking, a less predictable management philosophy. However, one needs to understand that they are also suitable at times and can be used to *leverage the characteristics that are specific to certain types of project conditions.*

### 17.7.3   Milestone Structure

While a lot has been spoken in favor of activity-based management, the milestone-based management philosophy is extremely popular in the 'neo-management' style of working. People who believe that the traditional way of project management has already exceeded its limits in the modern connected world often feel the need to innovate and improve upon their management style. The need for innovation stems from the changing technology and information cycles. With increased technology, the behavioral aspects tend to limit productivity. However, with the right skill sets in place, it is often the technology that seems to be causing trouble for the project manager. In any case, the concept of innovation is driven mainly by knowledge workers that are dominating the project management scenario. With such resources, it is possible to experiment a lot more as the dependence on equipment is a secondary aspect.

In any case, the reader must understand that there needs to be a *hierarchy of milestones* similar to those of WBS and activities if one has to use this methodology. In this hierarchy, one has to define certain main milestones. Then, each duration between the main milestones needs to be further defined using sub-milestones.

While doing so, it makes sense to incorporate different perspectives in each level. For instance, the first level could be taken from the business case. The second level could be taken from the handover or take-over points perspective. The third level could be taken from the critical path. The fourth level could be taken from the noncritical activities. In short, milestones could be plentiful.

The thumb rule is to define at least one minor milestone per path every 4–6 planning review cycles, which will enable to keep the pressure. So, if the weekly review or update is being done, the minor milestones should typically be a month to one-and-a-half month apart. This will ensure a good balance between the milestones and enable a proper control. Anything closer would make it more economical to go for an activity-based system. On an average, one should see one milestone every 2 weeks.

### 17.7.4 Monitoring

One of the main problems with milestones is that they need to be *defined* in the language known, in an objective manner. For instance, mere stating that the code is ready for user acceptance testing is difficult and 'stupid' too! It would be necessary to define rigorous criteria to enable the user understand the meaning of the milestone. This oftentimes is more difficult than defining the activity per se! Having a reasonable definition is a prerequisite for good monitoring.

I recall in a construction project that the handover was to be made with certain conditions. The question that lingered was the *exact definition* of the handover condition. For instance, the standard of cleanliness differs from one person to another. Similarly, the same phenomenon is observed in the milestone definitions. This is a cause of concern.

Defining a good milestone is definitely a challenge. And despite the best efforts, there is always a shade of 'gray' while making the definition. This has two fallouts: (a) It forces one to put more efforts in defining the milestones and (b) It makes the overall monitoring less robust due to the overbearing load on the project manager to define a state or a milestone rather than the action. However, at the end of the day, this method is more 'flexible' as the doer chooses the way he wants to meet the milestone.

### 17.7.5 Contract Types

Despite everything we said on planning, milestones still are common elements used in contracts. One often comes across a *payment milestone* that exists in every contract. Our milestone-based management philosophy is simply a 'backward' extension of this contractual condition. From a pure need-to-know perspective, the milestone-based management option is better for lump sum contracts and the activity-based management option is better for time and material contracts, Cost+ contracts, or others where the visibility of the resources is more relevant.

### *17.7.6   A Closing Remark on Activity Detailing*

Activity details are probably the most important focus areas for a project manager. The logic is plain and simple and summarized here:

1. The less you see, the lesser you know. Lesser you know, there is even lesser to monitor and control. Hence, any activity-based method is definitely better due to the potential visibility. More details imply more visibility.
2. Any slippage that occurs is instantly visible and the cascading effect too is visible.
3. An activity-based management option is prescriptive, and the verification is done by both sides fairly independently. On the other hand, milestones are often thrust upon without checking for the feasibility of the overall system.
4. Needless to say, the activity-based system gives better progress information
5. Control is much higher, and yields a more predictable nature of management.
6. Finally, consequential damage is traceable (cause and effect traceable)
7. Drivers of 'failure' are 'known' and 'visible' here unlike in the milestone methodology. And most importantly, one can use a combination of both. That is, because, *refinement in milestones requires activity definitions for finer detailing.*

## 17.8   Critical Chain Project Management

One of the more recent developments in Project Management has been the Critical Chain Project Management philosophy. Before we get to the bottom of the concept, we need to understand a few things about how the project is planned. We have covered a few of these aspects earlier; however, we are going to repeat them just to ensure a well-rounded perspective of the critical chain.

**A Quick Hands-On**

Back to the basics here, the reader now needs to focus on the following questions:

**How do you estimate project durations?**
**What is the probability distribution for the durations?**
**What is the expected 'free-time' in the duration?**

In other words, let us understand the estimation process and the buffers in them.

### 17.8.1  A Behavioral Genesis

In any organization, it is natural to assume that employees strive to meet the deadlines they commit to the senior management. Intrinsically, therefore, they tend to incorporate allowances for contingencies in case they feel they could fail. Management, most times, fails to appreciate the fact that 'free time' is being given to an employee. This fundamental approach used results in what we can call a *negotiating posture* between the management and the staff. This is pretty straightforward because most times the employee wants to play the game safe so that he does not enter into the bad books. So, if the manager is a 'tough task master,' he would not like to be appraised with a negative answer.

On the other hand, as a manager in a project, if an employee does come up and say that he has 'no work today,' the project manager would get restless because there *definitely are several tasks* that have limited resources. So, both these reactions are 'natural' and, often the employee is looked upon as someone who is doing 'less work' than what he is supposed to be doing. At the same time, the employee feels 'bashed-up' both ways! Choosing between the overload or between the perception that he is 'working less'! This is a dangerous situation.

The other interesting aspect about project management is the phenomenon of delays. Simplistically stated, *in a multitasking environment, it is easier to carry forward delays than advances in schedules!* This is because advances in schedules are often 'lost' because the subsequent resource is not available to harness the early completion. So, even if one were 'super-efficient', it would not help the remainder of the project because of the characteristics of the resource availability. The resource for the subsequent task might be on another project or another task within the same project and he would have to complete the same before coming to this one. Moreover, a resource completing early may have to wait, before he can start the next task… Management might feel that he is 'under-utilized'!

Moreover, business pressures often drive management priorities. Ironically, a project with delays is normally a 'higher' priority than another that is going on schedule. Depending on the contract and the business case, management often adopts different control philosophies across projects. This gives rise to a skewed portfolio management (also called *fire-fighting*), especially in multitasking environments.

A new school of thinking, where management formally accepts the need of buffers and encourages transparence among the employees, was, thus, created. This was the *Critical Chain Project Management*. The Critical Chain Project Management accepts the need for buffers and exploits a near optimal use of it.

### 17.8.2  Demystifying the Critical Chain

Critical chain is best understood from the way it is implemented. In fact, it is done using the following simple steps:

1. Complete transparence on time estimates (looks at the best times possible; also called optimistic times). This is also the 'target' time when you implement CCPM.
2. Formal buffers are provided at the end of each activity 'chain.' If the path is critical, the buffer is on the project and is called the 'project buffer.' If the path is not critical, a 'feeder buffer' is provided at the point where it meets the critical path.
3. The project manager ensures that delays are *within* the buffers at each stage. So, he only has to review activities from the 'buffer' perspective. This is an important difference to be noted as it affects the conduct of the project review meetings.
4. In managing the buffers, he has to ensure that the 'productivity rate' of the activities in the project is high enough to be as close to the realistic targets that one has.

Due to its innovative approach, many organizations are keen on applying this 'latest tool' for their projects. Just like the sensational predecessor on manufacturing, the Theory of Constraints has been readily absorbed and accepted by the corporate world. Seeing the new methodology, it has been on the radar of most organizations to go for an implementation, and rightfully so, as many organizations have truly been benefitted by this approach.

Despite all the popularity, the critical chain project management school does run into problems while being implemented. The first and foremost is on the durations itself. There could be 'unexplained' variances in the durations of activities. Despite the best intentions to have a good estimate, there can always be causes that are 'above and beyond' the routine evaluatory process used by the project manager. This creates a serious roadblock for the critical chain methodology. In such situations, the methodology starts failing and it leads to inefficient implementation.

Since durations are at the crux of the discussion, one always needs to ensure that they are scientifically verified. However, in most projects, the estimate is obtained from the 'gut' of the team member and is 'refined' by the planner/scheduler. In other words, the durations have not been scientifically verified. This lowers the confidence level on using the estimates for something that is so innovative. The transparence, though ensured, is meaningless when the durations are not scientifically verified. This is another major stumbling block for the philosophy.

The third aspect is that of change management. While the management is serious about implementing this methodology, the management philosophy continues to use buffer as 'free time.' This defeats the purpose of the implementation and changes the overall implementation quality. It also implies that change has set-in, but the top brass might have excluded themselves in the first pass! That puts everything back at square one.

The fourth problem is at the level of the project manager. The project manager is not always trained enough on managing the critical chain. The fallout of this situation is self-explanatory. At times, he is trained, but is not given adequate options to improve his productivity. This is a worse situation! It is like tying the hands and making them do things simultaneously! It just will not work.

The next aspect to understand is that although the company has a portfolio, multi-tasking of people across projects or even activities is limited. It is never the case that the programmer can also do the job of the DBA or the Solution Architect. So, multitasking of people is seriously limited to tasks that are less specialized. This invariably means tasks that do not require great know-how, in short, some kind of basal skilled labor. This seldom helps the project manager. In construction, the welders and the crane operators are often resource constraints. One cannot just take anybody and put them on the job. Hence, it is not very effective and feasible to use the critical chain methodology in these cases.

Still worse situations occur when the project is underway. Increased time pressure is directly transferred to team members making them work overtime over extended periods. This again becomes counterproductive. So, instead of ensuring a high productivity, the project manager indirectly makes the entire environment counterproductive! In such situations, the critical chain method would simply fail! And finally, sometimes tasks have a significant need to interact with people. Hence, a 'collective mess' is formed leading to 'chaos' and false reporting (politicking!).

## 17.9   Key Takeaways

In this chapter, the reader would realize that advanced planning methodologies are far too complex for straightforward application. In fact, a lot of the success of the project depends on what hinges the decision process. While baselines were dealt with in the previous chapter, the re-baselining process was the focus in this specific chapter. Most importantly, one saw how the re-baselining could solve issues and, at the same time, cause problems for the project manager.

The most important aspect in this chapter was the understanding of the project plans. Most planners are confused with 'schedulers' who understand little about the project, but are proficient in modeling projects in software systems. However, if done correctly, the plans reflect as robust schedules also provide valuable inputs for other areas like materials management, logistics, resource planning, etc. In doing so, one has to understand how the perspectives in the plan need to be developed, modeled, and used for analysis. This has been demonstrated using a few examples in the chapter.

As planners put it, integrating schedules as schedules is also a challenge for many large programs. A whole lot of conditions need to be identified and followed to ensure that this is done successfully. At the same time, limitations in information flow, project processes, and alignment issues on the timing of information are all vital parameters that are to be considered.

We also covered the two options on activity versus milestone-based scheduling methodologies. Though this is a matter of debate most times, it is observed that each has distinct advantages and can help the project manager better control the project in different circumstances. All said, however, they are not totally independent of each other, and, in general, a combination of both activity as well as milestone-based scheduling is preferred. While a pure activity-based scheduling system stands and supports itself, the same is not true with the milestone-based methodology. Hence, it is advisable to go for a combination of the two rather than a pure milestone-based scheduling methodology.

Finally, we touched one of the latest concepts in the Project Management faculty called the Critical Chain Project Management. We saw the simple process as it stems from the 'behavioral faculty' of project management. However, what was interesting to note for the reader was that this method is not always the appropriate one for the project environment. Therefore, the potential areas of concern were also covered in the chapter.

# Part VI
# Project Procurement Management

# Chapter 18
# Procurement: Commercial Risks Dimension

**Abstract** In this chapter, we focus on risk management exclusively from the commercial dimension in a project. In doing so, we begin by looking at the conventional approaches and define a Coin Model Approach to Risk Management. We also look at the Pricing Framework that is typically used for commercial perspectives of Risk. The chapter then covers some of the bidding and contracting strategies that are commonly employed in project management. We next delve into the contract types. We define a new framework to help the project manager develop a Cost Perspective to Contracting. We then describe the traditional Risk and Control perspectives from the supplier perspective. We finally discuss contracting types from the business case perspective. We conclude with a few pointers on innovations in the contracting domain.

**Keywords** Project commercial risk · Bidding · Contracting · Pricing framework for risks · Maturing views of risks · Firm-price contract · Fixed-price contract · Target-price contract · Cost-plus contract · Cost perspective to contracting · Monopoly · Oligopoly · Partnered sources · Competitive arms length · BOT model · BOOT model · BOO model

All along, our discussions have been focused on risk management from their intrinsic nature of occurrence and intensity and the management ecosystem. What we have not dealt with are the tools of risk management that use the vehicle of the commercial dimension in a project. In this chapter, we will focus on such tools, methods, and principles.

One of the best works on commercial risk management is by Boyce [15]. He defines two broad categories as (a) Technical Risks and (b) Commercial Risks. While this may be a view taken by many managers, our take on this is slightly different. We now present the Consulting Connoisseurs Coin Model of Risk Management. Both these risks are expressed in Fig. 18.1. In the figure, we define that the two risks are actually talking to each other, therefore, the two different sides of the same coin. In other words, they are inherently connected to the *risk itself*. The risk is like a coin. *Each face of the coin actually denotes its value in the given dimension, for the*

N. Gurjar, *A Forward Looking Approach to Project Management*,
Lecture Notes in Management and Industrial Engineering,
DOI 10.1007/978-981-10-0782-8_18

**Fig. 18.1** The Consulting
Connoisseurs Coin Model of
Risk Management

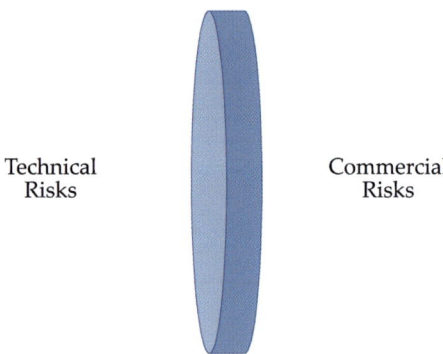

Technical                                   Commercial
Risks                                        Risks

The Coin of Risk Management Tools

*given project*. As we are considering the commercial dimension only, we have shown
them to have two faces, but we could also involve other dimensions like the legal
dimensions of risk management, the human dimension, etc., depending on what is
interesting to model.

One might choose to ignore a particular dimension, and that, is where the coin
model helps understand. So, a risk is like a 'coin toss' experiment…where one does
not always see all the dimensions in one go. At the same time, the dimensions always
exist. This is different from the traditional approach that tries to *delink one dimension
from the other*.

## 18.1   Pricing for Risks

While several of these areas are either directly or indirectly addressed in other chapters, one of the important aspects in risk management is that of the pricing strategy
for risks from a bidding perspective. Boyce provides a framework where he advises
the pricing of risks that are of moderate probability and moderate impact as shown in
Fig. 18.2. This is actually a more complex discussion than has been described. *With
changing/innovative contract models finding its place in the market, putting out a
figure often acts to the disadvantage of the bidder when they come for negotiations.*
Hence, the strategy is often a misleading one, though it does keep the bid competitive
through the negotiations phase.

The entire discussion is actually revolving around the possibility of bringing it into
the 'contract'. Hence, the generic view toward pricing also has suitable substantiation.
According to Boyce, the high probability-high impact risks are risks that could be
modeled using some notional value of five times the contract value. In which case,
the cost of inclusion is too high to bring in a small change in the quote given by
the bidding agency; in other words, 5–10 % are not going to suffice. So, the bidder

**Fig. 18.2** Pricing
framework for risks

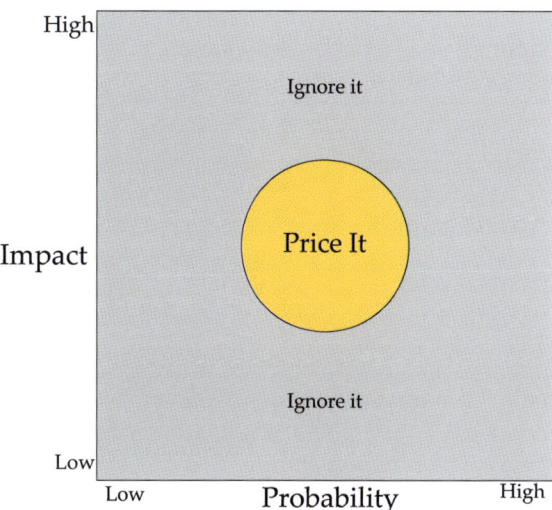

needs to have a concrete plan to shed or mitigate the risks. The other alternative is to choose not to quote for the project!

In the case where the risks are high impact-low probability, he again says that one can safely ignore it. The logic is that a notional value does not help if the risk occurs. And if it does not occur, there is no need to have a provision that only adds to the costs. As for high probability-low impact risks, he argues that the impact is very unlikely to be noticeable in the swings and roundabouts of prime contract performance. And low probability-low impact risks are anyway uninteresting according to him.

This is a good framework to start with, but seems to have certain serious limitations. It does not leverage on the strengths of the buyer or the seller in the project scenario, but tries to fix the responsibility on someone who is willing to accept it. The focus of the entire work is centered around the contract as a project instrument. While it is true that the contract is an instrument that needs to be designed and used well, it is equally important for the project manager to realize that the organization has a 'risk-exposure portfolio' that is one of the determining factors while determining the risk management strategy that is used in any project.

## 18.2  Bidding Dimensions and Risk Management

We will now focus on certain risk management principles that are applicable in the pricing strategy during the bidding phase of the project. A detailed treatment is also given in the procurement section. While quoting a project for pricing risks, one needs to understand a few important dimensions about the price. The first is the impact of the specific risk that needs to be considered.

Boyce talks about a progression of *maturing views of risks* in his work. He maintains four basic types of views that are interesting:

1. The Initial View: That takes a guesstimate on whether the risk is acceptable or not and works on the phase where one is deciding whether to bid or not.
2. The Considered View: That is at the time of tender preparation.
3. The Merging View: Which is the resultant of the contract and the negotiations. This view is important because here the risk frame changes focus from *winning the contract* to *completing the project.*
4. The Ongoing View: A view that is taken when the project is being executed aka contract performance.

Since the main transition is the actual bidding phase, the risk structure is further refined with respect to this phase as well. Boyce has given the following generic strategies for the project commercial manager to consider:

1. Pre-bid: Predict and Prevent
2. In-bid: React and Mitigate
3. Post-bid: Proact and Manage

The focus of the discussion is to constantly assess the bidding process. This has different connotations in different contexts. In large projects, companies have a proper team that can evaluate the risks. However, as the ticket sizes come down, the formal procedures come down drastically. Hence, evaluating the confidence level of a company regarding the outcome: 'winning the contract', is often tricky and not easily predictable. Boyce again provides for a norm in the confidence level and says there has to be a minimum confidence level, which he calls *worth bidding for*, to actually bid for the project.

In environments with high competition, it is often more complex than just having a 'confidence level'. Many companies try to manipulate the procurement processes at the buyer's ends by trying to get specifications, timelines, and budgets to their advantage. Hence, practically, this confidence level might actually look at a variety of elements rather than being a 'cleanly derived metric'.

## 18.3  Getting to Contract

The first focus of contracts for most managers is assignment of risks along with the scope and timelines. However, there are a lot of issues while dealing with multiple contractors on a project. Risk synchronization is a critical aspect that needs to be addressed by the project manager. In his book, Boyce has clearly stated that one needs to ensure that risks between various agencies should be managed in a synchronized manner contractually. He uses the term 'no less onerous', whereby the prime contractor risks should not be at the expense of other contractor performances. As an example, the procurement of equipment has been quoted. If the prime contractor ends work later, his own guarantee should be shorter than that of the other suppliers to ensure that there is synchronization. These are certain basic checks that the book speaks of, and the reader is advised to read the same from the book.

One of the interesting dynamics mentioned in the book is the pricing of risks. That is, a seller gives a discount for excluding a risk while a buyer gives a premium for their accepting the risk. Such a situation has a negotiation potential. He suggests the putting of a *price list of risks*. That is to say, how much would the seller require to accept the risk. This has to be seen in contrast with how much the buyer is willing to pay in order to assign the risk to the seller. This is an interesting concept. Unfortunately, systems approach would yield a different response to such a situation, the reason being that

1. Risks could interact with each other and not be totally independent and
2. The 'fixes' for the risks could be applicable for more than one item in the price list. It would mean a lot of 'double charging', that would be seen in the list.

Hence, though the methodology is powerful, it might not be optimal. This, of course, is dependent on the skills of the project manager who has to ascertain the competence of his purchasing team to ensure that they are negotiating the 'right values'.

### 18.3.1 Basic Contract Types: Cost Perspective

Looking at financial risks, there are four basic types given as

1. A Firm-Price contract: whereby the seller gets a firm fixed price, come what may!
2. A Fixed-Price contract: whereby the seller gets a fixed price, but that is with some riders. For instance, if the oil costs shoot up, the seller might be entitled to a price adjustment. This is typically used in most contracts where the owner is looking for reasonable 'fixity' of financial exposure.
3. A Target-Price contract: whereby both the parties have a target to meet. However, overspends above the target are shared by both the parties in a ratio that is different from the way in which savings are shared between the parties. Also called as a cost-incentive approach.
4. A Cost-Plus contract: where the seller gets money for using his resources and also gets a small predetermined percentage, of the cost, as his profit.

While a firm price is what the management expects, the seller may have different views. This is where the risk spectrum matches the financial spectrum. The variance expected in the project and business conditions plays a key role in this scenario. At the same time, there are also the financial cash flows that are required to keep the business moving. Hence, one has to be careful while evaluating a procurement perspective, as it would have multiple Cross-overs with each of these lines of thinking. Further, a lot revolves around the portfolio defined by the Buyer's Risks and the Buyer's Control.

Typically, a firm-price contract is one in which there is little control over the seller. This is because the entire risk is with the seller and, therefore, the seller requires to perform to ensure that the deliverables are met with the firm price. Here, the buyer's risk is seemingly low as the buyer has assigned his perception of risks to the seller.

However, in reality, despite the best efforts, the seller might not be able to contain all the risks. Typically, the sellers tend to 'load' their 'potential failures' into the price. So, if the seller is not expecting to perform to the contract, he would increase his price to ensure that he 'earns his part of the pie' despite any problems in the performance. In other words, if the contract does complete on time, it is an additional bonus *which is also worth fighting for!*. But the seller will not be unduly concerned about the performance in this case. Firm-price contracting is ideal for smaller scopes of work (duration and variability of scope are both low) and is also preferred in environments where the buyer has *limited experience* to execute projects. Such modes of contracting help the interests of the buyer the most. For other conditions, this might give highly skewed results. One has to be cautious while going for this methodology.

In a fixed-price contract, there is a good scope of price adjustments. The pricing is a little more fair in this case. However, a lot depends on the procurement process. Since one has parameters, in this case, for enabling price adjustments, there are two interesting fallouts:

1. Price Bid is valid for a longer time as the
2. Contracts are of long durations >1 fiscal year and input parameters are very strongly affected by Macro-economic parameters.

For instance, currency rate fluctuations are typically managed using fixed-price contracts. Boyce has also provided with strategies using derivatives to hedge potential currency risks. In any case, contractually, these are managed using such contracts. As a case in point, values of road contracts in Mumbai, India were escalated by around 30 times post contract award as they had to account for the fluctuations in the reference indices of the contract. While this was 'shocking' to the people, a lot of issues had to be factored in to ensure adequate correction. Establishing the actual sensitivity to each of the reference indices is of utmost concern for the project manager to ensure he is correctly identifying the contract risks. Further, one also needs to understand the multipliers given by the sellers. For instance, the price of bitumen may fluctuate. But the index chosen by the seller needs to be verified, assessed, and properly negotiated. Most times it is more complex than one thinks. A fixed-price contract too is similar to a firm-price contract. There is less buyer control and there is potential risk associated with it. For instance, if the contract is delayed and if the escalation in the contract cost exceeds the 'penalty', in the absence of the delay, then the seller continues to earn from the delay. This is a potentially dangerous condition.

A Target Price is one of the best arrangements from a risk-sharing perspective. It gives a higher incentive of the savings to the seller if he is managing to deliver within the budget (*Target Price*). However, if the costs go above the target price, the seller also has to part with some of his potential profit. This gives a reasonably good control due to better coordination between the buyer and the seller. However, since the risk is truly shared between the agencies, this mode intrinsically increases the risks of the buyer over the previous two options.

Lastly, in the Cost-Plus contract, it is like an extended resource of the buyer. Most risks are with the buyer in this case. The seller gains with *whatever amount he bills*. Hence, the seller would try to increase his billing, but this would automatically put

**Fig. 18.3** Consulting
Connoisseurs Model of Cost
Perspectives in contracting

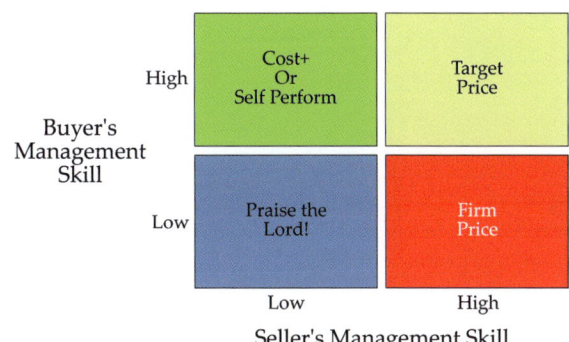

them on conflicting ends of the spectrum. The buyer will try to minimize the costs, while the seller is comfortable even with an increase in the costs. In short, the seller has it good in this case, where he minimizes his own risk exposure. However, the consequent fallout of such a situation is that the buyer has to have a greater control over activities of the seller.

The manager needs to understand which of these modes are ideal. This is given in the Consulting Connoisseurs Model of Cost perspectives in contracting in Fig. 18.3. The combination of the elements in the matrix helps as a quick guide. As one can see, if both have 'poor management skill/experience' in doing a specific project, it is time to start praising the lord for what happens! In all the other cases, there are good alternatives to understand how risk management can be contractually done from a price perspective. However, there are other dimensions in procurement that also need to be considered. We will also delve into those options in subsequent discussions.

### 18.3.2 Contract Types: Supplier Perspective

While negotiation is sought in project procurement, there are lot of concerns on how the seller environment looks like. Boyce again has defined a grid to characterize this as shown in Fig. 18.4.

1. A single source (monopoly): Here, the buyer has limited choice but to 'dance to the tunes of the seller'. Things get ugly when the seller does not have the buyer's project on his 'top priority' list. So, here, there is little control and huge risks that the buyer's appetite would command. While Boyce gives it a good score on the control scale, it has been witnessed in several projects, especially in the developing countries, that the monopolistic supplier, especially when he is an overseas one, does not always respond well to requests on better and improved project control.
2. A few sources: Again the situation is that of a noncompetitive nonmonopoly situation. Though there is a better possibility to bring in controls, the main problem

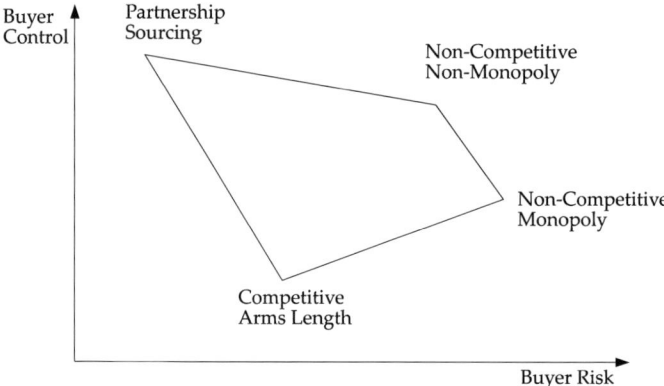

**Fig. 18.4** Risk and control portfolio

arises due to the fact that there is no competing pressure on the seller. In other words, it is very likely that there is some better degree of control due to the 'intrinsic trust' laid out on the seller and not using the 'competitive element'. However, this too is restricted. For instance, when one just picks a vendor from an approved vendor list, it could well be treated as a similar situation. The only incentive in this case is that the vendor would look for continued business and hence, he would be ok with allowing a little more control.

3. Multiple sources: Here the risks are low, but Boyce argues that the controls too are low in this case. However, it is difficult to completely subscribe to it. While the risk is lower, the contract itself is given on the premise that the project requirements (which include the controls) are met by the seller.

   That being said, the seller in this case knows that he is L1 and is operating with a minimal acceptable 'standard'. Hence, he does know that he is the 'best' that the agency has for the given project. There is a potential for misuse in this. The quality of negotiation is, therefore, critical in this case.

4. Partnered sources: Where there is a completely dedicated partnering posture between the buyer and the seller. This is a strong mechanism to do projects that are of very large value. A prime contractor, in a partner agency, would provide a cooperative stance to the project manager. This looks like an 'extended organization'.

### 18.3.3   Contract Types: Business Case Perspective

The business case perspective also helps manage risks in projects. The *business case perspective looks at engaging a contractor with the business interests of the project.* This is different from the traditional contracting models where the seller/contractor is linked to the objectives of the project deliverables, i.e., Scope, Time, Cost, etc. In short, these models require long-term engagement with the buyer.

Some of the basic variants in the spectrum include:

1. Traditional Procurement: Where the seller is engaged with the buyer for the scope of the project. In such a model, the critical evaluation is on the seller's experience to meet the scope of the project deliverable (be it a hardware product, a software one, or an equipment, etc.). The entire procurement process is based on an objective evaluation of the experience of the seller's experience. If the experience is not clearly established, at least the credibility to the capability of fulfilling the need, is to be established. A lot of weightage is given to the performance guarantees in this approach.

   In fact, the firm-price and fixed-price contracts are typical instruments used here. In short, this mode is used for products and for productized services that are typically required during the project life cycle. The procurement professional is a strong negotiator in this case, relying on *documents* establishing the basic experience and the technical abilities of the seller. The requirements are fairly rigid without much of a choice to change the 'content of the supply'. The procurements are typically handled using the two-bid system where the seller is to submit a technical and a price bid separately. Once the technical components are established, the procurement person is busy looking at the negotiation to bring in the 'best deal' for the company.

2. BOT Model: This model is called the Build–Operate–Transfer model. In this model, the seller provides the deliverables on a turnkey basis and then, continues to operate the facility for a fixed fee and for a limited period. Thereafter, the assets are transferred to the buyer. Clearly, the duration of typical BOT projects is much longer than the project itself (4–5 times the project duration). So, if a project is of three years duration, the BOT contract could extend anywhere between 12+3 = 15 years to 25 years. Such a long-term commitment to the operations is a challenge for most sellers.

   The BOT model has a different procurement orientation than traditional procurement. Here, the seller is typically taken for his turnkey experience and credentials. Since the project comes before the operations, though the operations period could be significantly longer than the project phase, these models are classified under 'projects'. Apparently, the project management community has gracefully embraced this as a 'project management innovation'. Although, perhaps more critical in such models is the operations experience of the seller. This includes all the dimensions of operations (labor, machinery and capital) that are verified.

   The main focus of the management is the derivation of the 'fee' that is to be paid to the seller. Usually the seller invests in a part of the project scope and has returns on his investments through the fee that is part of the revenue. Since the fee is to be paid to the seller, the scope of such projects is fairly well established as they are the main drivers of the investment made by the seller. For instance, land might be provided by the buyer, but the equipment might be provided by the seller. Since the scope is fairly well defined and outlined, the seller has a good 'fixity' in terms of the expected scope and costs. However, the discussion often revolves around the revenues. Dicey estimates often lead to escalations in the fees as it would affect the economies from the scale of the operation. In short, the bottomline during the discussions for both the buyer and the seller is that 'no one can foresee the future'.

3. BOOT Model: This model is very similar to the BOT Model. It is called the Build–Own–Operate–Transfer model. The main distinction between the two is that the BOT model runs on a fee. However, in the BOOT model, the seller will be taking a share of the revenue instead of the fee. The seller is given the status of an owner in this case, unlike the BOT model where the seller is an 'investor-operator'. As the owner of the facility, the seller has a significant share of the revenue. The BOOT model is useful in those cases where the buyer and the seller have jointly and agreeably established a market potential.

   Typically, there is a minimum market estimate that is used as a basis in these cases. After the period, the seller transfers the assets to the buyer. Problems in the BOOT model stem from the ascertaining of the market potential. It has been seen in most projects that incorrect projections cost both the buyer and the seller. Although this is a popular mode for infrastructure projects, and is meant to improve the risk-sharing equation between the buyer and the seller, renegotiations are often oriented toward increasing the risks/losses to the buyer. This is the unfortunate reality of many projects that operate in both the BOT as well as the BOOT model.

4. BOO Model: When the market potential is established, the better mode to go for is the Build–Own–Operate model. Typically, in such a model, the asset is not returned to the buyer. The seller operates it and retains the complete ownership of the project. In such projects, the seller uses his competence to use hidden capacities when the market grows and profitable operations ensue. In these projects, the life of the assets is quite similar to the duration of operation project. So, the remaining asset life can be a determiner for the choice of the model.

Business models are fairly complex and need a good handle of various dimensions. The contracting type is, therefore, dependent on these dimensions:

1. Asset Life: If the asset life is short, it is better to go for traditional procurement. If the asset life is almost the same as the operating life requirement, it is better to go for the BOO model.

2. Scope Engineering: The degrees of detailing and fixing of the scope are important parameters. Traditional procurement requires a fairly good or well-defined scope. However, when one goes for the BOO, the BOOT or the PPP modes, the buyer usually can specify the business requirement and allow the seller to detail out as per his choice. This gives the business advantage to both the parties as it can use engineering skills of the seller to ensure he delivers. So, this implicitly takes care of some of the risks for the buyer.

3. Supplier Experience: Supplier experience is an important dimension. However, it works both ways. A supplier might have experience, but still might not be used to working in environments with different expectations! Moreover, the lack of experience does not always lead to a conclusion that the seller is 'incapable'. A

cultural fit and a 'mindset fit' is often times also important to objectively evaluate the competence and the experience.

4. Market Potential: We have already covered this earlier.
5. Economy of Scale: The advantage with scope engineering being within the deliverable of the seller is that the seller can optimize it to varying levels of scale of operations. For instance, in a BOO project involving waste management, or even the setup of bloodbanks, the additional capacity can be sold by the seller to other buyers. Thus, it helps both the parties to optimize their business case.
6. Availability of Finances: The inclusion of the operations phase often requires the availability of finances. Many buyers try to reduce their financial exposure by inviting investments from the sellers. This is also a critical parameter that is used to choose the business model.
7. Buyer's strategic intent: Sometimes, the buyer doesn't want to include the operations of auxilliary facilities in their gamut so that they stay *lean and focused*. This is a strategic decision that many times calls for alternate business models. Despite the availability of finance and an established market projection, therefore, the buyers prefer to stay away from being the 'sole owners' of the facility. They, therefore, use contracts that promote seller participation.
8. Risk-sharing posture: Many times, the risk management directives from the senior management define the choice of the contract model that needs to be used.

The decision to use the business model within the contract is, therefore, dependent on several dimensions and the project manager needs to assess all these dimensions clearly. As one goes for *long-term contracts*, the decision is more strategic and the project manager is likely to have lesser *independent* control on the situation/decision.

### 18.3.4 Innovations in Contracts

The reader must be hearing several different options in contracting. In the current industry scenario, there have been two fundamental directions for innovations:

1. New models of ownerships, revenue shares, costs, etc.
2. Combination of existing models to help the buyers leverage from their current popularity.

Systems Thinking actually provides for inputs in both the types of innovations.

Some companies have opted for combinations of the models. There are models where the award is done on a target-price approach and with an option to later change to a fixed-price contract (within a period of 6 months). Such models are tricky as the seller is at a greater risk due to the possible change in the contracting of the buyer.

A lot of cues have been percolating from the new business paradigms used in outsourcing as well. This has also been indicated in our whitepaper on BPO, where several different levels have been identified.

The next challenge is to define the outsourcing relationship. This is typically a mixed bag of management thought, and it often holds the key to the extent of benefit from the process. At the lowest end of the chain, one has the outsourced professionals as 'contractors'; people who have little say in the relationship and the management of the parent company. The next level is that of a 'client-service provider' relationship. Here, there is a little more autonomy for the entity performing the outsourcing, especially at the operative level. As one goes up the progression, it is the case of a 'business partnership'. Here, both the businesses interact at the tactical level as 'partners' and operate in sync with each other. The highest level of progression is the 'strategic' level, where a steering committee operates at the strategic level of the two companies. Depending on the nature of the relationship, one can identify additional benefits from the interaction of the two business entities. Needless to say, a higher level of relationship requires a strategic assessment of the capabilities of the providers, a criterion that is often used during the selection of the entity. It is always beneficial to have a relationship at the strategic level, as this would help both companies to optimize their operations and investments, and ensure smooth transitions with the demands of the markets. This also ensures a substantial minimization of the risks involved, as per a recent research publication of mine. If you think you are already operating at that level, ask yourself if you have the president of the outsourcing entity as a 'virtual member' of your board. The other issue that one faces is the strategic-operational disconnect in most companies. Middle managers and line managers often view outsourcing as a potential risk to their existence. This factor tends to distort the working relationships and the synergy of operation of the two entities. Do you think you have the right relationship with your outsourcing partner? A simple test would be to ask your managers if they would like to share more responsibilities with their counterparts in the outsourcing entity.

The above message is taken from our article on outsourcing [19]. Clearly, though traditional contracting has been static in terms of its business model, there has been a transformation at the operational interface. A large part of this is driven by the popular 'partnership approach' that is currently in vogue in the outsourcing industry. In short, the acceptance of this philosophy has improved over time.

All said, there exist several companies that are, simplistically speaking, project averse! They prefer the acquisition strategy where they go for buying existing operational facilities and assets rather than risking the project route. However, that is a separate discussion point and beyond the scope of this book.

## 18.4  Key Takeaways

In this chapter, the basic focus has been on the procurement processes as they link with commercial risks for the organization. The pricing of risks as individual units is a common practice for the procurement professional. However, it skews the price as it leads to double calculation of risk control measures. Moreover, every such price equation also carries with it some *lead time* for risk identification and control. In other words, if the risk identification mechanism is absent, then the control will be delayed due to the lead time of the control mechanism. This means that the project suffers. Project managers need to understand this dynamics of the negotiation process in order to be sure that there is meaningful negotiation of terms and conditions. In many organizations, the contract is formalized at the leadership level, detailed at the procurement level, and 'passed on' to the project manager. In such organizations, there is often a huge risk borne by the process of risk management. The bidding dimensions mentioned thereafter, need to be necessarily mapped with the project management methodology so as to ensure good synchronization.

The next focus was on understanding certain basic types of contracts. Most companies have their own procurement processes. These are normally driven by their own management framework, the competence of the management, and the legal issues in the business environment. Hence, instead of focusing on generic procurement practices, the focus has been on understanding certain key variants in the procurement contracts. These types have been drawn from the cost perspective, the business case perspective, the supplier perspective and lastly, from the innovation perspectives. Due to the increase in outsourcing globally, certain cues from this area have been taken up for the project manager's consideration.

# Chapter 19
# Account Management

**Abstract** This chapter touches the unconventional area of Account Management in Projects. We begin by touching upon the conventional key account management framework. Thereafter, we describe an innovative account management framework used by Consulting Connoisseurs. In doing so, we first define the critical areas. We first define the Communication Philosophy and help the reader understand the Account Management Philosophy Grid. We then move on to the Partnering or Engaging Principles and describe the Framework on Account Management Partnering Principles in detail. The third critical dimension is that of Posturing and we also define and describe the Account Management Posturing Framework for the reader. Finally, we define the critical dimension of Content and describe the Account Management Interactions on the Content Framework. We conclude the chapter with a case study that evaluates the Communication Leverage in a Project Procurement Scenario.

**Keywords** Key account management · Key account champion · Communication philosophy · Account management philosophy grid · Absorption philosophy · Repository philosophy · Fact-based philosophy · Qualified hands philosophy · Common baseline philosophy · Partnering principles · Gaming principle · Power play principle · Support Principle · Collaborate principle · Strategic intent principle · Account management partnering principle framework · Posturing · Random posturing · Selective posturing · Incidental posturing · Need-to-know posturing · Free-for-all posturing · Account management posturing framework · Content interactions · Account management interactions on the content framework · Communication leverage

One of the most ignored areas of project management is the post-procurement account management. This is a critical area and it often needs closer introspection by the project manager. Most times 'procurement dynamics' cause a lot of 'storms' in the project. Some of them could actually be 'storms-in-the-tea-cup' but it is up to the project manager to identify and control these issues.

**A Quick Hands-On**

The reader needs to understand the following phenomenon:

> **The manager has taken over a project and the kick-off meeting with the contract agency has gone really well. However, over time, the manager is witnessing a change/potential misalignment between his organization and the contract agency.**
> **Why do you think this happens?**
> **How should one control it?**
> **How should one leverage from it?**

The reader needs to give at least two reasons for each.

## 19.1  Key Drivers

Apart from the realities on the progress of the project, account management is very strongly driven by the nature of communication that exists between the parties. This is true because the skill of account management lies in the way in which a project manager deals with his 'client' agency. Now the client agency and the account manager could have several different outcomes on account management.

The objectives of the two agencies are often important to elaborate:

1. The buyer is interested in managing the cash flows of his organization and believes that the project would save more if lesser is given out.
2. The seller is interested in not just managing the cash flows, but also wants to ensure a smooth and a cooperative environment for the remainder of the project.

Account management is a critical activity as elaborated in the work by Ryals [20]. This is reiterated in the definition of key account management that is popular today. Account management is a systematic process for managing key interactions and relationships with most valuable customers. It focuses on the creation, implementation, execution, and evaluation of an overall plan to guide the account team in developing new relationships at the C-level, aligning the best resources to the most profitable opportunities, and delivering what was promised. This is from the 'sales' perspective. In most literature, the fundamental focus is on *multiple sales*. However, in a project environment, there is a *multi-stage sales process* that operates throughout the life cycle. In other words, despite being a single order, the account manager/project manager needs to understand that there is a need to ensure smooth relationships throughout.

As Ryals describes it, KAM is a radically different organizational process used by business-to-business suppliers to manage their relationships with strategically important customers, and it produces measurable business benefits and mentions that there are failures due to improper implementation.

She goes on to describe the seven steps to ensure success as follows:

1. Recognize that KAM is an organizational change, not a sales technique. So, a successful KAM needs the alignment between operations and the sales team to ensure that they are giving it the same priority! Best-practice companies choose to train their operations and supply chain people in KAM, along with their sales people.
2. Get high-level buy-in; preferably C-Suite and Ryals mentions Rolls Royce and Siemens as examples.
3. Appoint a KAM champion. As Ryals puts it, once the organization has accepted that it is embarking on a major change, and senior managers understand what KAM is and have bought in to it, the next step is to find someone who is going to champion the KAM program and drive the implementation. Usually, this will be someone high up the organization, and it helps if they report directly to the top management, at least for the duration of the project. This way, KAM gets onto the top team agenda and the champion gets the support they need to make changes. Your KAM champion should be passionate about KAM and needs to have good influencing skills and great energy levels. Tetrapak has two KAM champions who travel the world to 'sell' the message about KAM within the company.
4. Identify your key accounts carefully. To get the KAM program started, you need to identify some key accounts, and you need to develop an offer that differentiates them from the rest of the customer base. Generally, the number of key accounts should be small. Be clear about what defines a key account and stick to that. We will touch upon this aspect as we delve into topics in our subsequent sections.
5. Appoint and train your key account managers. Many organizations make the mistake of simply moving their best sales people into key account manager roles.
6. Set the right metrics. What gets measured gets managed. If you have tasked your key account managers to build long-term relationships with their customers, do not carry on rewarding them as though they were doing a standard sales job.
7. Benchmark and build. Your key account program should not be static over time.

The critical understanding for the reader is that *account management for projects is similar to that of sales key account management* and one needs to understand the commonalities to exploit the research and development that is already done in this area. We will now touch the key dimensions of Communication in the context of account management. Each of these has a quantifiable effect on the maturity of the account management.

## 19.2  Communication Philosophy

The term philosophy is like strategy, often misused in several contexts. Since it is already abused in literature and everyday management sciences, we take the liberty to use this term as one of the key dimensions. Technically speaking, philosophy is defined as the following:

The study of the fundamental nature of knowledge, reality, and existence, especially when considered as an academic discipline.

A theory or attitude that acts as a guiding principle for behavior.

We now present the Consulting Connoisseurs Framework on Communication Philosophy. This has 5 sub-dimensions. In our context, this dimension essentially has the following categories:

1. *Optimizing Efforts in Absorbing Information rather than Generating Information*
   This is the lowest level of account management. Here, the philosophy is simplistically an 'answer when asked' kind of a situation. Needless to say, problems in such projects could be plentiful and one needs to be cautious as a project manager who does not receive information from the other partner. It is the focus or approach of the agency to react to the requirement.
   The information is often unavailable. Typically, such an agency comes to the meeting without adequate documents and is totally unprepared for their reviews. The proactive project manager many times gives information to such agencies, but is unable to get information from them.
   The situation is subject to the chaos in the organization. Usually, staffing issues are abundant and the lack of processes also gives rise to discontinuity in the information on both sides. Availability of information too could be compromised in such cases. Thus, there is a possibility of getting a mix of both truth and untruths in the project.
   Many times, account management also leads to escalations that go up to the CXO levels in these cases. This situation is not very uncommon in present-day project scenarios. Much of these escalations are oriented toward fact-finding rather than any decision or action.
2. *Handling Sensitive Information must involve 'qualified' hands (esp. with external agencies)*
   Another popular philosophy that is used by many organizations. Here, the general trend is to deem all that is shared with the client as 'sensitive', requiring a smart communicator who needs to quickly guage the situation and react accordingly. Such environments are typically those involving 'doctored' reporting that many times leads to phenomena like the 'schedule chicken' that was discussed earlier. However, this could also be due to a lack of skills in the organization, leaving the qualified hand to manage all the information. In this situation, the advantage is that the single point of contact is more likely to give the information desired. While in the first case, the process and the system lead to an overall 'break-down' in the account management; in this case, it is more of a person-dependent process. However, potential use or misuse is always a risk that one needs to consider.
   Most projectized organizations also have their project managers working as their account managers. And it is the project manager who keeps a tight control over the entire communication process by attempting to be a 'single point of contact.' In an environment where integration is less or where the overall management system is highly fragmented, yet the communication is forced through a single person, this kind of a situation is fairly common. Again, this leads to a mix of both

the truth and untruth in the project. Both the previous alternative and this one are risky and similar in their profiling. However, the second one has a 'designated' person to handle the information and a directive within the organization on how this information would be processed.

Escalation of issues, though present in such situations, does not help much because everything depends on how the organization 'plays its cards.' Hence, the escalation frequency is relatively lower in this case. Escalations are often meant for discussion and high-level resolution in this case.

3. *Fact-Based Communication*

Fact-based communication is one of the most popular paradigms. Here, the two parties discuss facts rather than their own impressions and feelings, thereby making it clear to each other that the real world is the key driver to account management. Such a philosophy is useful for both the parties as there is a fundamental element of trust established between the parties due to their fact-based communication. However, it does not always mean that the agencies are proactive. This is another challenge posed by such environments.

The biggest advantage of this system is that there is no need to 'doctor' information. A free and fair information flow can be allowed. However, it is very likely that the information is not always available at the right time and specific to the requirement.

Escalation of issues is for resolving the 'facts and the subsequent plan of action' based on the facts presented to both sides in order to take the project forward.

4. *Leveraging of Repository*

A fact-based system gets leveraged when the repository of information is wisely used. Here, past information is used to ensure an adaptive and a learning environment that helps identify systemic issues in the project. It works as a double-edged sword by providing sufficient information and, at the same time, 'expose' the agencies that are not doing the job: Trends, analysis, and a lot of management inputs into understanding the 'behavior' of the *progress of the project*. The main idea of account management is to create sufficient management 'ammo' for taking things forward.

Using the repository is typical in claims management. Hence, if a project environment does believe in using this philosophy, they often have trained professionals to manage their projects and these persons have the ability to create a 'temporal map' of the developments in the project. A lot needs to be understood about how one handles uncertainty. However, when a company does go for this philosophy, usually a proper data capturing and evaluating mechanism is put in place. This helps smooth out the account management process. The thrust is in the development of escalation of issues, that is usually done to ascertain responsibilities and safeguard risks from a claims perspective. It is used as an evidence to check the 'behavior' of the project account/issue.

5. *Communication for a Common Baseline*

The highest philosophy is to communicate to ensure a common baseline. Successful account management involves a proper alignment of the interests of both the sellers and the buyers in the project environment. The focus is always on moving 'forward' rather than on ascertaining the responsibilities. Usually, any discrepancy in understanding the risks and the responsibilities is handled at the C-Level. This philosophy is very popular in environments where both the buyer and seller are large organizations. Usually, if one of the agencies is smaller than the other one, it is less likely to have a common baseline philosophy in operation; though, when done effectively, it works wonders.

Escalation is more at the operational level since the strategic alignment is relatively static. There is a high degree of trust and there is free sharing of the truth. The common baseline ensures that there is little or no ambiguity between the two parties and they move forward.

All these philosophies are essentially characterized by two dimensions in the Consulting Connoisseurs Communication Philosophy Grid Model: (a) *Truth and Honesty* and (b) *Proactive, Accountable, and a Responsible System or Individual*. The grid, therefore, is simplistically defined as shown in Fig. 19.1. The escalation issues also differ with the fundamental philosophy in each case.

An interesting aspect about the choice of the philosophy is that it depends on the competence of the procurement professional as well. In many organizations, the procurement negotiates contract terms and is unable to truly verify the authenticity of the selling organization. Therefore, orders are committed to, without actually knowing the reality. In several projects, therefore, despite engaging expediting agencies, the entire timeline appears to be shifting. One of the key and difficult areas of the account manager is to ensure that both move toward a healthier philosophy and relationship.

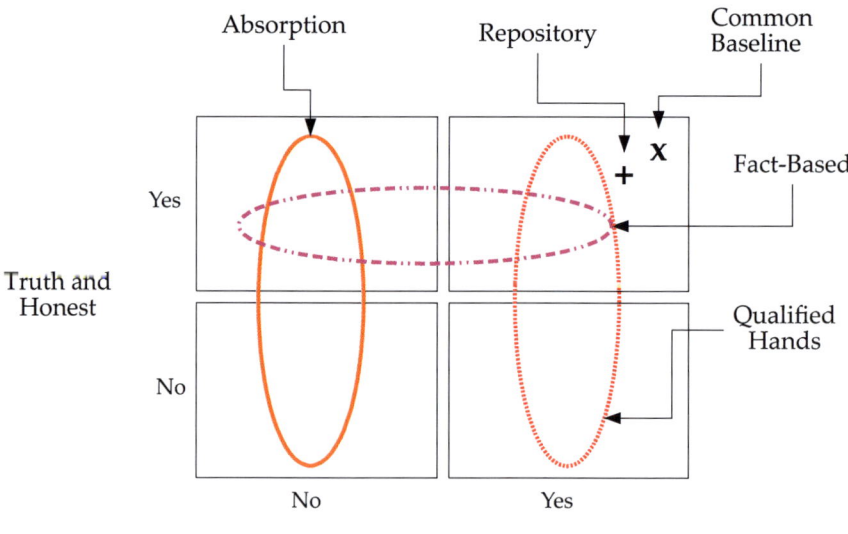

**Fig. 19.1** Consulting Connoisseurs Communication Philosophy Grid Model for Account Management

The other aspect is that the philosophy, as we are using in this case, is a 'global phenomenon.' In other words, it is independent of how the 'individual' employee of the organization plays the cards. This is, therefore, more intricately linked with the organizational culture rather than individual behavior. The reader needs to be careful while making this distinction.

To put it in perspective, let us look at a few situations that would give a better perspective to the reader.

**Situation 1:**

*Owner agency*: What is the status of the project?

*Supplier agency*: It's on …

*Owner agency*: Do you know that the other contracts are nearing completion. Yours is the only one pending!

*Supplier agency*: We are on it!

*Owner agency*: If there is a further delay, the subsequent contractor will also have issues. This is serious for the project

*Supplier agency*: There wouldn't be necessary, we should be ok

*Owner agency*: The maximum delay I could accept is 2 days, because our supervisors are coming in later

*Supplier agency*: In that case, let the invoiced amount be released. With the delay, there should be less cost to be attributed to our end.

The above case gives a good understanding of how the 'absorbing philosophy' works.

**Situation 2:**

*Owner agency*: What is the status of the project, Tony?

*Supplier agency*: Harry, our senior manager, would know.

*Owner agency*: Okay, the boss is upset about the progress.

Tony calls up Harry and briefs him. Harry then 'doctors' the response with the given information and takes it forward.

The above case is one where a 'qualified hand' actually handles sensitive information.

**Situation 3:**

*Owner agency*: What is the status of the project?

*Supplier agency*: It's 30 % complete.

*Owner agency*: When is it likely to complete?

*Supplier agency*: With the current schedule, it looks like the project would take another 3 weeks to complete.

The above case is a typical 'fact-based' environment.

**Situation 4:**

*Owner agency*: What is the status of the project?

*Supplier agency*: We are 30 % complete at this stage.

*Owner agency*: Well, that is not good enough. Last week, you had indicated that you would be 40 % complete by now. It looks like you have lesser programmers at site. That department is consistently understaffed. What are you going to do about it?

*Supplier agency*: We have a shortage as this is the festival time in India. We are trying to fill it in by overtime. However, approvals from your side are also getting increasingly delayed now. On an average, the turnaround time has increased by 1 day each month! That is severely constraining our productivity as we are short staffed and have a further issue of a 'long waiting period' that is increasing our 'downtime'!

*Owner agency*: Let us know how we can resolve both these issues …

This is a case of using the repository to try and enhance the account management.

In the last case, there is a better convergence on how the issues have to be taken forward with the other company. It relies more on providing a foundation to understand each other's position and take things forward; hence, the common baseline philosophy. Every project would have a different philosophy for account management. Of course, the effects of each of the philosophies are going to be drastically different, but it is definitely a choice that needs to be made by the project manager. Often times, there is a management directive to the manager on the nature of the philosophy that needs to be adopted. Such situations cause a lot of confusion and the project manager needs to understand how to handle them correctly.

## 19.3   Partnering or Engaging Principles

The next important dimension for the project manager is to understand the partnering or engaging principles. While the previous section looked at broad-based philosophical and systemic components, this aspect looks at the interface or the interaction between the various parties. Again, there is a wide range of alternatives, for the companies to choose from, in this situation. We will now delve into the same.

1. At one of the extreme ends of the spectrum, one has the 'gaming' client who is always into *power play*. In short, such a client typically wants to set the 'traps' on the other party, appearing to be pulling the other party down. This posture creates either significant conflict, pressure, or a disturbed relationship at the end of the experience. Power play postures are used to create a political 'clout' where the other agency is constantly being 'picked' on any slippages (not just projects, even plain language slippages are caught!). Negotiations are typically power play based where one of the parties tries to have the 'upper hand' and make the other 'toe-the-line.' However, this does not always yield meaningful results. Though this posture is preferred by those wanting to 'play boss,' it is far from the truth in the project. Power play often leads to distortions of the truth due to its intrinsic nature. The agencies often witness an implicit abuse from the other party, due to various reasons.

   Many conventional government contracts often run on the power play engaging model. Traditionally, the 'competence' of the project manager has been measured using his skill to 'keep the upper hand' with the contractors. This was nothing but 'power play.' With the advent of the knowledge worker, management principals have become more sensitive toward effective partnering as it is witnessed that 'power play' does not always give the intended results. Therefore, though the traditional managers still advocate this as an effective partnering or engaging principle, the reality is changing and power play tactics are becoming less intense in the modern era. The power-play results in the lowest form of account management.

2. The first improvement over the power play principle is the *support* principle. Here, the agencies are open to understanding each other without having the need

to power play. Of course, the needs of information are met by both sides. In this posture, there is better engagement between the two agencies. The interactions are more dignified, as the organizations now are treated as essential partners for project success. Usually, this mode is popular with agencies from different cultures (east-meets-west) in a given project environment. Further, in projects where the environment is cordial, the support principle is seen at work. Both the agencies understand what is expected out of them and none of them go overboard to make the environment counterproductive.

In the support principle, however, the delivery streams of both the agencies (and therefore, their schedules) are not directly linked with each other. In other words, there is a tie-in, but there are adequate floats in each case. Even if there were no floats, the 'independence' and the 'freedom' along with a strong 'compassion' are valued higher than the interlinking. For instance, an LSTK contractor typically operates on a support principle. He has his ways of handling his own issues and is fairly clear about how he needs to go about the project. At the same time, the contractor does not want too much interference from the owner and is willing to provide information to the owner on a regular basis. This is, therefore, a slightly improved version of account management.

3. An improvement of the support principle is the *collaborate* principle. In the collaborate principle, both the agencies tend to share their information without being explicitly asked. They are, in some sorts, 'open books' and 'open information' at work. The flow is better. In this case, both the agencies understand that it is important to exchange information to ensure that the project objectives are met. While the former two were ways of partnering, this alternative is like a close partner. A lot of emphasis is given to the operational efficacy that is generated by the exchange of the information.

   A 'collaborate' option is more common when the activities of both the agencies are intricately 'tied' to each other throughout the course of the engagement. In other words, decisions from one agency could influence the other and vice versa, throughout the course of the project work. And unlike the support principle, here there is scope for discussion and decisions throughout the course of the project. Though the organizations work in sync here, their stakeholders still belong to separate entities. In other words, the steering committees might have little representation from the contractor side. This version is commonly seen in many 'knowledge-intensive' projects and 'large budget' programs.

4. The final engagement principle is that of the *strategic intent*. In this case, the owner stakeholders of both agencies are together and share the same amount and sense of priority and importance, followed by commitment and management backing to ensure the success of the project. Both the agencies work as partners in this case, meaning more like extended organizations of each other. This could be either as embedded organizations, where the smaller team becomes a part of the larger team, or as two close-knitted departments that are working on the project. The key distinction is that the partnering is at the CXO level in this case with a possible steering committee comprising members from both the sides to review the situation. This is the best form of account management as it ensures syn-

chronization and a positive work ethic and environment between the partnering agencies. Here, the concerns are shared by both the stakeholders and they try to work in unison. As an account management strategy, therefore, there is little gap between the directions of thinking and working between the agencies.

The reader can appreciate that the partnering principles can also be derived using psychological models such as the PAC model of Transaction Analysis developed by Eric Berne. However, unlike the PAC model, an organizational model is more stable to the type of interaction. And since the organization has a large group of persons that are interacting, it is always required to understand the overall 'sense' of the partnering principle that is in action.

The above parameters are modeled using the Consulting Connoisseurs Partnering Principles Grid for account management. If one is to take cues from the transaction analysis models, this can be defined using two dimensions in our case: (a) The level at which the commitment is fixed in both the agencies and (b) The objectives of the interaction. The former actually identifies the minimax condition of commitment. For instance, if agency A has the CEO involved in making the commitment while agency B has the project manager involved, the level of the commitment for the combined interaction is operational. However, if the CEOs of both the agencies are involved in the commitment (and this is through the life cycle of the project and not just the placement of the orders), the level at which the commitment is fixed becomes strategic. Similarly, the objectives of the interactions are important. One could have a control-oriented objective (involving gaming at the operational level), or could look at the progression as shown in Fig. 19.2.

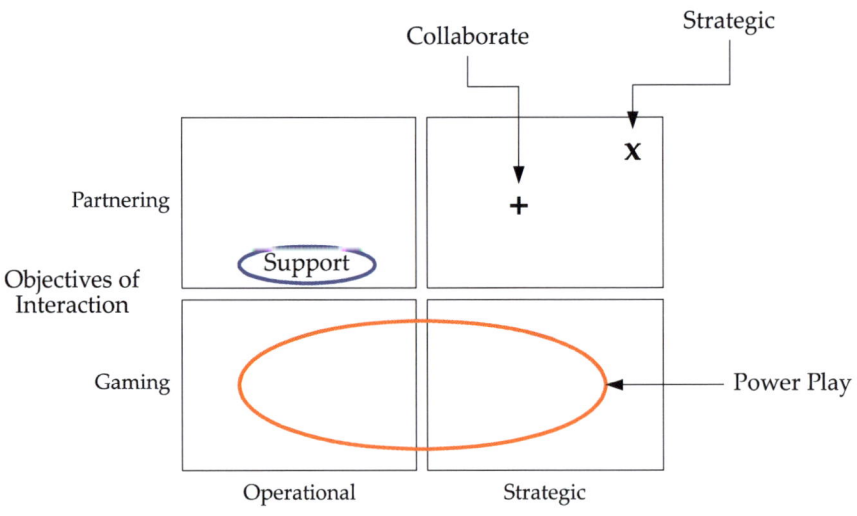

**Fig. 19.2** Consulting Connoisseurs Partnering Principles Grid Model for account management

One could also understand these variants using a few examples, just as we did before.

**Situation 1:**
*Owner Agency*: What is the status of the project?
*Supplier Agency*: We are at 30 %
*Owner Agency*: You are not doing well then
*Supplier Agency*: We are as per our plan and the agreed baseline.
*Owner Agency*: But you are slipping. There is less staff at site! Quality issues.
*Supplier Agency*: Though we are short-staffed, we are managing to keep the commitment. The bugs are on the colors of the fields and not on the programming logic.
*Owner Agency*: You are not delivering. The project is not being managed well. I can see problems coming up …

This is a typical power play discussion. Here, the owner is not interested in understanding the objectives and the actuals in the project. It is more of a situation where the owner agency wants to have the 'upper hand' and it appears to be a 'witch-hunt' in any case!

**Situation 2:**
*Owner Agency*: What is the status of the project?
*Supplier Agency*: We are having a delay of 1 week currently and it is likely to go up to 4 weeks.
*Owner Agency*: Why is it?
*Supplier Agency*: Due to a delay in supply, our office in Florida hasn't received the material. Hence, there is a delay.
*Owner Agency*: That is serious. How do we solve it?
*Supplier Agency*: If you have the material in any of your warehouses, we could use that. Alternatively, if you know of any other supplier who can ship it readily, please let us know.
*Owner Agency*: We will check and revert back. How do you propose to catch-up?
*Supplier Agency*: We have booked additional capacities at other workshops as we expect some spikes in the workload.
*Owner Agency*: Okay, let's review it again …

This is a typical situation of support. Here, the idea is less of 'pressurizing' the supplier agency. It is more toward supporting his endeavor of delivering to the project. Small quantity procurements are often in this bracket.

In large-scale programs, a lot of information is continuously exchanged with the partner agencies. In other words, it is more of a collaborative environment rather than a support environment. Most times, information is exchanged as 'packages' between the agencies. However, when the quality of collaboration is good, the agencies have a continuous stream of information 'bytes' that keep providing the teams with the inputs they require to ensure they deliver for the project.

The fundamental distinction between the collaborative principle and the strategic intent principle is the nature of the project plan. In a collaborative environment, the plan is fairly static between the agencies. When the strategic intent is involved, the plan is always a subset of the intent. In other words, in order to meet the strategic objectives, the management might discard the existing project plans and come up with new task lists to move ahead. For instance, temporary solutions due to delays are common. This is a situation where the strategic intent supersedes the project plan.

Technological alternatives too could be explored. In other words, one could expect larger changes in the environments with strategic intent-based partnering. In fact, the entire business case, and therefore, the project scope could be altered in this kind of partnering. Everything is a subset of the strategic objective and the partnering at the CXO level is to ensure that the strategic objective is not compromised. In a collaborative environment, however, the plan is fairly fixed and one may not change the dependencies to a huge extent. In other words, the business case is seldom changed.

## 19.4   Posturing

The next important dimension in account management is the posturing toward communication and information flow. The philosophy typically focuses on the beliefs and the intrinsic priorities of the management system and the principles provide guidelines for the relationship. The posturing tells us how it operationally overlaps over the information flow processes and systems. There are essentially five types of posturing used in account management:

1. *Random*
   In this method, the account management process is initiated, developed, and worked on a random basis. In other words, there is no definite way in which information is generated, stored, evaluated, and distributed. This is the 'surprise' mode or the 'dis-organized' way of working. It involves chaotic information flows. So, even if a company has strategic intent and has a philosophy to develop a common baseline, the random posturing would severely compromise the account management activities.

2. *Incidental*
   In this posture, the communication is based rather incidental. For instance, information is 'generated' because there was a slip in the timeline. Again, this is a posture whereby the priority of communicating is lower than expected. Hence, there is ample room for speculative working.

3. *Selective*
   The selective working process screens out information and shares only selected bits and pieces of the same. There is a clear demarcation of what needs to be shared and what does not need to be shared. Thus, there are pockets of information that are often missing in this case.

4. *Need-to-Know*
   A more refined selective form is a need-to-know form. In this posture, information is selectively available at various levels depending on the potential purpose and the use of the information. The previous posture only has selective information. Most corporates today implement the need-to-know posture. They share information as per the 'pay grade' of the employee. This often causes potential misalignment in the organization as there could be a difference in perception on what the 'need' and the 'use' of specific information could be.

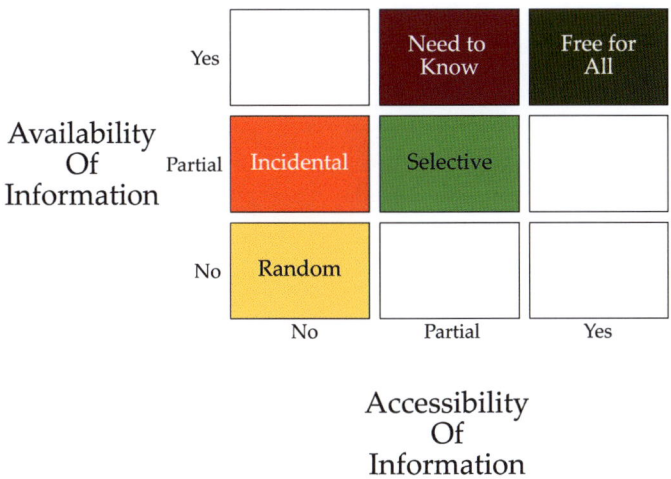

**Fig. 19.3** Consulting Connoisseurs Posturing Grid Model for account management

5. *Free-for-all*

The free-for-all is an open posture that believes in leveraging the true value of information by simply making it available to everyone involved in the project. Needless to say, a gaming partnering principle could misuse this posture drastically!

It is important for the reader to delink the dimensions and understand that they are different components of account management. Communication dimensions are typically combinations of each of these and they lead to certain specific results. This is defined in the Consulting Connoisseurs Posturing Grid Model for account management. A near approximation is found from the portfolio given in Fig. 19.3. Structured availability of information and a structured system of accessing the same are important in this case. The more the structured availability and access, the better the posturing from an account management perspective. However, the project needs to be a combined effect of all these elements.

## 19.5 More Dimensions

There are two more dimensions that are interesting in the context of account management. These two dimensions additionally affect the efficacy of the account management process.

## 19.5.1   *Content of Information*

One of the key aspects in account management is the content of information that is being 'managed' by the account manager. This again has different dimensions. The content of information helps the partnering agencies understand the nature of the issues. At times, there are several issues that are relevant.

1. Technical
2. Schedule
3. Quality
4. Organizational Dynamics (People)
5. Commercial/Financial
6. Legal/Contractual

One needs to rate the communication of each of these dimensions and put them in perspective. This is done using the Consulting Connoisseurs Interaction on Content Grid Model as shown in Fig. 19.4. While technical issues and discussions are rated as top priority in most organizations, the subsequent ones have lesser and lesser importance. However, it is often the overall connection that is of interest to the project manager. Hence, we don't subscribe to a direct ranking/subordination approach.

## 19.5.2   *Network*

The nature of the communication network is also a key parameter. In most literature, it is said that there has to be a single contact person who plays the role of a key account manager. Unfortunately, in projects, especially the larger ones, there are

**Fig. 19.4**   Consulting Connoisseurs Interactions on Content Grid Model for account management

multiple entities that keep communicating with the customer's team. Hence, it is often necessary to understand how this communication unfolds in the realm of account management.

The key aspects are the following:

1. Who Communicates (Number of Points)
2. With Whom (Number of Points)
3. When
4. How
5. Recording/Storage
6. Access/Disbursing

All these dimensions put together give the project manager inputs to manage his accounts better.

## 19.6  A Case Study: Communication Leverage

At a turnkey water solutions provider, there were significant problems with respect to account management. The company had a turnover of approximate USD 8 million and was into manufacturing ultra-high purity water solutions for various clients in the pharmaceutical as well as food and beverage industries. The water treatment solutions included ultrafiltration, reverse osmosis, and UV treatment plants that had subsequent ozonization and other processing for ensuring water of usable quality (FDA grade) for their intended applications.

Despite being a fairly established name in the pharma domain, the company was facing severe problems. For instance, during the project phase they had the following:

1. Completion Issues at Sites: whereby the projects were witnessing inordinate delays in the PQ (Performance Qualification) and handover. Hence, their cost issues were plentiful as the average cost of personnel at site was huge and the problems were plenty.
2. Payment Issues: which were not just related to the projects having the completion issues! Rather, these had now overflown into other projects that were being negotiated with the client agencies. Hence, not just the current projects were having problems, the subsequent ones too had serious issues.
3. Guarantee Issues: the company was not able to fulfill what it had promised in the purchase orders.

Unfortunately, the problems were not just during the current project phase, but also had a bearing on incoming/subsequent contracts.

1. The market was maturing. In other words, a need to improve client engagement model was definitely a top-priority issue. The earlier purchasing strategies were more toward 'bills-of-materials.' However, over time, the engagement models were gearing up toward an 'uptime guarantee of the equipment' and a 'cost per liter' BOO model.

2. There was a need for a new project management methodology. With the large requirements at client locations, traditional management methodology was failing to provide the company with good results.
3. The bleeding margins also meant that there was a dire need to have a renewed account management methodology.
4. And finally, a new control regime was desired. And interestingly, this was not just felt by the company, but also by its customers.

In short, this was becoming an ideal configuration for an account management re-engineering case.

The first step in the process was to understand how the company was doing with respect to its clients. Four key clients were taken and their accounts were evaluated. In each case, the dimensions of the client (of our consulting company) were compared with his key account customer. The dimensions where the client fared better than the key account customer were those that gave the client a *communication leverage* as far as the account management was concerned. Needless to say, when the customer scored better, he had a communication leverage when compared with the client. This is shown in Fig. 19.5. In this case, the client was doing well in some areas, but was not doing as good in some others. As we moved on, there were other three cases which showed distinct differences as shown in Figs. 19.6, 19.7, and 19.8. Each interface had a distinct and a different behavior. Communication leverage was a strong focus while negotiating with the clients, particularly because the negotiations were more dependent on the interface rather than the technical and commercial terms per se.

The most difficult issue for the client was that Customer D was the one giving the maximum number of orders. They were also the ones giving high-value orders. Moreover, they were keen on taking the account management to higher levels of collaborative and strategic intents. Yet, the outstanding from the current orders was extremely high. Hence, it was a complicated situation for the client to take things ahead and gain good strategic advantage as they did so.

As a first step, therefore, the reasons for the outstanding payments were explored and understood. This exercise gave many pointers in the direction of the account management and sales strategy. On closer introspection, it was observed that most of the steps in the sales processes were not controlled as shown in Fig. 19.9. In the figure, all the activities above the process line were done with the customers. All the activities below the process line were actually internal to the company. The amount of clarity this exercise brought to the company was amazing. Every function head, as well as every team member, was able to understand and appreciate the nature and the process by which business was being done at the company. The deep blue activities were those that were always controlled. The gray activities were those that were never controlled and the gradient colors were those activities that were sometimes controlled (depending on the customer). Next, the impact of not controlling the activities was again important for everyone to understand. In fact, the configuration in the situation actually led to far-reaching consequences. For instance, the implications of not controlling the first two activities in the sales processes are shown in Fig. 19.10.

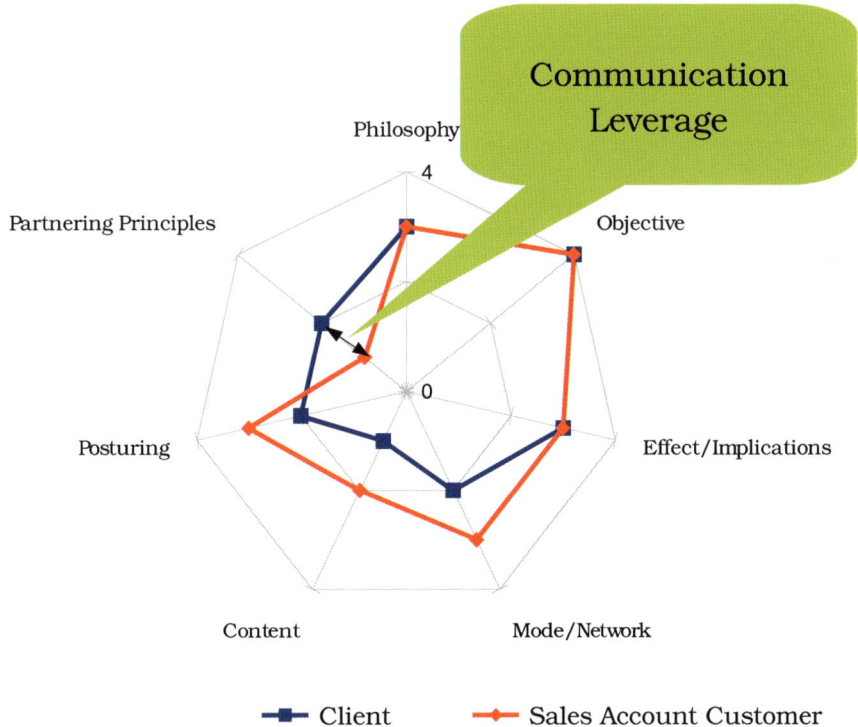

**Fig. 19.5**  Evaluation of the account management interface with customer A

This activity of process mapping, followed by technical, commercial, and legal assessments, was repeated for every engagement model and every business case that the company was operating in, and the detailed assessment was then used to redefine their overall account management strategy. In the process, a lot of their *delivery processes for the projects had to be realigned to ensure that they were in sync with the account management requirements*. In other words, a complete realignment of the organization was worked out to ensure that the performance was improved and that customer satisfaction was also improved in the process.

However, as a case in context, Customer D was to be addressed on priority. Customer D had indicated that they were keen on a strategic tie-up with the client *if the client met with certain conditions*. However, given the bleeding margins of this otherwise profitable customer, it was necessary to explain the situation to the customer to ensure that they are on the same page as far as the overall strategic initiative was concerned. This was a complex task. Yet, it needed to be done. An *account management maturity model* for the client was then developed. This was mainly required because most clients were moving away from the traditional contracting models to the BOOT or the BOO models. Though this was a gradual step, it did have several

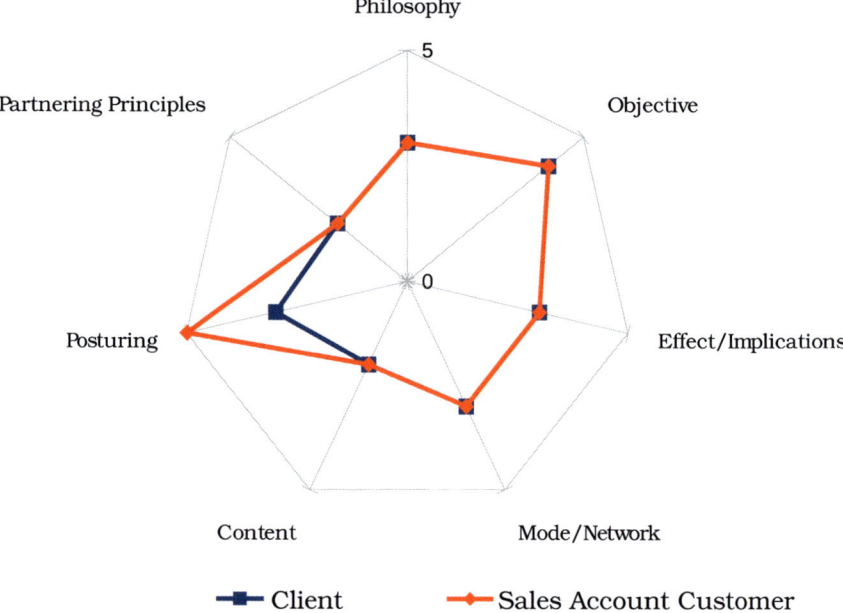

**Fig. 19.6** Evaluation of the account management interface with customer B

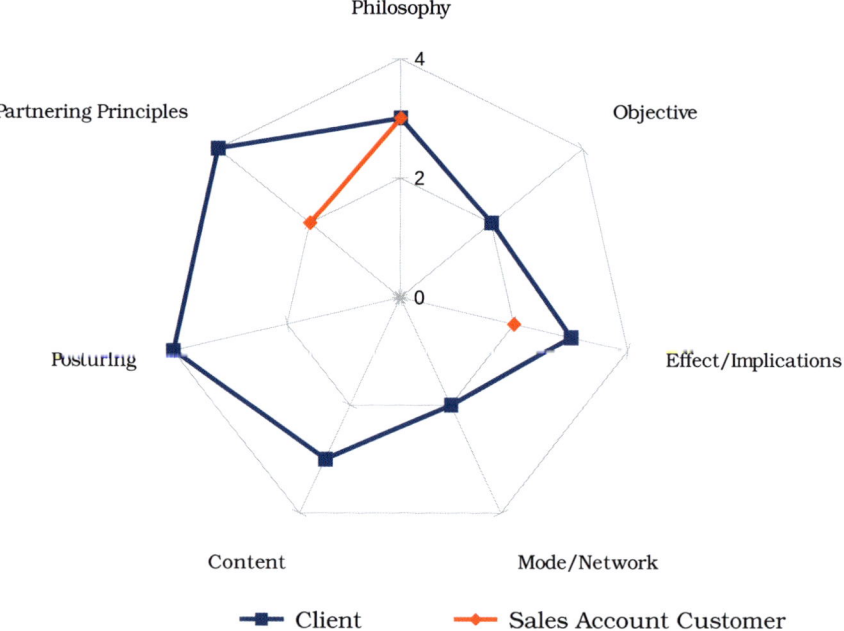

**Fig. 19.7** Evaluation of the account management interface with customer C

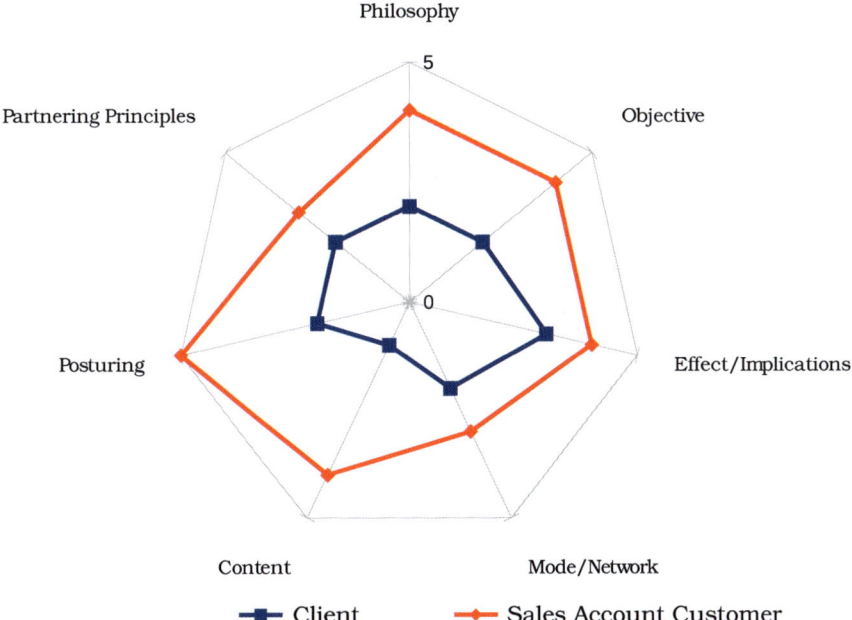

**Fig. 19.8** Evaluation of the account management interface with customer D

implications that were to be considered by both the client as well as their customers. The general progression is shown in Fig. 19.11.

In order to take the customer engagement to a higher level, it was essential to clarify the strategic drivers that affected the relationship. The first fundamental business relationship for the client to understand was that a higher availability also meant a higher cost-per-liter situation. This is described in Fig. 19.12. In fact, most customers failed to understand this transition, especially when they came from the traditional contracting background. Every customer was willing to pay for the 'components,' but they were not willing to pay for the 'risk cover' that needed to be factored in while talking about higher availability. Hence, it was essential to redefine the account orientation by explaining this key factor to the customer.

The next step was to explain the key drivers of the availability. These are shown in Fig. 19.13. In this exercise, the first objective was to 'defuse' the ammunition that the customer was creating through his communication leverage. Thus, the drivers of availability were clearly defined and identified and it was made amply clear that the communication leverage that the customer seemed to enjoy was not truly giving them any benefit. The key determinant was the technology that was to be used. The appropriateness of the solution in terms of the technology that was to be used was determined using three parameters: (a) Input Water Quality, (b) Production Requirements, and (c) Output Water Quality desired. The technology then drove

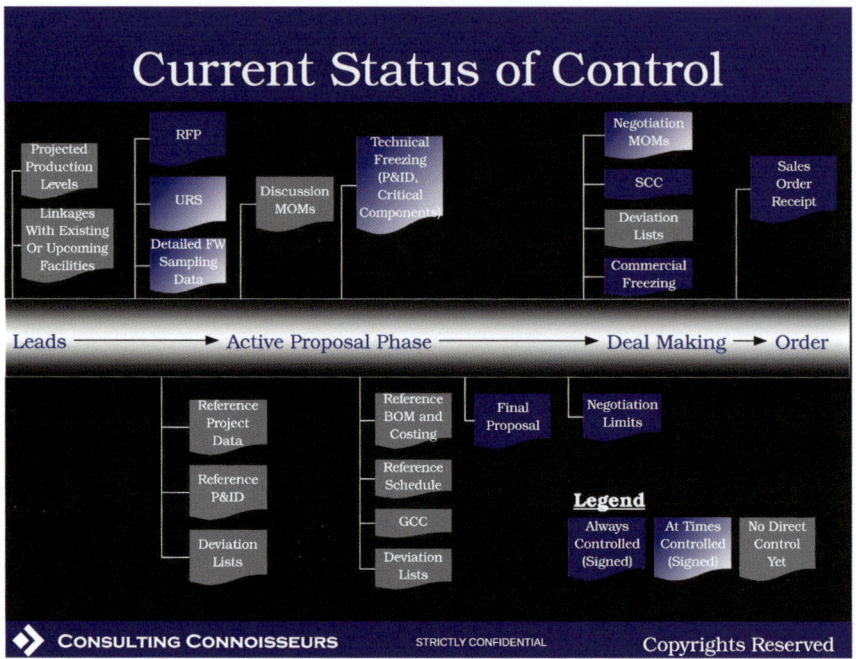

**Fig. 19.9**  Sales process: temporal map showing controls

the operations (including the standard operating procedures), the servicing, and the maintenance.

In order to meet the requirements of the client as well as the customer D, the key optimization parameters included

1. Minimization of the cost of investment: both current as well as future
2. Minimization of the cost of operations: both operational costs as well as life cycle costs (operations and maintenance)
3. Minimization of Downtimes: both costs/durations of downtimes and the frequency of downtimes

The parameters that determined the availability also determined the cost. The relationship is shown in the table given in Fig. 19.14. The figure actually shows the degree of influence of each of the individual parameters on the costs. As one can see, the investment is actually the project cost in the conventional contracting model; however, there were also additional considerations with operational costs and servicing costs. Also, the cost of the downtime, beyond the scope of the client, is also highlighted (not all was within their scope). The most important driver of the costs is the input water quality. It was observed that many of the outstanding payments and the inordinate delays were due to the input water quality that was conveniently dropped off the contract, thereby causing an unsurmountable risk in most cases. In

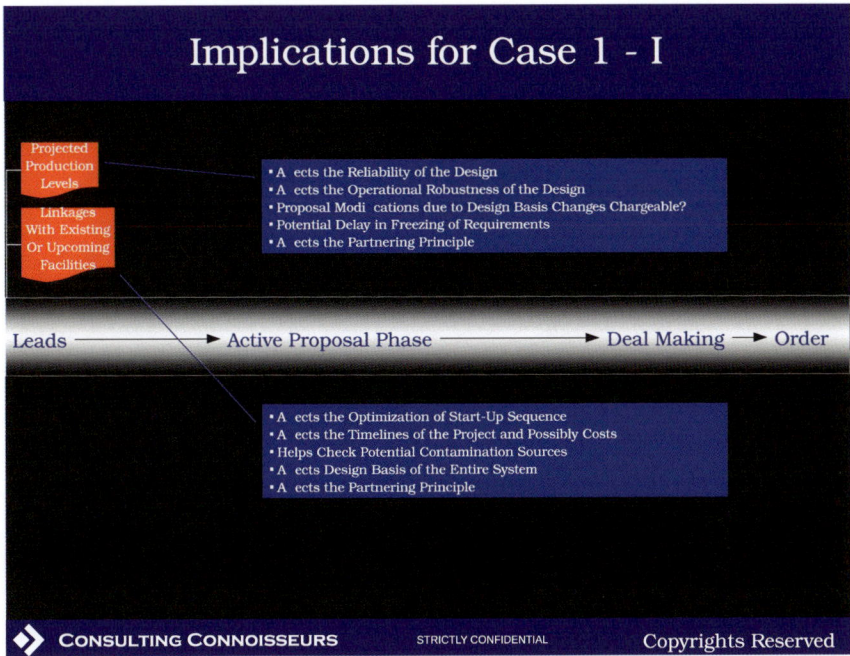

**Fig. 19.10** Implications of non-controlling of activities

the enthusiasm to get things going to the 'higher revenue target' and the 'large company,' the management over-promised the customer based on their past experiences (particularly of water quality and also on the site conditions). This decision costed the management a heavy price in terms of the payment schedules and the overall cash-flow situation of the business. In any case, without delving into that aspect now, the focus was on clarifying the situation to the customer so as to ensure a balanced approach to take things further.

The availability progression was then defined with the customer. This was meant to clarify the drivers of costs in the context of availability. The main factors of downtime have been highlighted in Fig. 19.15. This figure clarifies the understanding of higher costs for the change in the model to the typical BOO or BOOT model. The overall factors were then used to define a smooth progression into the next level of account management.

After customer D understood the critical parameters and the factors that influenced the success of the contracts, they appreciated the need to manage the drivers in such a way as to ensure that their own strategic objectives were met. Therefore, they agreed on a resolution plan for the closure of all the open issues in the existing site and described a method (with strong due diligence) to ensure that the client was able to identify the risks and make suitable amends to the contracts to ensure better availability of the equipment. After all, the downtime severely affected the profitability of the entire pharmaceutical plant. Though the water system unit costed much lower

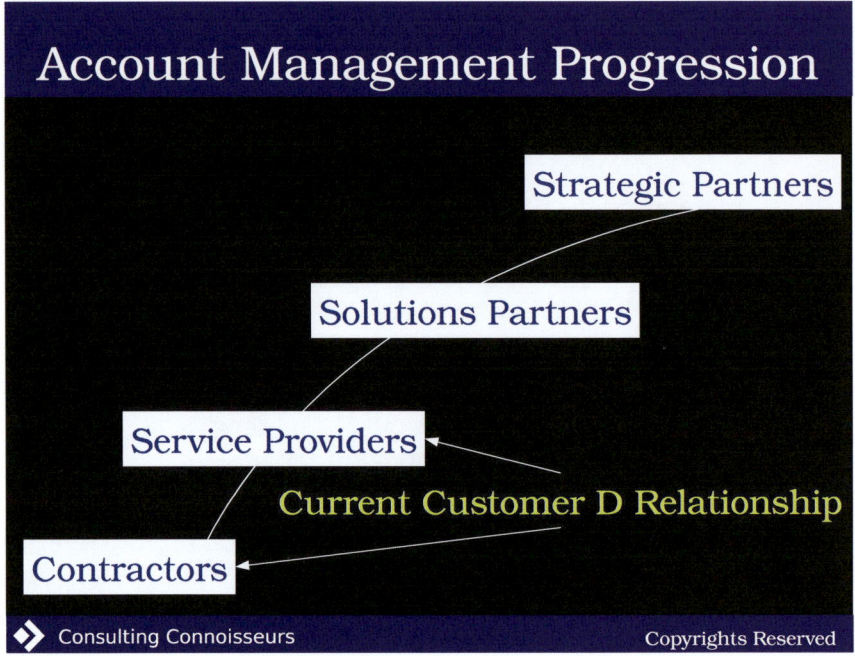

**Fig. 19.11** Progression of account management

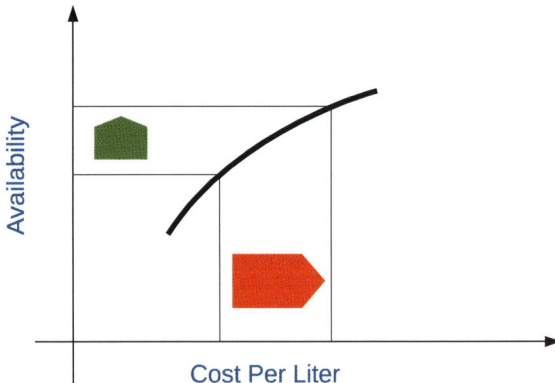

**Fig. 19.12** Availability–cost relationship

than the remainder of the process plant, the downtime of the unit implied that the remainder of the plant too suffered from a 'consequential operational downtime' or 'damage' leading to an overall loss to the customer. A path was, therefore, defined for taking the relationships to newer heights.

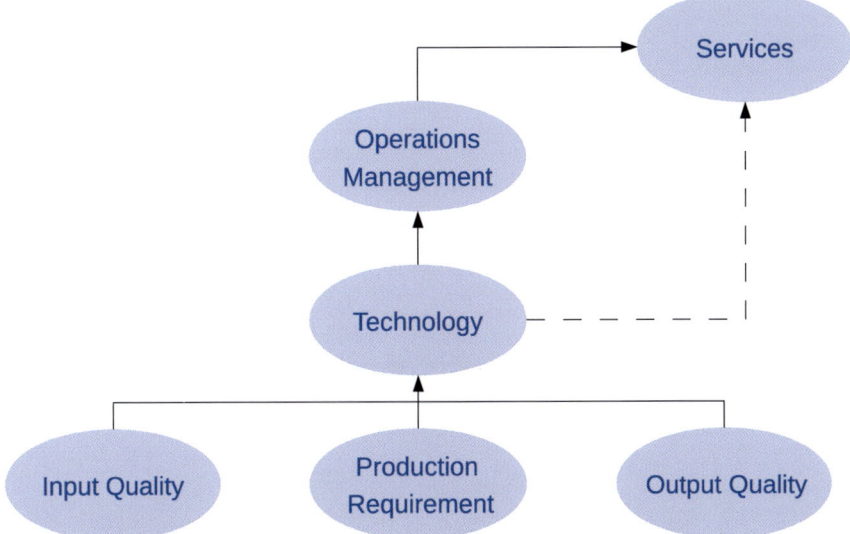

**Fig. 19.13**  Tree of drivers of availability

| Cost / Type / Driver | Investment | Pure Operations | Maintenance & Services | Opportunity Cost of Downtime |
|---|---|---|---|---|
| Input Water Quality | +++ | +++ | +++ | +++ |
| Output Water Quality | +++ | + | + | ++ |
| Production Requirement | +++ | + | ++ | N.A. |
| Standard Operating Procedures | N.A. | +++ | +++ | ++ |

**Fig. 19.14**  Drivers of costs

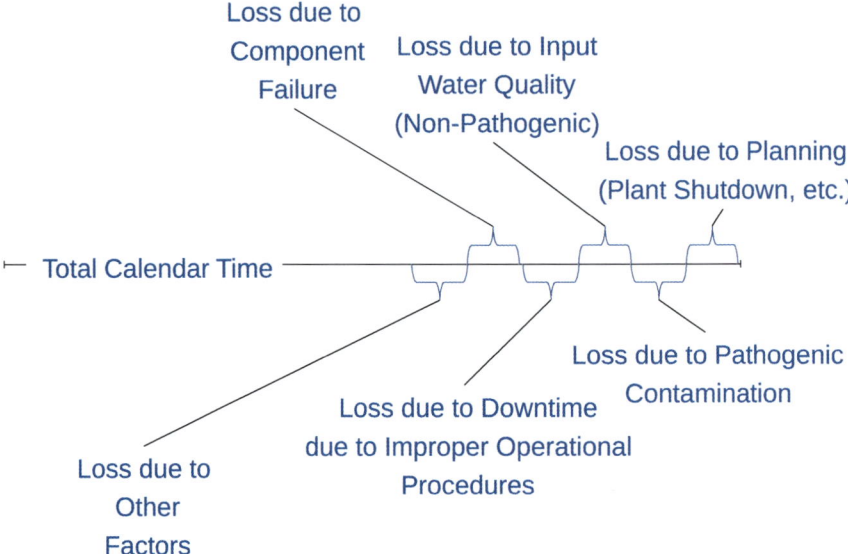

**Fig. 19.15** Availability progression

## 19.7  Key Takeaways

The present chapter has clarified the need to have a robust account management system. However, unlike the key account management in 'operational sales,' the project account management is more complex due to the duration of the engagement with the client on the same project order and the value of the contract. Since a project sale also triggers, in most cases, succeeding operations and maintenance contracts, it is essential to ensure that the process is done correctly.

One of the key differentiators between operational sales and project account management is the fact that information is generated and collated from across the organization. Unlike key account management in traditional sales environment, the role of the account manager is less pivotal in the project environment. In other words, project environments tend to be more diffused than the sales environments. Further, as the key stakeholder, the project manager is also many times involved in the account manager process.

The chapter then goes on to describe the importance of communication in account management, stating that communication is the single factor that many times decides the fate of the account management process. The various dimensions were then looked into in greater detail. To put things in perspective, a detailed case study of a water treatment plant supplier for the pharmaceutical industry was also discussed.

# Chapter 20
# Public–Private Partnerships in Projects

**Abstract** This chapter exclusively covers the Public–Private Partnerships aka PPP mode of contracting in projects. We first touch upon the existing models and literature on PPP and then move on to define a new conceptual model for PPP. In doing so, we begin by understanding PPP by way of analogies. After refining the basic concept, we conceptualize PPPs in the project management domain to enable a better understanding of how the framework is developed. The chapter next covers the key dimensions and challenges in four phases of PPP projects, viz., Environmental Readiness Phase, Initiation Phase, Realization Phase, and Post-Implementation Phase. In the Environmental Readiness Phase, we define the Bureaucratic Competence Framework for PPPs that often drives the course of the project. We also define a new Pricing Scenario Framework that incorporates the environmental readiness factors in PPP. In the realization phase, we define a new Quality Framework for PPP projects.

**Keywords** PPP · Defining PPP · Environmental readiness phase · Initiation phase · Realization phase · Post-implementation phase · Bureaucratic competence framework for PPP projects · Pricing scenario framework for PPP projects · Quality framework for PPP projects in realization

Traditional contracting has undergone a metamorphosis over the past few years. Today, especially in large-scale infrastructure projects, a paradigm of increased partnering is being advocated. One such paradigm, where a government agency comes together with a private entity, is the PPP or Public–Private Partnership model. This model has been the topic of discussion in most recent research papers on contracting. In this chapter, some of the aspects revolving around PPP will be discussed in greater detail.

The literature available has serious limitations as the perspectives show a heavy bias toward the faculties of project management, financial analysis and, looks at restricted macroeconomic drivers to understand these initiatives. Therefore, despite satisfying the conditions in their framework, they fail to provide any predictive and usable perspective; resulting in significantly different results at ground level from those in the literature. Moreover, the delicate balance between public interest and private gain seems to be violated in many cases. This is, therefore, an ongoing area

© Springer Science+Business Media Singapore 2017 341
N. Gurjar, *A Forward Looking Approach to Project Management*,
Lecture Notes in Management and Industrial Engineering,
DOI 10.1007/978-981-10-0782-8_20

of management research and many publications in this area are trying to 'nail' the various elements so as to ensure a more holistic understanding. All said, there is a lot of scope for improvement under the current knowledge levels and practice.

## 20.1  Introducing PPP

In this section, we will formally define PPP. Public–Private Partnerships, as we know today, comprise of a special mode of development in which the government and private parties come together to maximize the value provided to the society. It has been defined across literature in many different ways.

---

**A Quick Hands-On**

**What is PPP as we know it today?**
**How is it different from traditional contracting?**
**Why is it becoming increasingly important to research on PPP today?**

---

Charles [21], in his master's research paper has provided a few definitions that are worth looking into:

1. Paoletto (2000) defines partnerships as 'collaborative activities among interested groups, based on mutual recognition of respective strengths and weaknesses, working toward common agreed objectives developed through effective and timely communication'.
2. For Brinkerhoff (2002), partnership is a 'dynamic relationship among diverse actors, based on mutually agreed objectives, pursued through a shared understanding of the most rational division of labor based on the respective comparative advantages of each partner'.
3. According to Venkatraman and Bjorkman (2006), the common elements that determine partnership are 'beneficence, (–) non-malfeasance, (—) anatomy (of each partner), jointness (shared decision making) and equity (benefits to be distributed to those in need)'.

Kalidindi and Singh [22] have also described numerous modes of PPP in their research publication as follows:

PPPs, in the broadest sense, can cover all types of collaboration across the interface between the public and private sectors to deliver policies, services, and infrastructure. The term PPP refers to a wide range of arrangements with simple arrangement such as management contract on one extreme of the spectrum, while arrangements such as full privatization or divesture remain on the other extreme of the spectrum. Various approaches are in use to classify the arrangements between the two extremes of the spectrum. One of the approaches is to refer to the wide variety

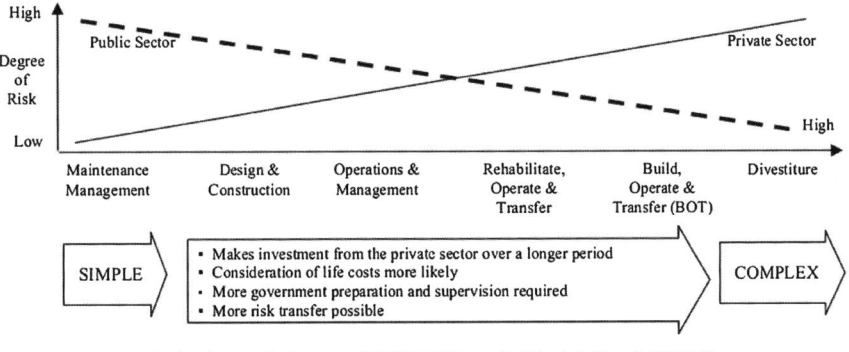

Ref adapted from ADB[2000] and World Bank[2004]

**Fig. 20.1** PPP spectrum as given in Kalidindi and Singh [22]

of arrangements based on the involvement of the private and public sectors in the various phases of the project life cycle (Pakkala 2002). However, the most common way of referring to the different arrangements is based on the extent to which the responsibilities and risks are transferred from public sector to private sector.

They go on to explain these with a figure adapted from the ADB report that is reproduced in Fig. 20.1. The figure has multiple modes that are defined across the spectrum. The key parameters in most definitions (and partnership documents) are identified as

1. Duration
2. Investment involved
3. Governmental Policy
4. Risks and Bailout.

Unfortunately, none of these dimensions actually allow due considerations for a predictive study.

Most of the definitions are vague. In fact, the Rajasthan Government, Indian Union, on their website, have provided with another definition for PPP [23]. This is an interesting definition as it covers both *What-Is* and *What-Isn't* PPP. According to them, PPP is defined as follows:

The Public–Private Partnership (PPP) project means a project based on contract or concession agreement between a Government or statutory entity on the one side and a private sector company on the other side, for delivering an infrastructure service on payment of user charges.

They also go on to clarify what-isn't PPP as follows:

1. PPP is not a simple outsourcing of functions where substantial financial, technical, and operational risk is retained by the public institution.

2. A PPP is not a donation by a private party for a public good.
3. A PPP is not the 'commercialization' of a public function by the creation of a state-owned enterprise.
4. A PPP does not constitute borrowing by the State.

The comparison of keywords between the different definitions reveals interesting levels of understanding of the concept. Paletto uses the term 'collaborate' that is defined by the Merriam-Webster dictionary as: To work jointly with others or together especially in an intellectual endeavor or to cooperate with an agency which one is not immediately connected. Brinkerhoff calls it a 'dynamic' relationship, where the word dynamic is defined as: always active or changing, having or showing a lot of energy, or relating to energy, motion, or physical force. He also uses the words shared understanding and comparative advantage. Comparative advantage is also seen in the other definitions. Paletto in fact speaks of timely communication. All the definitions seem to include a lot of strategic elements that are essential in the consideration of the PPP model.

## 20.2  A Better Understanding of PPP

To take this further, it is essential to understand how PPP differs from traditional contracting. The simplest way to understand this is using the power of analogies. Once one gets to understand the common thread in the analogies, one will have a better sense of the PPP paradigm and how the project manager must orient himself/herself toward the same.

### 20.2.1  Common Analogy

In our training programs, we often use the power of the analogy to help initiate a brain-storming session among our participants. Hence, it is advised to work on this analogy with a group of your colleagues or friends.

**A Quick Hands-On**

**How would one define the distinction between a housewife and a maid?**

Both the housewife and the maid perform pretty much similar types of tasks. They both perform household chores, yet they are distinctly different. The housewife is called a *life partner* and shares many more privileges than a maid. At the same time, the relationship also has a lot more mutual dependence. If one can understand the fundamental difference between the two, it would help understand the context of PPP better.

### *20.2.2  Another One*

We all enjoy our ways of working. And let us assume that there is a simple household that one is looking at. One is looking forward to renovating the house. This means a lot of furniture is to be made.

**A Quick Hands-On**

**What would be the difference between getting a carpenter to do the furniture and self making?**

In both the cases, the objective is to make the furniture, the process is similar. However, would the carpenter have the same status as a 'partner'? This is an important aspect to understand.

### *20.2.3  A More Complex One*

I am sure many of us have heard, seen, or even taken nutritional supplements.

**A Quick Hands-On**

**What is the fundamental distinction between a nutritional supplement and a medicine?**

We take to the web encyclopedia or Wikipedia for more details.

A drug is any substance containing a chemical which binds with a receptor in a cell membrane or an enzyme which produces some biological effect by altering the cellular functions as a result of that binding. It is usually synthesized outside of an organism, but introduced into an organism to produce its action. That is, when taken into the organism's body, it produces some effects or alters some bodily functions (such as relieving symptoms, curing diseases or used as preventive medicine or any other purposes). Note that natural endogenous biochemicals (such as hormones) can bind to the same receptor in the cell, producing the same effect as a drug. Thus, 'drug' is merely an artificial definition that distinguishes whether that molecule is synthesized within an organism or outside an organism. For instance, insulin is a hormone that is synthesized in the body; it is considered a hormone when it is synthesized by the pancreas inside the body, but if it is introduced into the body from outside, it is considered a drug.

It is a substance which is not food, and which, when ingested, affects the functioning of the mind or the body, or both. However, under the philosophy of Chinese medicine, food is also considered a drug as it affects particular parts of the body and cures some diseases. Thus, food does satisfy the above-mentioned definition of drug so long as ingestion of it would alter some bodily functions.

So, the reader needs to understand the distinction very clearly. The following is taken from a recent case paper from India:

> The word medicine means, any substance/chemical which when used in a disease will lead to cure or improvement. Dietary supplement means any substance (usually a food item/ingredient) which helps in maintenance of health.
>
> The clear distinction between health and disease is not there. What is healthy for a particular person/sex/ethnic group is a disease for someone else (for example, being slim is healthy for a model but is a sign of malnutrition for a pregnant woman or for doctors). Since the distinction between disease and health itself is not clear, dietary supplement versus medicine is also not clear in many circumstances. In some way, we need to draw a distinction for various reasons (for rigorous preclinical and post-marketing testing and for taxation purpose).
>
> If GLA-120 is used to treat diabetic neuropathy, then it should be considered a medicine. If for some reason it is used in normal healthy people (which it is not supposed to) as a tonic to improve the general well being, then it is a dietary supplement. If GLA-120 is a medicine, then the question still remains whether it has undergone the same rigorous stage I (testing in healthy volunteers), stage II (testing in diseased people), and stage III (randomized trials) clinical trials and whether its entire side effect profile and harmful effects are known in large group of people. If it is a dietary supplement, it may not require such rigorous clinical testing and it may be available over the counter without doctors' prescription.

This opinion of the court changes the overall mindset involved.

## 20.3   A Simplistic View

Simplistically speaking, much of the PPP literature relies on defining 'effective contracts'. If one has to contractually define a 'wife' and then, look for a fitment of the definition with the potential 'candidates' around, one would see a 'maid' qualifying for most of the terms and conditions desired (no pun intended !?! :-)).

The fundamental distinctions are the following:

**A partnership relies on doing things together. In other words, the fundamental premise is that both the parties believe and agree in doing the activity together. This means that at every stage there is operational involvement and support to both the sides so as to ensure that they are able to achieve the objective.**

**The second important aspect is that a partnership does not include those areas that involve getting those things done that we do not want to do. In other words,**

**as far as a maid is concerned, she does those tasks that the employer would not want to do by himself/herself. In the case of the carpenter as well, the employer does not want to do it. Rather, he would like to get it done through the carpenter. Though both might be doing things 'together', the choice of the activity is on those things that one agency would not like to do.**

**The most important characteristic is that the 'termination or separation' of a partnership is more expensive than those of traditional contracts. In other words, in a separation, a partner will have a share of the 'values or benefits' in addition to mere costs.**

After looking at these variants, the reader would probably be able to appreciate the fact that most projects and contracts in the PPP mechanism are misnomers! They are more of traditional contracts put under the 'garb of PPP'. This has many fallouts as we would soon see in the later part of the chapter. Traditional contracting models are more oriented toward the costs, the concept of having an outsourcing agency to do some task in the public domain and tries to give them more 'independence' and therefore, more 'responsibility'.

Traditionally, just as the world has changed, economic models too have been changing (in the context of PPP). For instance, today, one finds a lot of couples that have working partners. So, the traditional housewife is now transitioning to a working woman who earns money for herself, and in some case, her family as well. This could be compared to a strategic partnership variant to the case with the housewife. Here, the roles are shared between both the parties. On the other hand, in a bank, one is seldom interested in the way funds are applied. Hence, the relationship is more of a strategic stakeholder rather than that of a strategic partner. It is important to note this distinction. In a strategic partnership, *both* the parties would be responsible for both the strategy as well as the operation. However, in a bank, there is no partnership in the true sense. It is more of a 'stakeholder' perspective. Though the relation is more toward the strategy, the operations are independent. Hence, the reader needs to understand this perspective from the emerging trends in project management scenarios today.

## 20.4  Why Is PPP So Important?

The Project Management scenario has become more and more challenging today due to the following reasons:

1. The average cost of projects has increased over the decades
2. The business cycles have shrunk over the decades
3. Technology has given more options to explore into 'data' and make more meaningful decisions
4. The project execution methodology has changed from the traditional project method to 'bundled execution methods' like BOO, BOOT, etc.

5. The project environment has become more cross-bordered with expanding markets and geographies.

With increase in the value of the projects and shrinking business cycles, most projects today also include a piece of the operations to ensure a better balance between the ups and downs of the business cycles and enable a kind of risk sharing and also partnerships of the associated agencies involved. The PPP trend has, therefore, caught up in most countries.

No project manager can ignore this aspect. However, most of them are still trained in managing projects using their conventional approach to a PPP project. They look at the contract as an instrument of robustness. It is, often times, comparable to looking at ensuring that one has a great *key* without looking at the 'great car'. Hence, the entire premise is to ensure that the contract is honed to the specifications of ensuring that there is a 'wonderful clarity' on the roles and responsibilities from both the sides. When a project is a long-term commitment as in the case of the PPP projects, this approach fails miserably! For instance, in India, as of 2011, USD 7000 billion was the approximate value of the PPP projects that were ongoing in the country. Making a contract robust is like trying to specify the *perfect wife!* Every married reader would know that this is a utopian target!

Hence, as a manager, one needs to be clear of how the project is envisioned and how the basic premise of working differs from a typical traditional environment. We will cover these in the subsequent sections. In the world bank website (worldbank.org), one can find many case studies of PPP projects. Each of these also speaks of the 'value-for-money' of these projects. Despite everything put on the site, the fact remains that most infrastructures in developing countries still have huge problems in giving good quality services. Yet, the analyses speak of numbers that can often seem a little distorted from the reality.

## 20.5  Key Dimensions and Challenges

In any PPP situation, there are two basic dimensions that are interesting. They are (a) the Phases of the Project and (b) the Entities in the Project. We will look at each of these in greater detail.

As far as the phases are concerned, there are four basic phases that are of interest for the project manager. Each of these has a cascading effect on the overall 'Project' that is being considered:

1. *Environmental Readiness Phase* In this phase, the basic ground work has to be clear. It is a pre-PPP phase, where one actually checks and verifies if the environment has the necessary infrastructure to accept and use a model like the PPP project. Depending on the environmental readiness, the overall project conditions would change.

For instance, availability of financing agencies for long-term projects is a crucial aspect. If there is no clear agency available, the overall project would need to raise funds from international agencies. These have their own challenges and considerations.

2. *Initiation Phase* This phase involves the actual initiation of the project. It involves the precursors of the projects (typically governmental schemes), posturing, and the overall tendering process. The phase continues till the kickoff meeting post award of the contract/partnering agreement.

3. *Realization Phase* This phase involves the actual execution of the 'project' in the overall 'project'. It looks at the realization of the essentials of the project.

4. *Post-implementation Phase* This phase involves the operational phase post completion of the 'project execution'.

The other key dimension involves the entities in the situation:

1. The Agency/Partner
2. The Competition
3. The Political Pressures
4. The Legal Pressures
5. Conditions from the Funding Agency
6. Conditions of the Implementing Agency
7. The Contract
8. The Scope of Engagement
9. Fair Practices
10. Miscellaneous Entities

As one can see in the list, the contract is a 'miniscule' part of the equation. The overall picture, therefore, gives a large number of challenges in the various phases. We will cover these aspects in the subsequent discussions.

## 20.5.1 Environmental Readiness Phase

The principal challenge in the environmental ecosystem is that of knowledge or *bureaucratic competence* as far as the PPP model is concerned. Most bureaucrats continue to treat the 'partner' as a 'contractor'. They also fail to understand how a partnership works. Rather than being a transformation from traditional contracting, the bureaucrat might be treating this as a renamed version of traditional contracting. This has far-reaching consequences in the subsequent phases.

The overall scenarios are highlighted in the Consulting Connoisseurs Bureaucratic Competence Grid Model shown in Fig. 20.2. The different bureaucratic competencies actually reflect as different styles, depending on how the bureaucrat is oriented toward power play. The reason why the power play aspect becomes important is that many bureaucrats fail to appreciate that they need to learn, owing to their comprehension of how power play works. For instance, even if the bureaucrat does not have the

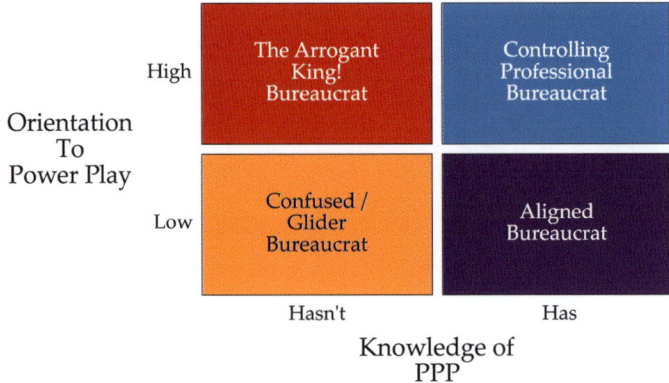

**Fig. 20.2** Consulting Connoisseurs Bureaucratic Competence Grid Model for PPP Projects

knowledge, he would continue to keep his 'upper hand' if he wants to be 'high' on power play. This is a rather difficult situation.

The second biggest challenge is the *availability of market players.* Most times, there are very few market players that are interested in a specific kind of PPP project. This also depends on the criteria used to determine the suitability of a market player and the admissibility of exceptions to the criteria. In many cases, newer areas of exploration often lead to very little experience from both the parties, resulting in a total experimental approach. This is a rather dangerous obsession and the project manager must understand how to steer clear from such situations without an abnormally high exposure.

The third challenge that is documented today is the *clarity in the definition of the operating model.* Most times the operating models are not very clear. Therefore, the packaging gets divided into subsequent elements. This is similar to the 'housewife versus maid' trap. It is similar to the forms of truth with the elephant that was discussed in the risk management section. Most times, the government has a clear direction and SOP-based methodology when it comes to self-performing the project. However, this is seldom realized in the execution stage because a lot of 'overhead' costs are not factored in correctly. This is where the definition of the operating model becomes important. Furthermore, quality checks and their correlations with payouts are often a bigger issue.

The last challenge that is crucial is the *stability in the development mechanism.* It is often seen that the bureaucrats do not have a proper understanding of the development mechanism. In other words, there are many instances where there have been price revisions (usually upward revisions) because some factor had significantly changed. Any such change is usually more expensive to the end user than to the agency, but it does not help anyone else other than the contracting agency (that too …to a limited extent).

**A Quick Hands-On**

For the project manager, each of these parameters would yield a different kind of an opportunity or an impediment, depending on how the overall project environment characterizes itself. In doing so, there are two basic issues that the reader needs to understand.

**How detailed should the plan be?**
**How should one determine the adoption and the *transition* from one strategy to another?**

While these issues are often critical and relevant, they do not always feature in the project manager's perspective of evaluating the environmental readiness. In fact, most times, the focus is restricted to a 'Tender/RFP/RFQ' that is released and the success factors associated with the award of the 'Tender'. Therefore, the typical strategies used by the players are oriented toward the type of contract agreement that is attempted at being entered into. The main determinants of the strategies are:

1. *The level of maturity of the environmental readiness factors*
   Each of the factors would have varying influences on the perception and the maneuverability of the project manager. There are two broad categories: High or Low, depending on how one would evaluate the outcomes of the assessment. The more rigorous and clear the evaluation and the prediction of the 'expected future' is, the better is the handle to understand and negotiate a balanced contract.
   However, if the maturity is found to be on the lower side, it would be a 'strategic' discussion with very little insight into the 'driving' realities of the future. In other words, the uncertainty component actually increases in this case.

2. *The nature of the contract*
   Traditionally, long-term contracts have always had their set of challenges. And it is usually a question of negotiation when it comes to identifying the actual factors that would and should truly drive the change management process. In long-term contracts, this has far-reaching consequences. For instance, if one talks of a toll bridge project, a change in the toll rates would affect multiple factors. Therefore, one needs to understand and identify these handles. Now, while defining this as a determinant of the strategies, one is trying to evaluate the kind of parameters that are used to trigger this kind of a change. These factors could be global or local. For instance, if one were to use a global oil price index to adjust the toll rates, it would be a global factor. However, if one were to look at domestic cement price at the time of construction, it would be a local parameter. In other words, a local factor gets directly factored into the cost equation while a global factor needs to be derived using some management logic that may or may not be reflective of the true reality. An astute bureaucrat would always strive for good local adjustment triggers over global ones.

In the current context, our interest is in determining the potential of a contract to deviate from a 'pure' form of a *fixed firm price* contract. In other words, the parameters that are used to trigger, their relative sensitivity and the ease to implement the 'change' in the economic premise of the contract. Historically, global factors were always preferred to individual ones as they have been believed to help average out the nuances of more direct measures as well as the fact that they have been considered as premises that were 'beyond' the control of both the parties involved.

In any case, the number of such variables and the nature of these variables are what could be categorically classified as either 'global' or 'local' variables.

We now introduce the Consulting Connoisseurs Grid Model of Broad Pricing Scenarios for PPP considering the environment readiness factors. Figure 20.3 shows the possible broad pricing scenarios in the combinations of these two parameters. If the environmental readiness factors are not too mature and the contract handles are localized, then the project is actually a very risky deal for both the parties. What it essentially means is that the project itself is a black box which no one has real insight of. This mode is a risky one and, often times, to cover the risks, the contract would be awarded *with a huge risk factor*! Depending on how both the parties choose to 'swing', the known unknowns result in a huge premium in the contract price finalization, in which case, the actual problem in the contract, reduces to one in which the private entity is gaining more than the public value it provides. However, if both the parties are keen on the contract award, this could also result in an extremely risky 'undervaluation' of the project costs, in which case, the private agency would bleed in the initial years before a massive change in the contract is initiated and executed. Therefore, in this quadrant, there is a risk of the outcome going both ways.

In the second quadrant, when the parameters are global and the maturity is low, the private entity has more means to express gains and would try to 'fortify' its own profits. In other words, there are now more means of 'modifying' contract conditions and getting profits through changes that are external. So, this kind of a contract would

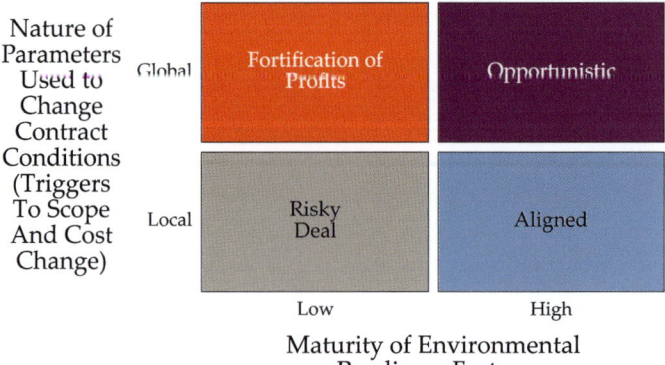

**Fig. 20.3** Consulting Connoisseurs Broad Pricing Scenarios considering environmental readiness factors

invariably result in a higher price of the contract. In other words, there is going to be a lot of extra money that will be shelled out when compared with 'self-performing'. Instead of leveraging from the size/value of the contract so that the margins are reasonably controlled, the tendency would be to safeguard the profits in a higher percentile of cases than normal, due to the fact that the changes in the contract will not be directly related to the local factors that are more 'realistic'; rather due to some global parameter that may or may not be translatable to the current situation. Hence, the tendency in these cases is that the costs are much higher than required and it is only the profit motive that drives the overall contract situation.

In the third quadrant, when the maturity levels are high and the factors are local, the contract is typically very well aligned and realistic. Only in these cases would one truly have a contract that is comparable with 'self-performing'. In other words, both the parties are clear on what they are entering into and understand the key drivers that need to be considered along the entire project cycle and have an efficient mechanism to address the issues that could arise in future.

The last quadrant where the maturity levels are high and global factors are being considered is the one that has an opportunistic pricing mechanism. In other words, the tendency would be to gain more from this contract. However, due to the maturity levels, it would not be a 'windfall' gain as with the other cases. Rather, it would be a case where there exists an opportunity that has potential, although limited in many ways. Hence, the contract structure leans toward an opportunistic posture for the private agency whereby, the agency is looking for a means to look at possible premium.

## 20.5.2  Initiation Phase

The initiation phase is a far too complex one in reality. The reader needs to understand the *constraining factors and the specifics of the situation*. While a major part of the constraining factors does come in from the environmental readiness domain, there are several factors in the initiation phase that exert their influences on the outcomes of the project. In this subsection, we will look at some of these in greater detail.

In most countries, PPP implementations are administered under schemes. The first factor of significance is the understanding of the basic principles that drive the PPP schemes. In other words, such schemes are designed and defined to ensure the implementation of the most appropriate option. For instance, in India a lot of PPP projects have been defined under the JNNURM scheme. The scheme acts as a guiding principle to the fundamental implementation structure of the PPP model. These schemes are usually 'hard-coded' and are prescriptive in nature. An agency (government department) that initiates the project, therefore, is bound by the premises of the scheme. Often times, one needs to check the scheme for completeness and potential errors. These considerations manifest themselves as two other factors, viz., the extent of adaptation permitted and the ability to factor in the learning. As the scheme rolls out, corrections for 'local conditions' need to be considered and incorporated in the right perspective. If the basic principles are: (a) missing or not elaborated, (b) unclear, or (c) discontinuous in their logic, the scheme would tend to begin with a handicap to start with. Again, if the contract cannot incorporate issues based on the learnings from previous applications, it would once again reflect in serious issues for the administrators. This is, most often, a case of the policies and the way the flexibility is perceived and implemented (in practice) in the specific government agency.

The most challenging issue in the initiation phase is that of corruption. While corruption has been accepted as commonplace in most countries, it severely skews the entire balance between public value and private gain. While many economists tend to describe this as a cost to the country, the implications are far beyond the 'pure costs' in a project. The reader needs to carefully understand and evaluate this parameter while modeling the project scenarios.

Lastly, in the initiation phase, the focus of the key bureaucrats as well as the leadership of the private entities are important considerations. Their individual appetite toward both the risks involved and their focus on complying with guidelines is a critical measure. Usually, a lot of contractual discussions are based on these two parameters.

### 20.5.3   The Realization Phase

The realization phase is typically the 'physical realization phase' or the 'asset building phase'. However, since the project is not being executed in isolation, each critical factor needs to be studied in the context of the *relative* influence of the environmental factors. After the initiation phase, the realization phase has two main challenges from the perspective of the private agency:

1. The Quality of Implementation
2. The Protection of the Business Case

For instance, most road projects have been delayed due to land acquisition issues, right of way issues, lack of geotechnical information, etc. In such cases, when the delay is within a threshold, the private agency tends to 'bleed' as they are required to absorb the 'shocks' in the project. In other words, the difficulty in hip-hopping along the various areas is not a factor that is generously considered by the public entity most times; essentially due to the fact that they continue to think of the PPP model as an extension of the traditional contracting model. Under these circumstances, the private agency is often forced to adopt 'cost cutting' measures to enable holding the fort in the interim. Some of them include looking for alternative solutions to maintain the servicing of the requirements, changing the engineering specifications to 'lower standards', or even compromising the quality of the implementation in the same standards. Once the delay exceeds the contractual thresholds, the project business case gets suddenly altered to incorporate a cost escalation. But until then, there is a high likelihood to have altered situations in the project.

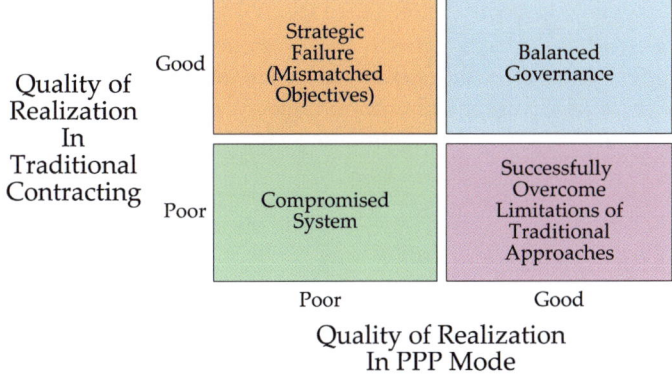

**Fig. 20.4** Consulting Connoisseurs Quality Perspective based Grid Model for the realization phase

The evaluation of quality during the realization phase is done using the portfolio shown in Fig. 20.4 as the Consulting Connoisseurs Quality Perspectives Grid Model. If the quality of traditional contracting approach is superior to that of the PPP approach, it would actually be a compromised initiative. Such a PPP project is likely to face many life-cycle cost adjustments. If both the quality of the traditional contracting as well as the PPP model are poor, this is an indication of a compromised system. In other words, corruption and poor management practices are plaguing the system. If the quality of the PPP realization is far superior to that of the traditional contracting approach, though it would minimize the overall life-cycle costs, the competitive forces put a lot of pressure on the profitability of the project. In other words, the latter two would potentially require a change in the business case assessment. The ideal situation is when both the traditional and the PPP modes have the same qualities; both of which are good. Such a situation is that of balanced development.

The bigger issue is the historical development of the quality situation. If a project starts having problems that are linked with poor performance, it is difficult to set it right. For instance, if the bitumen/asphalt layer is laid too thin, a rework would cost the private agency (usually) a lot of additional costs. This could be triggered by the fact that the private player wants to safeguard the profits in the event of an unfavorable event that is forcing the reduction of the cost base. Once 'done', it would be difficult for the private agency to redo the work even if it gets a cost-adjustment amendment later in the construction case. Hence, the historical development often clouds the conceptual clarity that governs the normal decision process.

Like the initiation phase, the realization phase too is plagued with corruption issues in most PPP projects. In the parlance of PPP, the factor of corruption rephrases the model as *personal gain before public interest or private gain*. While most countries do face these problems, cultural influences often overshadow the decision process and the ability to do justice to the business model. A 10% increase in the project realization costs could impact the IRR by as much as 2.5% depending on how and when it occurs. Hence, even if the governments tend to 'ignore' the aspect of corruption as economically trivial (1–2 % of the GDP), the impact from a PPP project perspective could be enormous. Hence, it is necessary for the governments and the private entities to recognize this aspect and ensure that it does not skew the results of the project. In fact, corruption is also identified as one of the largest reasons why the quality of PPP projects gets compromised and why many of them fail.

### 20.5.4  Post-implementation Phase

The post-implementation phase is the operational phase in the life cycle of the PPP project. This phase is interesting because it actually puts the subsequent 'operations' into the ambit of 'project management'. And if one were to stick to the traditional thought process that tries to demarcate the two faculties, it would call for multiple layers of advanced skills.

The post-implementation phase has several characteristics that are interesting:

1. It is often in this stage that there is culmination of the effects of errors during various stages of the project life cycle; in that, it becomes visible and evident. In other words, the project sunk cost is incredibly high and, most times, irreversible.
2. The traditional focus of projects that restricts itself to the 'project completion' becomes a misplaced priority! In a PPP project, the deemed completion has little significance, due to potential ramifications on the profitability of the subsequent operations.
3. The worst 'nightmare' for the traditional project manager is actually the absence of a 'handover' where someone else is to inherit the 'headache/baby'. On the contrary, it is like the ghosts of the project's past that come back to haunt the company by severely impacting the bottomline. In other words, one does not have the privilege to *pass the buck* to the operations as both are internal to the company.
4. One of the more important aspect in the post-implementation phase is that the risk exposure of any company is expected to start reducing once the project is in the post-implementation phase. However, due to the longer duration of the contract, it is not always easy for a company to ensure that their risk exposure is contained. In many cases, the risk containment starts getting attached to certain contractual modifications that may not always be immediately approved. Hence, exposure control is a critical parameter. For instance, changes in toll charges are, often times, seen to affect the 'populist' image of the government. Due to this fact, there is a tendency to delay the changes. This has its impact on the overall project profitability.
5. The biggest change is that the account management cycle in PPP projects is longer and, most times, not a one-time earning. In other words, instead of having a good big pay-in at the end of the realization phase, the pay-in comes in piecemeal flows extending over the next 10–20 years. This changes the overall sense of account management for both the agencies.

In the world of modeling and simulations, however, the differences between traditional projects and PPP projects are *downplayed* as cash flows are just 'parameters' in the overall scheme of the framework. That being said, a modeling and simulation approach has a mechanism to provide with seamless integration of the project phase with the subsequent operations phase. Traditionally, the use of modeling and simulation has been elaborate in the financial domain. However, the versatile nature of the technique makes it easily amenable to applications in the operations domain as well.

## 20.6  Key Takeaways

In the present work, a holistic approach is adopted to understand the entirety of PPP, bringing in multiple perspectives from the faculties of macroeconomics, governance, specific deliverables (projects as well as operations), systemic dimensions, legal studies, partnering principles, and political sciences. To apply these parameters it was important to evaluate PPP initiatives from a phase perspective and analyzing them, thereafter, for first-order challenges from each of these faculties. The current publication explores this methodology by dividing the PPP cycle in the following four phases, viz., (a) The Environmental Readiness Phase, (b) The Initiation Phase, (c) The Realization Phase, and (d) The Post-Implementation Phase. After characterizing these challenges, an evaluatory framework is proposed to assess the various relationships between the entities in a typical PPP framework. The evaluatory framework provides us with four critical dimensions, viz., (a) Appropriateness of PPP modes for the given situation, (b) Potential modes of failure for a given case, (c) Areas that need to be addressed, and (d) Management tools that would be recommended to augment the needs in each case.

# Chapter 21
# Key Mantras in PPP Projects

**Abstract**  This chapter is, again, a unique compilation of the Key Mantras in PPP Projects. In all, five mantras are discussed and described for the reader. The first mantra is on factoring the experience of the partnering agency (notably the Public Agency) in PPP Projects. The second mantra speaks of how one needs to incentivize a PPP Model in the contracting. This again is a different paradigm from the traditional approach. The third mantra provides room for a new concept called Independent Contracting that is very essential in PPP projects. The fourth mantra talks about Integration and the Balance of Trust in PPP projects between both the parties involved. The fifth and the last mantra talks about corruption and its impact on PPP projects and how these need to be circumvented.

**Keywords**  Mantras · Experience of public agency · Incentivizing PPP models · Independent contracting · Independent specifications generation · Integration · Balance of trust · Corruption in PPP

The success of PPP contracts has been function of various parameters as seen in the preceding sections. However, it is amply clear that despite these variances, there are a few mantras/basic strategies that need to be followed. In this section, we will touch upon these basic strategies.

## 21.1  The Experience Dimension

Oftentimes, it is observed that the governmental agency has little experience of doing a project that they intend to do using the PPP mode. In such a situation, one often sees skewed pricing in the overall project. Most road projects, for instance, are based on a good projection of the traffic they would carry. It is, however, observed that most of these projects go in for course corrections in terms of the monetary equations involved. The common problem cited is that of poor projections. However, this is a more serious issue than the projections alone.

© Springer Science+Business Media Singapore 2017                                         359
N. Gurjar, *A Forward Looking Approach to Project Management*,
Lecture Notes in Management and Industrial Engineering,
DOI 10.1007/978-981-10-0782-8_21

1. In most cases, instead of revising the durations of collection, the DoTs often resort to modifying the rates of the collection. This 'northward' movement actually creates the potential for windfall profits for the players.
2. Despite everything done, it is also believed that the projections are 'never good enough' for what they were supposed to be. Hence, there is a constant bias in the business case. Unfortunately, this bias often works to the advantage of the lowest bidders who later 'adjust' their prices and become more expensive than the others who had lost the tender because their factored prices were higher.

In India, in the state of Gujarat, there was a drive to promote solar power. Now, the government came up with an incentive model that enabled private players to literally take full benefit of their technologies. In fact, the incentive of the produced electrical unit was so high that most installations were making great deals of profit in these contracts. This is the other extreme end of the experience dimension.

To avoid an overly high or an overly low 'profit' scenario, it is necessary to understand that one does not go ahead and define the PPP contract in haste. In other words, it is necessary to be realistic without being a party to favoring the 'wrong/crony mode of capitalism.' In other words, what is truly essential is to create a kind of a threshold.

The easiest way to do this is to gain *first-hand experience*. This has two fallouts. It expects the user to spend time in knowing the dynamics of such projects. At the same time, it delays the projects in order to gain meaningful experience before going down the PPP route. Proponents of this approach claim that it is the best way to avoid inappropriate exploitation from the private partner. At the same time, the opponents of this approach claim that smaller version pilot models that would be used to get the first-hand experience would probably be very different from those of the larger version models. In other words, the dynamics of the scale are always difficult parameters to assess and to control the extent of 'private gain' that these contracts could give.

The other important aspect is that the repayment of the project to the private entity is through the 'public value' that the project is expected to bring with itself. Traditional projects do not often go into the speculative mode to quantify the amount of public value. This is one of the pre-requisites in the PPP mode, and despite having the experience of the project, the governmental agencies seldom try to quantify the public value through a phased monitoring approach. This is largely because the governmental projects are treated to be a 'pool' from the taxpayer's money. Hence, there is no significant effort normally undertaken to enable the calculation of the actual number of beneficiaries from a given project. This is particularly true in projects where there is no specific monetary mechanism directly linked to it. For instance, a bridge constructed by the government could have a subsequent toll collected. Under these circumstances, there is a mechanism to monitor and track the usage. However, if the government decides not to have any toll or charge levied, there is virtually no mechanism to estimate the number of beneficiaries.

Simple economics says that the *Price* has a relationship with the *Quantity Demanded*. Hence, the same service would have potentially different demand patterns if the service/benefit would be chargeable or not chargeable. In the event one is not able to ascertain the quantity demanded appropriately, there are repercussions.

### 21.1.1 The First Mantra

The discussion above actually helps us understand the first mantra viz.

**When no experience is available, create a threshold before advocating the PPP Mode.**

The threshold should clearly help one understand the economic relationships as well as give a first hand experience. In other words, it is necessary to have an experience that is representative enough of these two parameters.

## 21.2 Incentivizing a PPP Project

Most PPP projects have a fixed frame of business. For instance, a particular system may be designed for the benefit of 10,000 households. However, when the system starts getting overloaded with time, it might have to cater to 20,000 households. Recognizing the PPP agency as a separate business entity is important, and more often than not, this is also contractually done. However, giving it the flexibility to leverage from the changing conditions is an important aspect to understand. Most PPP contracts fail in achieving this. To elucidate this in a better way, let us look at a case study.

### 21.2.1 A Case in Point

At a municipal corporation, a PPP tender was released to enable operators run specific bus routes on a partnership basis. The government agency was to provide them with the land for their maintenance and the terminal for the buses while the private agency was to maintain and operate the fleet. The key metric used in evaluating public transport is the *cost per kilometer*. This perspective was extended to the operator using three different conditions:

1. Royalty per Kilometer operated
2. Blanket Penalties
3. Fixing Service Conditions.

These three conditions are actually not aligned with the PPP model per se. Let us look at a partial list of recommendations based on an alternative perspective that we provided as a part of our engagement with the municipal corporation.

1. *Number of Trips*
   The tender specified the number of trips required to be made by the operator. This was an implicit disadvantage for the government agency as change in passenger requirements would mean expensive change orders and perceived lower service

levels. Same was the case with buses. The tender specified the number of buses required in the fleet of the operator.

**Recommendation:** To make them as the *Mandatory Number of Trips* that is the bare minimum. A time schedule/time table was needed to be attached. The number of buses specified was dropped.

2. *Treatment of Fuel Costs*

The tender specified the fuel costs were in the operator's scope. This had an implicit disadvantage for the operator as the fluctuation in fuel prices had been significant over the past 10 years. This would, therefore, result in cumbersome price escalation/adjustments. What is worse is that there could be an *undue advantage* in the process.

**Recommendation:** Fuel costs need to be treated separately. This will ensure a better response to the tender. Operator can pay the government agency for fuel as part of the agreement with a fixed surcharge (as the government agency used to get fuel at a negotiated rate with the fuel companies).

3. *Traffic Operations Management*

The tender did not specify the use of GPS to track operator performance. In order to have better quality of services, the government agency may implement a GPS-based system for the traffic operations. The government agency was already gearing up to the implementation of GPS for their remainder fleet.

**Recommendation:** Operator was to implement GPS-based tracking system for traffic operations. This could be used as a pilot/test facility to extend across the remainder (entire) of the government agency operations.

4. *Route Information PIS (Public Information System)*

The tender did not specify the basic infrastructure for PIS for operator's buses. In order to improve the usability of bus services, route-wise information, and tentative timings should be provided at each bus stop of the operator. This would help develop a pro-active PIS-based system.

**Recommendation:** Operator to provide route information (schedule, timetable, fares) at each bus stop for their routes.

5. *Royalty Calculation*

The tender required payment of Royalty on a per kilometer basis. This had an implicit disadvantage for the operator as it did not provide any incentive for greater profit in operation. Further, certain elements like fuel need significantly higher level of risk-hedging that would add to the operating cost of the operator.

**Recommendation:** The tender should split the Royalty as a Fixed Charge + Fuel Cost + Fixed Charge per Bus.

The last point is depicted in Fig. 21.1. Clearly the split actually helps the private agency *gain more only when the public value it delivers is higher*. In short, it creates a win–win for both the agencies.

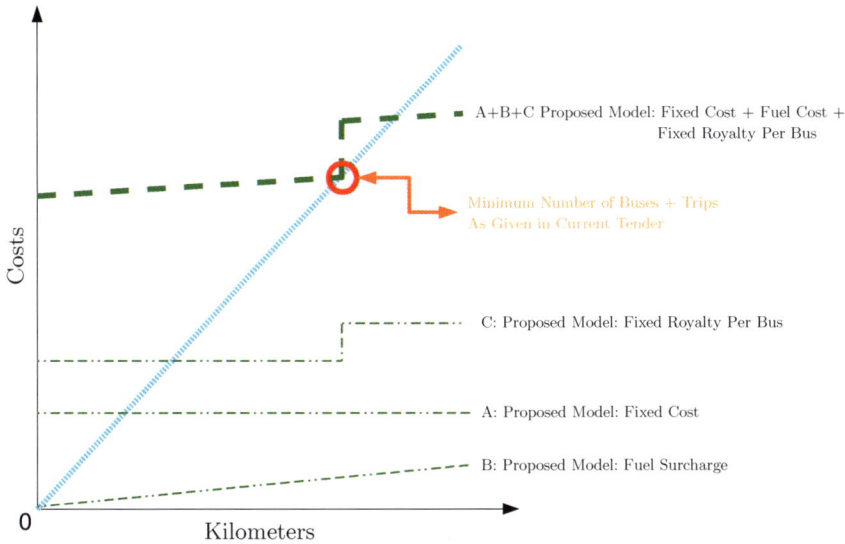

**Fig. 21.1**  Incentivizing the royalty calculation

## 21.2.2   *The Second Mantra*

The moment one gives the 'complete' status to the business entity formed, to operate as a full-fledged economically savvy business, the fundamental premise of the PPP project changes. In the above example, we have seen that the earlier version of the tender was more of an outsourcing model. Including the partnering principle has helped incentivize the project.

Hence, the Mantra:

**When a benchmark is available, define the profitability as a proportional function to the company/agency performance.**

To make a project successful, one needs to ensure that the contractor derives his gains using both *his efficiency* and *his ability to leverage on the scale.* The ability to leverage on the scale would imply that he is able to serve more customers, in the case of market changes; or even go beyond the general ambits of a fixed service contract. Thus, the excess revenue generated would yield excess profits for the same assets that would be used.

Without incentivizing a contract, one actually ends up losing the perspective that the PPP tender is all about maximizing public value and also private gains. In other words, if a contractor has a cost structure of 80 % costs in his 100 % price, the current concept of tendering actually puts the onus on contractor efficiency. In other words, a contractor would earn if he 'meets/fulfills' the tender requirements and is able to convince a value proposition. In other words, it is a battle between 75 % versus 80 % versus 85 %. However, when one says that it is possible to stretch upto 130 % of the overall revenues/price, the battle is a different one. Incentivizing does precisely this particular aspect. It allows the PPP entity to independently design and optimize a portfolio to a set of boundary conditions.

## 21.3   Unorthodox Views on Independent Contracting!

In many governmental institutions, the idea of going for a PPP project is to bring in some new systems and ideas. This is a logic that is strongly supported by the research done in considering PPP as a viable option. However, when it comes to contracting, the entire logic changes! In fact, the resulting condition is, what I refer to as *independent contracting,* due to a lack of finding a better term for the same.

What happens in independent contracting? In independent contracting, the directive is to try to implement a proposal on the PPP mode. It typically follows the steps given below:

1. An initial idea is developed with the help of the bureaucrat.
2. This idea is then verified for its feasibility. This may or may not be done rigorously.
3. A directive is passed in the governing council (house of representatives) approving a detailed project feasibility.
4. Typically, an agency is lined up for doing this work. The role of the agency is to conduct the DPR and optimize the implementation.
5. The report is then discussed and approved in the house/governing council.
6. Next, an agency is lined up for the tendering and the subsequent supervision.
7. Since this is a 'new' implementation, the typical tendency would be to finalize the PPP as an 'independent' project. In other words, independent contracting.

There are many reasons why the independent contracting paradigm is popular:

1. Since the project is implementing something very 'new,' the tendency is to *keep it simple and avoid complications* by making it independent of the existing setup.
2. Most times, the governmental agency is interested in the *proven track record phenomenon.* The application of this phenomenon usually implies that the implementation agency is one that may be operating overseas. The resulting procurement mechanism becomes fairly complicated due to the distribution of liabilities and the local partners that are involved. Hence, an 'independent' mode is preferred.
3. Finally, the implementation requires complete engineering of a system. Today, the technology curves are so dynamic and fast-changing that it is a practical constraint, most times, that makes independent contracting a more lucrative option.

The effects of these situations are simple to understand. Typical PPP Projects are argued and justified as if they are *Greenfield Ventures* rather than an upward extension of the existing system. This might give a good handle on the 'private gains' expected out of such ventures, but it definitely puts huge questions on the 'public value' involved. Any Greenfield activity actually needs and includes huge costs in terms of 'getting the site ready'. These could be avoidable costs in most cases. In other words, independent contracting actually increases the costs of the resulting/intended services, thereby lowering their potential in 'public value'.

While the overseas players might have very good know-how, there is limited experience of the players in the domestic market, thereby making the criteria of the proven-track-record literally go for a toss! This is very true, especially in the developing countries. Even in the developed nations, the proven-track-record needs to be understood well by incorporating the variances in the local conditions. Therefore, if the ecosystem is known, it would be a matter of safeguarding the profits of the private agency. If the ecosystem is unknown, there is a likelihood of the agency increasing its price to cover up for potential unincorporated costs downstream.

Finally, in order to save time and leverage from the proven track records, the implementation agency often tries to *copy-and-paste the specifications* from other tenders. This is an observed trend and most PPP projects are plagued with lengthy procurement cycles due to inordinately long technical clarifications in the specifications involved. Like other governmental contracting projects, there are also cases of over-specification that might favor a particular vendor more than others. While this may not be uncommon in traditional procurement, the implications in PPP environments reach out through a much longer timeframe than the traditional procurement model.

## 21.3.1 The Third Mantra

Due to the various reasons given above, the resulting specifications and the projects are, often times, a total mismatch with the reality. This problem, therefore, needs to be addressed at all levels. As a preliminary checklist, one needs to understand the following aspects:

1. The rigor of the feasibility study that is expected in terms of both time and cost.
2. The characterization of local conditions and the comparative assessment of how these differ from the existing implementations elsewhere.

Both these aspects give us two different mantras that are being clubbed here:

**It is always advisable to have an independent specification generation activity instead of an 'independent contracting' activity.**
**While going for independent contracting, it is always advisable to go for the RFP model rather than the RFQ mode of tendering to ensure a fair chance to incentivize the business case.**

The flexibility and the falsely perceived notion of 'accelerating projects' in the case of independent contracting force one to think that it is a powerful method to advocate. However, in practice, it is often a bane rather than a boon, because it has little relation with reality.

## 21.4   Ensuring Integration

The next mantra is an extremely complex and an important one to understand for most. In our earlier treatment, we had mentioned that a partnership needs to identify and differentiate between outsourcing, traditional contracting, and PPP. We tried to understand this through a few examples (maid versus. wife, etc.). We will now see where the systemic issues come into play.

### 21.4.1   Contrasting Philosophies

Just imagine the case. It is difficult for a marriage to survive its test if the husband and wife have strikingly different views. What's worse that could happen? Imagine a condition where they both have strikingly different values! Nothing more disastrous, right???

According to the 7S Framework of McKinsey and Co, Shared Values are an important aspect to the success of an organizational entity. In the case of PPP, this is an amalgamation of two entities that come together for a specific reason viz. the public entity and the private entity. Clearly the shared values principle is violated in any normal kind of a PPP arrangement.

**To understand this aspect better, one needs to realize that the PPP project is stressful for the public entity because its own organization is based on an intrinsic distrust.**

In other words, the public entity designs its systems and its checks and balances based on these principles that translate as given below:

1. Never trust an individual
2. System should take care of its issues (rules and more rules)
3. Two people can not be unfaithful to the system at the same time
4. Accountability virtually ends at the officer level (i.e., doesn't translate down)
5. Layers of checks and balances to ensure there are no loose ends whatsoever.

This results in a lot of overhead costs for the same activity. While on the one hand, the beauty of the system is in the way it manages, it also slows down the overall processes.

And in the PPP Project, the posturing of the public entity needs to be different.

**From a distrust-based value for their own management, the entity now has to demonstrate excessive trust on the PPP private partner agency.**

In short, the outcome is always disastrous, if one considers the value perspective. In other words, if one is subject to an environment that closely (at least technically) scrutinizes the work, and all of a sudden, this entity is set free, we would have nothing short of a *Roman Holiday* in store!

Therefore, ensuring appropriate integration is key.

## 21.4.2 The Fourth Mantra

In order to ensure that your PPP model truly leverages on the partnership angle as well as maximizing the overall gains, the following mantras are to be used:

**Ensure integration to leverage strength**
**Ensure integration to protect the business case and**
**Ensure integration to maximize values.**

Each of these aspects is performed in different ways and there are too many variables to consider. However, they are all looking at different levels of integration. From a project management perspective, this actually means multiple rounds of business case modeling and simulations, as well as an elaborate mechanism to model and simulate strategic partnerships. Since each of these exercises is extremely elaborate, we would restrict ourselves to the understanding of potential issues involved. As a project manager, one needs to go through these rounds if one wants to ensure that the PPP project is successful.

**A Quick Test**

As we know the framework is too elaborate to discuss here, let us at least understand a few points that might force one to consider a detailed treatment. I am presenting a small questionnaire/check-list that might be useful in ascertaining the facts.

1. On a scale of 1–10 where we have 1 (very easy)-10 (extremely difficult), how would you judge the ease of terminating the PPP project (in terms of expected cost and time)?
2. If the demand on the PPP project increases by 100 %, how could the project work out to be (in terms of service level to the public at large)? Would there have been a cheaper and more appropriate alternative than the one chosen now?
3. How many members of the public entity will be operationally involved in managing the facility full-time?
4. If the price of the service to the people is reduced by 5 %, how much would it affect the viability or the profitability of the project?
5. Under what circumstances, other than price adjustments, would the project earn more profits than projected? Who, other than the private agency, are the beneficiaries of such unnatural gains?
6. What specific strengths of both the parties are used in a given project? How do they translate into the execution? Is there a better alternative?
7. Is the project incentivized to ensure that both parties are aligned to the same thinking?
8. Is the management control uniformly distributed among both the agencies at the operational level? (Normally PPP projects have just a steering committee that has little operational control).

These questions would give one a general feeling as to how the overall integration is. In particular, like we have discussed incentivizing of the overall project, we could also discuss about incentivizing the integration. This is rather tricky, but feasible, and requires expert consultants to derive the same.

## 21.5  Curbing Corruption

One of the biggest problems in working with governmental entities/institutions is that of corruption. Corruption, however, comes in different forms in the PPP projects. It has varied effects on the project such as

1. Inflating the Cost of the Project
2. Stripping the Revenues of the Project
3. False Reporting
4. Compromises in Quality
5. Misappropriation of Funds
6. Compromizing the Business Model
7. Delay in Bureaucratic Procedures
8. Vested Interests that are against the Public Needs, etc.

As a project manager, one needs to understand that corruption might be *considered to be a necessary evil* by many professionals across countries. Over the past few years, there has been enactment of complex legislative machinery and systems to curb corruption. However, the low rate of conviction, coupled with the long periods of trials in most countries have hardly made a dent in the existing practices. In developing countries like India, pay-offs are in the range of 20–40 % and the concept of *passive income* is gaining a wider acceptance globally.

Despite all these aspects, PPP projects are particularly sensitive targets to the issues of corruption. Moreso because the population at large, does keep raking up the issues concerning PPP projects through their elected representatives. Hence, the *political risks* in such projects are very high. Therefore, the project manager must define innovative measures to steer clear of potential corruption issues. This is particularly emphasized here as political risks are often much greater than economic risks.

### 21.5.1  The Fifth Mantra

To have a successful PPP project, therefore, one needs to innovate on a model that disables the key operational drivers in the 'corruption value chain.'

The mantra is, therefore, simple:

**Reduction of bureaucratic red tape and ensuring robust payment models.**

This can be achieved in a variety of ways such as (a) Using public involvement for project appraisal, (b) Management reviews made transparent, and (c) Accelerating resolution processes through timebound mechanisms. Of course, a lot of systems extraneous to the 'core' project need to be considered while incorporating this mantra and it would usually require expert consultants to work it out.

## 21.6   Key Takeaways

This chapter has tried to bring out some of the mantras that need to be used in the context of PPP projects. Most governmental agencies as well as private agencies have a lot of difficulty in getting the PPP project balanced. With the approaches provided for in this chapter, by way of the mantras, it is possible to take the projects to a higher niveau. In doing so, we also have to understand that many of these mantras need expert involvement to ensure careful consideration at the time of application. The modeling and simulation methodology provides for a detailed/meticulous treatment of most of these factors. However, it is not a one-size-fit-all model, and hence, needs to be treated on a case-to-case basis.

Regardless of the complexity in implementation, the reader needs to understand these factors and diligently identify them in a given project situation. A 'first implementation' at a particular governmental institution is often risky and it would make sense to do a rigorous case analysis involving some direct data collection exercises such as surveys, culture studies, etc. These would help the project manager to design and position his project appropriately and ensure success for all the entities involved.

# Part VII
# Other Project Areas

# Chapter 22
# People Dimensions

**Abstract** This chapter covers two important aspects, viz. hiring and attrition. In Hiring, we look at the cardinal question from the HR perspective, viz., 1-2-3 of HR. The question is whether one has to recruit 1 super performer or 2 mediocre performers or 3 average performers. A modeling and simulation framework is presented for the reader to understand and evaluate his own independent situation. The second aspect that is covered in this work is that on attrition management. We begin by delving into the basics of attrition management strategies that are common in projectized organizations. The chapter then moves on to show the limitations of traditional frameworks and presents the Neo-Attrition Analysis Framework. The remainder of the framework describes the classical approach and develops the model and simulates to show results in a typical organization.

**Keywords** Project human resource management · 1-2-3 of HR · Attrition management · Strategies for attrition management in projects · Neo-attrition analysis framework · Focus of the classical attrition framework · Perspectives of impact as per the classical attrition framework · The black box neo-attrition model · Simulation

In our earlier chapters, we spoke about the organizational design. In this chapter we will delve into two crucial aspects of people dimensions that every project manager needs to know: (a) hiring and (b) attrition. Projects are typically high-pressure environments and it is often difficult to find the right person at the right time. Hiring, therefore, becomes an important and 'time consuming' activity for the project manager. Attrition is another area that adversely affects the project manager's role. Hence, it is necessary to get a firm handle on both these issues.

The subsequent treatment is taken from our white papers, publications, and proprietary frameworks.

© Springer Science+Business Media Singapore 2017　　　　　　　　　　373
N. Gurjar, *A Forward Looking Approach to Project Management*,
Lecture Notes in Management and Industrial Engineering,
DOI 10.1007/978-981-10-0782-8_22

## 22.1   The 1-2-3 of HR

This section is taken from our Website www.consultingconnoisseurs.com, where we have described the overall model [26].

Human resource management, as a faculty, had its genesis from the complex elements of Productivity Dynamics. Even today, the focus of HRM continues to be improving business gains and practices from the perspective of the fourth factor of production, viz., labor or man. Do you often see that you recruit superlative 'talent' from a salary perspective, only to realize later that there is a need for economic adjustment? Do you see a floating population that swings from super achievers to separation candidates? If yes, it is time to look into your HR practices. Today, the lack of the knowledge element (as shown in Fig. 22.1) to enable a reasonable integration of concepts for an objective decision at the strategic level has made most HR professionals shy away from this focus. In other words, they seem to be distant from this principal focus and have not quite understood how this works. Fortunately, systems thinking through modeling and simulation provides an answer to these situations today. Let us understand it in greater detail.

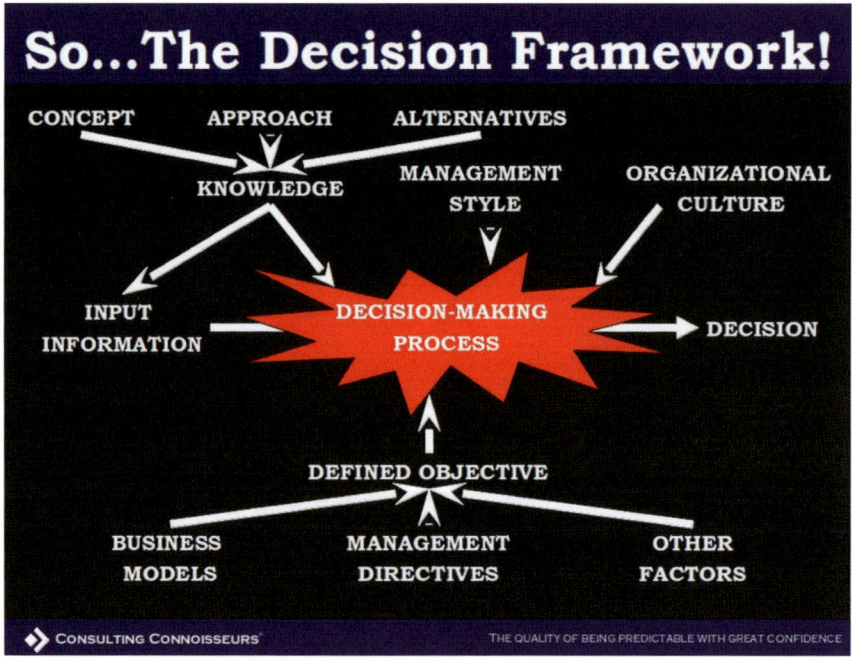

**Fig. 22.1**  Typical decision framework (®Consulting Connoisseurs)

### 22.1.1 The Cardinal Question

Ever since its origin, HRM has had a very cardinal question that has driven most research. In today's business environment, the question can be translated as what I call *THE 1-2-3 OF HR* rather than the A-B-C of HR!

**A Quick Hands-On**

In other words, the cardinal question is

**Should one recruit 1 high performer or 2 mediocre performers or 3 low performers for a job?**

The question, though simple, has profound implications on how an organization is designed, how it works, and how profitable it would be. Different companies use different strategies when it comes to leveraging productivity dynamics. And while the answer to many HR professionals would appear simple, it is fairly complex to derive at. What we often hear from them is a best guess rather than a rigorously defined strategy. Let us try to understand some of the factors that drive the answers to the 1-2-3 of HR in a typical organization.

### 22.1.2 Factor 1: The Individual Productivity or Performance

Since we are speaking of productivity dynamics, the first fundamental factor is that of the individual productivity. It is important to understand the performance of the individual objectively. Most HR practices/employers have individual productivities or performances defined on the job. However, for recruitment practices, it is often difficult to rigorously evaluate the performance of the individual before the recruitment process. Nevertheless, this is one of the crucial factors in the 1-2-3 model of HR. Most times HR managers use implicit approaches such as recruiting from top colleges, top employers/competitors, high academic scores, a possible aptitude test score, etc. In any case, there are multiple methods to correlate these indicators to the expected productivity. While we are not debating on which is better, *the actual individual productivity is a known unknown to many.* In fact, one might even assume that it leans toward a reasonable known more than an unknown.

**Is the individual productivity of your employee reasonably known to you?**

If not, you might have to derive the same.

While individual productivities are important considerations, the productivity dynamics from the 1-2-3 model requires to understand the variance in the individual productivities that are being talked of. This is most times significantly large (non-trivial) and drives the outcome considerably.

### 22.1.3  Factor 2: The Group Productivity or Performance

Excellent individual performance does not always translate to excellent team performance. In fact, this varies greatly with the organization. For the perfect organization, this actually creates a synergy (i.e., as an example, 2 members doing a job of 2.5 members). In most organizations, however, this is not adequately leveraged and results in a coordination loss (i.e., 2 members doing a job of 1.5 members). The strategic implications could be complex. For instance, when individual productivity is high and coordination losses are also high, an organization might prefer using the intrapreneurial approach of management rather than a close-knit culture to minimize the coordination exposure. Do you often see managers spending more time with certain employees than others? Do you also often find 'confused' employees in your company? Some of these symptoms are indicative and critical aspects for group performance. Do you know the extent of the synergy or the coordination loss component in your team performance? Again this is team specific and the HR department needs to actually understand (to say the least), if not evaluate, the extent of this parameter. While team building training has become exceedingly popular, objective measurement of team productivity at the operational level is seldom seen in any business. Global parameters often blur the picture as there are many different 'compensatory actions' by specific individuals that come into play. The supervision effort is also a function of the group productivity and it can safely be assumed that the high performer works with lesser supervision. However, the extent is important to understand and measure.

In addition, what is also important in the framework of productivity dynamics is to understand the group productivity function in both stable and transient conditions (when some employee leaves).

### 22.1.4  Factor 3: CTC as a Function of Performance

The economic reality translates into a management decision factor when one gets a handle of the premium to be paid for higher performance. The more the premium, the more sensitive is the market condition. And in such cases, the HR is really at the cross roads: trying to balance the budgets, the availability of candidates, and the perceived performance potential! However, the problem is that this often ends up being a 'localized' decision rather than a strategic one. To make the best of it, one needs to have a strong handle of the CTC ranges as functions of performance.

Moreover, as the old school might argue, the sum total of the CTCs (total man-power cost) as a percentage of the total costs also plays a decisive role in shaping the HR strategy. While this is true for many organizations, it does not really hold for high performing companies or companies that are in high/innovative IP offerings.

### 22.1.5 Factor 4: Attrition Rates

The rates of attrition are critical factors in any HR plan. The attrition rates could change with business cycles, productivity of employees, productivity of teams, and several other factors. The variances in the attrition rates are important to understand. An important statistic that is used to 'justify' many HR policies, attrition, is rarely used in tandem with other factors to see the cumulative effects in the context of productivity dynamics. The critical point, therefore, is to look at interrelationships of the various factors rather than treat them in isolation, and this is provided for by the systems thinking approach.

### 22.1.6 Factor 5: Cost of Replacing

Critical to the economic perspective is the cost of replacing a resource. In the current business environment, this is a shared cost for most companies with a network of placement consultants along with in-house efforts.

The in-house component of the costs are, in most cases, hard to track. And they are significantly high. One, therefore, needs to have a good handle of this cost. This is important for every HR to clearly define.

### 22.1.7 Factor 6: Time to Replace and the Time to Restore

Any case of separation/attrition is followed by a period of instability (transient behav-ior). To characterize any transient, two parameters are fundamental, viz., the time taken to replace and the time taken to restore. The time taken to replace is the time taken to get a successor for any task or role. The time to restore, on the other hand, is the total time taken to restore the operations to the previously stable state. In other words, it includes the time to replace, the time to train the replacement and the time to establish practices as in a reasonably stable environment. While the time to replace is characterized by a phase of 'lower' manpower costs, the time to restore actually helps understand the increased supervision effort as well as the potential loss of performance due to the interim transient state.

### 22.1.8   Factor 7: Cost of Under-Performing

While all the first six factors are aimed toward understanding the explicit costs and the hidden costs associated with the factors of under-performance, the actual cost of under-performing assets to a business is primal to understanding the criticality of this factor. The question:

**What happens if one is not able to achieve what he is supposed to?**

holds the key to this situation. And the effects, in monetary terms, at the local as well as the global (business) level need to be carefully understood and interpreted.

While it is relatively easy to understand this cost at a business level, the same becomes complex to translate at the function/team level. This is another situation where a model and a simulation can assist the HR to evaluate and do a reality check of their own assumptions. This is, however, the most critical parameter to define from both a business as well as an HR perspective.

### 22.1.9   The Integration Model

While it is intrinsically trivial to know the direction for each of the factors, the important aspect includes the sensitivity and the interplay of the associated parameters. As an HR strategist, it is essential to know the integrative aspects of these factors. Fortunately, modeling and simulation provide for a good answer. A reasonably simple integration model can then be used to define and simulate the results. A simplified example model is shown in Table 22.1 to demonstrate some of the concepts indicated.

Of course, a real-life model will be a little more complex as it needs to reflect the current scenarios in the organization to enable providing a basis for decisions involved in the HR strategy.

### 22.1.10   Conclusion

The beauty of modeling and simulation is that it provides one with a very concise interpretation of management requirements and enables the factoring in of complex concepts at the ground/operational level, making it a practical tool today.

The 1-2-3 of HR is among the most popular as well as complex problems in the HR fraternity and there has been little effort done to understand, evaluate, and answer the question. Yet, it holds the key to the success of high-performing teams and organizations. For a confidential discussion on HR strategies and assistance with modeling and simulation based tools to enable real impact, you could contact systems consulting experts like us.

**Table 22.1**  HR integration model

| Time Bucket | 1 | 2 | 3 |
|---|---|---|---|
| Job Allocation per Unit Time | 2.2 | 2.5 | 2.5 |
| Cumulative Work-Load | 2.5 | 5 | 7.5 |
| Coordination Loss in Stable Condition | 0 | 0 | 0 |
| Coordination Loss in Unstable Condition | 0.1 | 0.1 | 0.1 |
| Productivity of Person | 1.6 | 1.6 | 1.6 |
| Salary per Person | 2 | 2 | 2 |
| Number of Persons | 2 | 2 | 1 |
| Stable Month 1 Unstable 0 | 1 | 1 | 0 |
| Work Performed | 3.2 | 3.2 | 1.44 |
| Cumulative Work Performed | 3.2 | 6.4 | 7.84 |
| Cost of Recruitment | | | |
| Salary for Time Bucket | 4 | 4 | 2 |
| Work Performed less than Work Planned | 0.7 | 1.4 | 0.34 |
| Cost of Delay (where negative) | | | |
| Cumulative Cost | 4 | 4 | 2 |

## 22.2   HR in Projects: Attrition Management!

The role of Human Resource Departments in projectized environments is more often operation-oriented to provide support to their internal customers. However, this posture affects the optimality in the decision-making, the strategizing of the organization, and the resultant efficacy and the profitability of the business.

Challenges in HR management for project environments include:

1. Effective matching of customer expectations (*and not just requirements*)
2. Profitability from the Human Factor of Production.
3. Agility in the strategy focus to ensure appropriate response of the human assets.

The cascade effects of these strategic issues manifest as serious challenges, as shown in Fig. 22.2, in the areas of:

1. *Attrition Management*
   How an organization should deal with situations involving employee attrition
2. *Productivity Management*
   How an organization should ensure its productivity at the customer end isn't affected
3. *Systems and Process Implementation*
   How an organization develops its ecosystem to leverage its strengths
4. *Management Skills for better Performance*
   How an organization should improve the productivity of the workforce.

Although these issues have direct implications on the revenues of the firm, most HR professionals remain 'busy' in using conventional exercises like Exit Interviews,

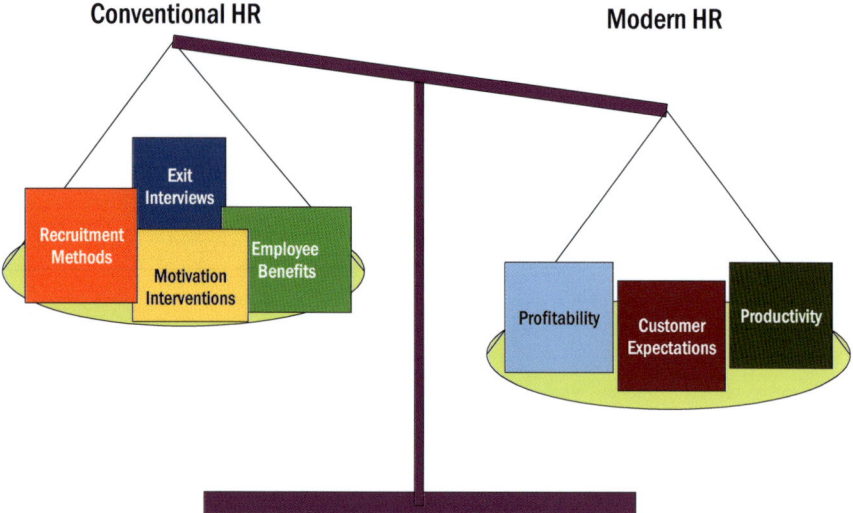

The Changing Balance of Attrition Management

**Fig. 22.2** HR focus today

Employee Benefit Re-design, Motivation Interventions, and Recruitment Budgets. In other words, most HR departments have serious inadequacies in dealing with the true challenges before them.

### 22.2.1  The HR Dilemma Revisited!

In most organizations, HR department is aligned to focus on the *human perspective* of the organization, thereby missing the *business perspective.* In other words, the professionals are trained to orient with the employees due to the *bottom-up* methodology in HR, but are expected to protect the management, viz., the *top-down* expectation creating a serious dilemma for the professional. The end-result is most times, a strict 'budgetary orientation' for possible 'HR Activities.' To illustrate this further, let us consider the simple case of attrition management.

If one were to ask a good HR professional about the most popular HR Tools used in the context of attrition management, one would find the following:

1. Exit interviews
2. Revising monetary and non monetary benefits
3. Interventions for Employee motivation
4. Better recruitment methods, etc.

These need closer introspection to establish their suitability in today's market conditions. In order to do that, one could have a quick diagnostic to establish the fundamentals as shown in Fig. 22.3.

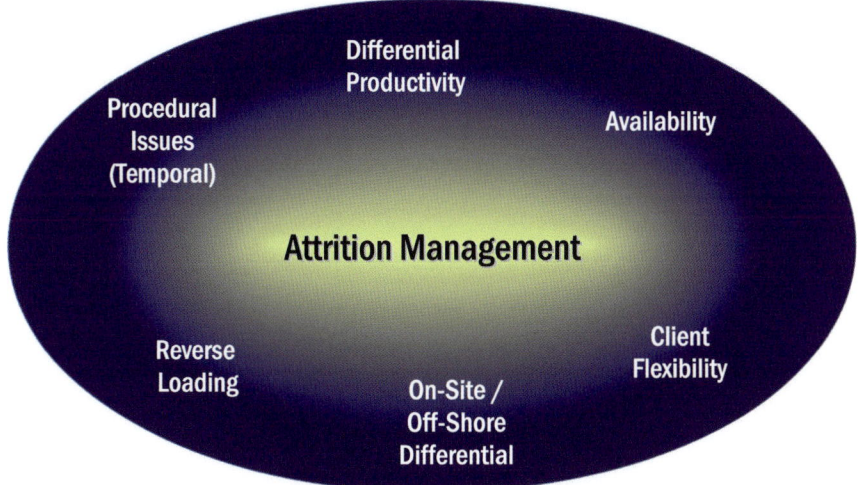

Determinants of Attrition Management Strategies

**Fig. 22.3**  Strategy determinants: attrition

## 22.2.2  *The Quick Business Test!*

*The First Symptom*, from a management perspective, a poor attrition management strategy gets implicitly indicative by the following symptoms at the executive level:

**Spending too much time in the recruitment of existing positions,**
**Extreme time-pressures, quality issues, and manpower adjustments when the organization is approaching contractual deadlines,**
**Aberrations in manpower costs vis-a-vis projections across project lifecycles**

If you see any one of the symptoms active, it would be due to a poor attrition management strategy.

**A Quick Test on HR**

At the HR level, therefore, one needs to revisit the HR activities:

**Do you see a reduction in attrition-related issues through your current exit interview processes?**
**Have measures used for boosting employee motivation truly increased the profitability? What is the RoI of these measures?**

Has revising monetary benefits for employees improved the performance in terms of fulfilling client expectations (and not just client requirements)?
Is the culture leveraging team-work by showing an increasing trend in the marginal profit-per-employee figures in recent years?

If your answer for any of the above is '*no*', then you definitely need to revamp your attrition management strategy. Fortunately, there is a solution for this.

### 22.2.3  Need for Innovation

The classical model has always been one that looks at functional control by trying to correlate between the causes and the phenomenon. This is elucidated in Fig. 22.4. Unfortunately, this approach forces the manager (and his HR staff) to go overboard in trying to understand the causes. This forces the manager to focus on the 'non-trivial' and discard the 'vital.' This is, therefore, a very serious situation.

## THE CLASSICAL ATTRITION ANALYSIS FRAMEWORK

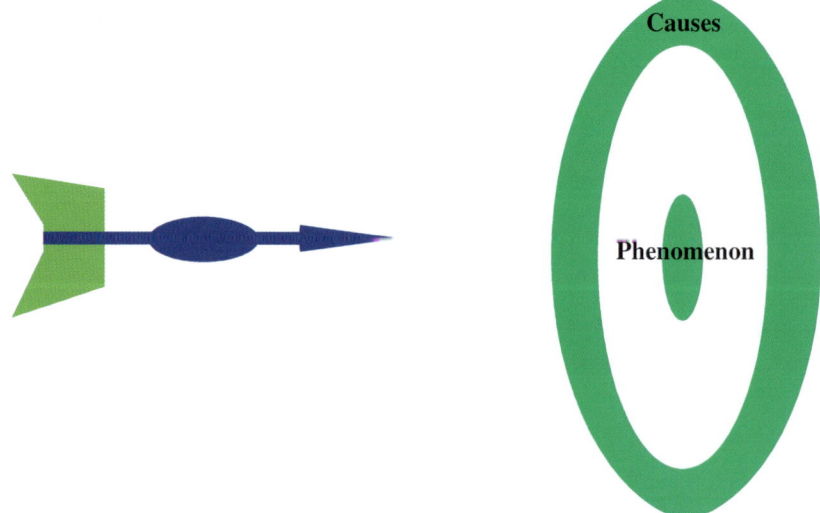

**Fig. 22.4**  The traditional approach

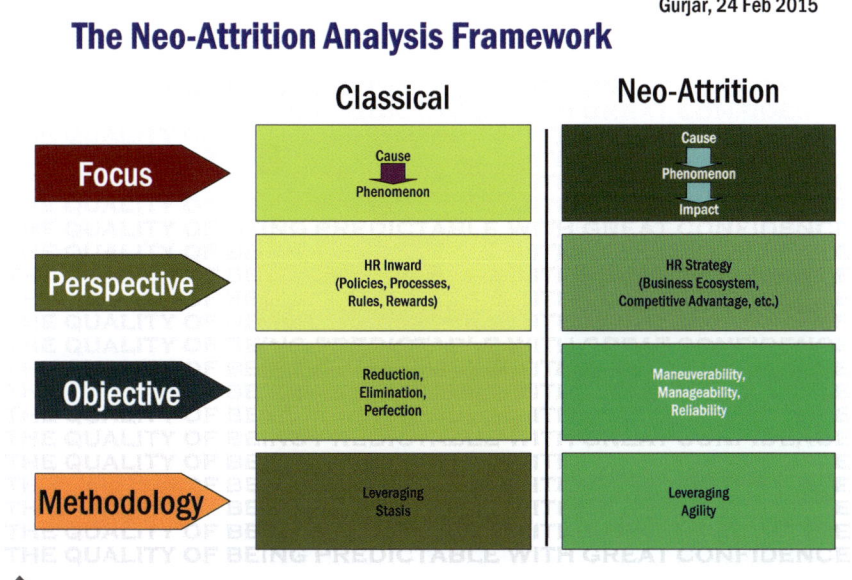

**Fig. 22.5** Comparison summary: traditional versus neo-attrition

The *Neo-Attrition Analysis Framework*® that I have proposed is radically different from the conventional approach. The main differences are highlighted in Fig. 22.5. The four dimensions where the Neo-Attrition Analysis Framework differs fundamentally from the traditional approach are elaborated below:

1. Focus
2. Perspective
3. Objective
4. Methodology.

We will touch upon each of these in greater detail.

**Focus**

The basic difference is that of the focus. As mentioned earlier, traditional methods focus on the causes and the phenomenon. This focus actually has a bigger issue involved. The point is depicted in Fig. 22.6. As seen in the figure, it is evident that there are certain interesting points to note:

1. There are certain visible factors and certain invisible factors that cause attrition.
2. It is often seen that the visible factors are not the accurate ones in many organizations. In other words, the factors are often *masked* behind the real reasons.
3. Most HR initiatives try to make the *invisible factors* the *visible ones*.
4. In addition, they are also trying to ascertain the accuracy of the visible factors.

Gurjar, 24 Feb 2015

# The Transition in the Approaches: Classical

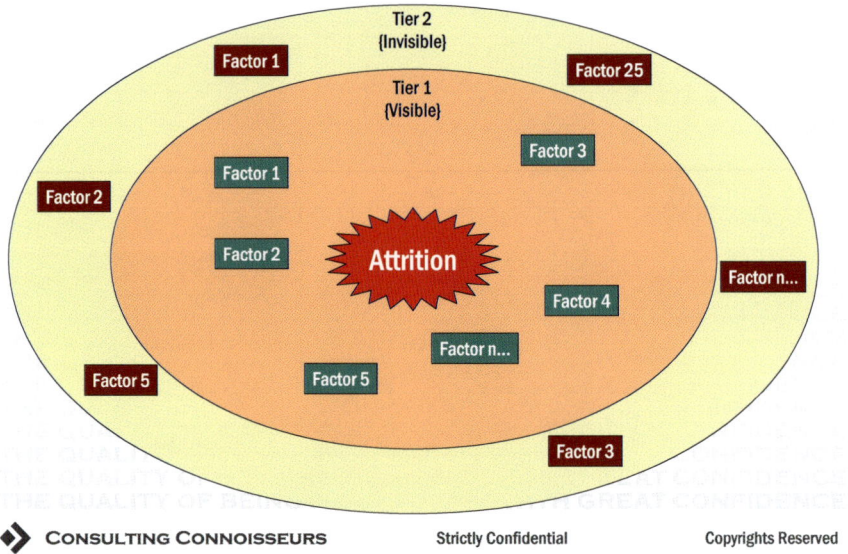

**Fig. 22.6**  Focus in classical approaches

This leads to a conflicting situation, especially considering the political under-currents in the organization. Worse, it is trying to force out a virtually impossible task. It is like the *perpetual motion machine* dream that triggered the engineering revolution. However, at some point, realization dawned that the Carnot Cycle could not be breached by natural means. However, HR has still been trying to work on this area.

This carries forward into four interesting scenarios as shown in Fig. 22.7. If the visibility is high and the causal correlation is perceived to be high, then the company needs to overhaul its system. This is, most times, the objective of the HR professional, if the traditional HR approach is to be believed.

If the causal correlation is perceived to be low and the factors are visible, the project manager and the HR manager would believe that the system is well managed. But even when the factors are invisible, as long as the causal correlation is perceived to be low, the management would be safe by 'ignoring' the area. However, if the causal correlation is perceived to be high and the factor is seemingly 'invisible/unexplained', the management is bound to panic!

Now, the entire philosophy, therefore, hinges on the one assumption, viz., *the factors obtained are reasonably accurate*. However, one is also aware that this is not the case. Hence, the true management indicator is the statistic that one has at the end of the period. All the other factors are interpretations to explain and control the same.

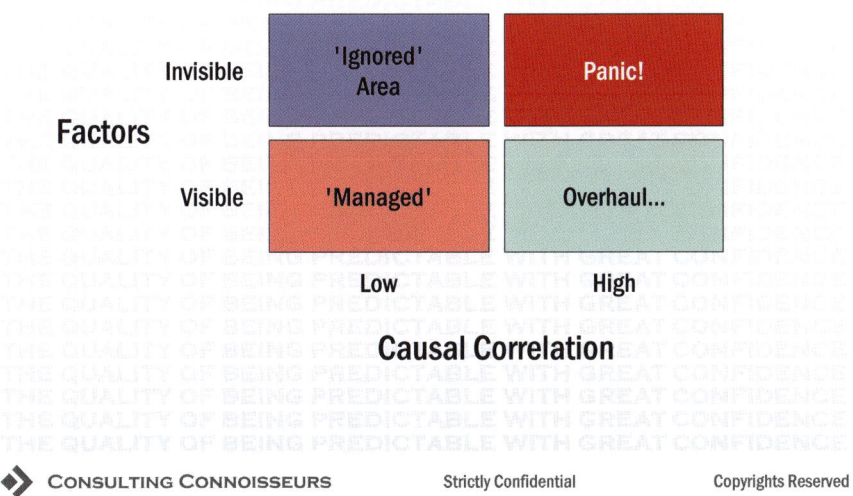

**Fig. 22.7**   Scenarios due to focus considerations

In the Neo-Attrition Analysis Framework, the cumulative effect of various parameters is taken into consideration. This allows one to overcome the narrow focus on known unknowns and is able to paint a more realistic picture for the user.

**Perspective**

The traditional approach is internally focused. In other words, the idea is on HR parameters like policies, costs, procedures, rules, rewards, compensation, etc. This is elucidated in Fig. 22.8. The focus swings between two extreme positions: 'assigning blame' or 'saving the management.' Both the perspectives are not adequate enough.

- Cost
- Time
- Effort
- Culture, etc.

**Fig. 22.8**   Traditional perspective for impact

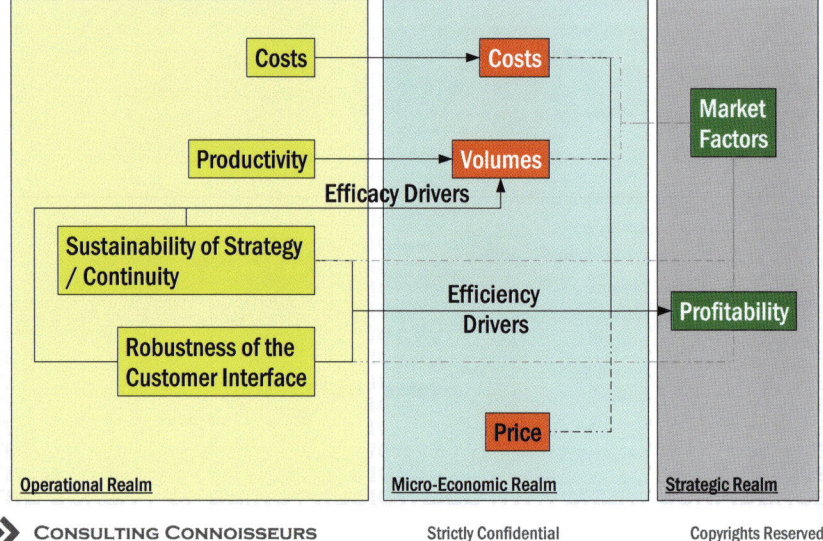

**Fig. 22.9** Neo-attrition analysis framework for impact

In most cases, such an approach is actually applied to justify budgets, training programs, or buy allowances for the project manager. And if the HR is a separate entity, it provides for ample room to start a well-laid scenario for a 'blame game.'

The modern approach, on the other hand, relies heavily on understanding the translation of this phenomenon in terms of the business eco-system. Here factors such as the overall project profitability, customer expectations, productivity, etc. are central to the project manager's concerns. The simple change in considerations is actually represented using a three realm approach as shown in Fig. 22.9. As one can see, the fundamental premise used in the traditional perspective has completely metamorphosed into newer parameters/factors that are more relevant to the project manager.

**Objective and Methodology**

While the traditional approach focuses more on rejection/elimination, the thinking is good yet wishful! It is comparable to the question in the mid-90s where managers were asking if one has to go for an IT investment or not! In the present days (or even back then), it was not about whether one has to consider IT, it is more of accepting that IT is there to stay and one needs to ensure that one acknowledges and leverages this style of thinking. Likewise, the modern thinking is that attrition is going to stay/continue. Hence, it is more on the management and the maneuverability of the phenomenon.

Moreover, the traditional approach is all for 'stable' organizations and employees. The modern approach is one that leverages agility. It accepts the fact that there are going to be changes. Hence, the manager needs to understand how to factor in and *create this agility.* We will explain this in greater detail as we discuss the model that would be used.

### 22.2.4 Environments Requiring Such Models

Project environments are very strong candidates for Neo-Attrition Analysis Frameworks. Especially in the context of large programs, the following factors seem to have a high correlation with the attrition phenomenon:

1. *High growth environments*
   Large projects involve massive ramp-up of personnel. Such situations are often pretty stressful for any business and the stress, when transmitted to the project team, has its own fall-outs. So, high growth environments do witness a lot of turnover.
2. *Improper Ramp-Up*
   Many project managers agree to the fact that they fall short of manpower, especially during the early stages of the project. This is true, but the other side of the story is equally interesting. Hiring too early makes the company burn through cash and fails to meet the expectations. A parallel study in sales force planning has been published in the Harvard Business Review [25]. This is particularly so because there are a lot of dependencies that cannot be simplistically compensated by just adding the number of personnel. Just as the article speaks of inflexion points in the learning curve, one could also define inflexion points in the ramp-up curves.
3. *Unexplained Outcomes with the Same Strategies*
   This is one of the best indicators of something going wrong with the overall management strategy. As mentioned earlier, the drive to identify the knowns is often too high in such situations. However, it is easier to work around using the modeling and simulation approach.
4. *Fast Changing Postures and Tactics*
   Most project environments have a reasonably well-defined strategy when it comes to implementation. However, these are not as static as most people think. As the execution progresses, changes are expected. These changes could be simple, tactical, or even strategic. With changing business conditions, the 'strategies' could be changed on a seemingly regular basis. This calls for a very *flexible* approach among the project personnel. Most project personnel find this to be a 'crazy' environment. The result is invariably a high attrition rate.

Apart from these factors, there are also auto-correlative factors. For instance, one employee leaves and takes his team along with him at his new workplace. Such factors too are important and need to be understood in the broader context of work.

### 22.2.5  Preparing the Model

While there are several ways in which this situation can be modeled, we advocate the GPSS approach. For those who are new to the world of simulation, this is called the General Purpose Simulation System that actually connects the dots between various events and flows. To describe this condition, I have shown a simple state diagram for attrition in Fig. 22.10. Any leaving employee would move through these various states. In doing so, there are incidence rates that are interesting to understand and the entire chain. For instance, some employees would agree to continue if they are given a raise. Others might not! So, such conditions are also interesting from the modeling and simulation perspective.

Most companies actually publish only the attrition statistics that capture the number of employees that leave an organization. For the project manager, it is also important to know the number of employees that have tried to leave and the number of employees that were retained through a change of role or a change in salary. In any case, once these aspects are clear, company-specific details need to be incorporated in the model. These look like the factors given in Fig. 22.11. Most companies fail to incorporate the team size in the equation. This leads to misleading figures and strategies. If you recall the discussion on the overloading of employees that we had in the previous chapters, the overall equation gets distorted. Once the model is fairly

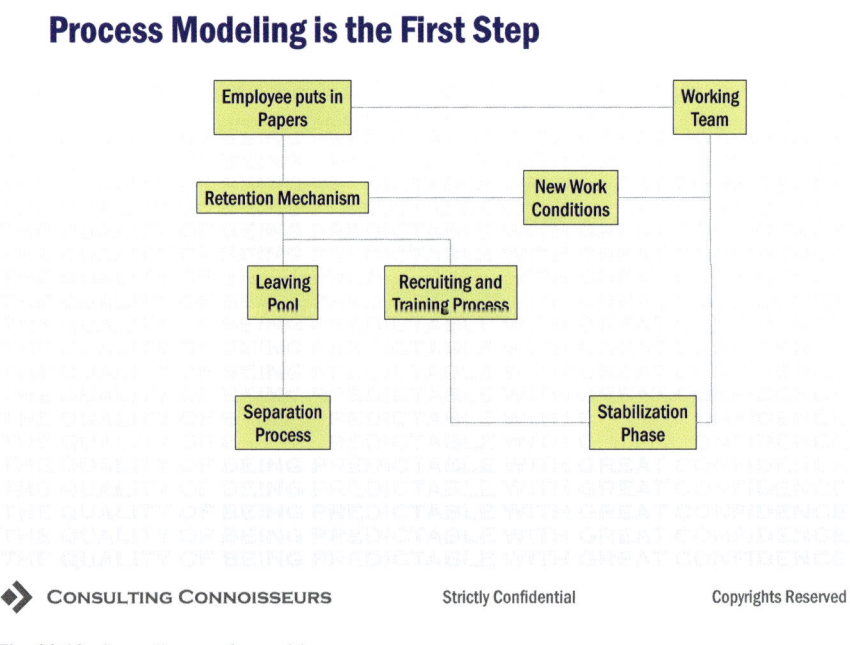

**Fig. 22.10**  State diagram for attrition

# Parameters...

- Notice Period
- Working Team Size
- Business Demands
- Continuity Planning
- Productivity
  - Of Leaving Employee
  - Of Joining Employee
  - Of Resigning Employee
  - Of Fully-Trained Employee

Company Specific Scenarios can be Built-In

 CONSULTING CONNOISSEURS          Strictly Confidential          Copyrights Reserved

**Fig. 22.11**  Numerical parameters in the model

## Predictability of Results...

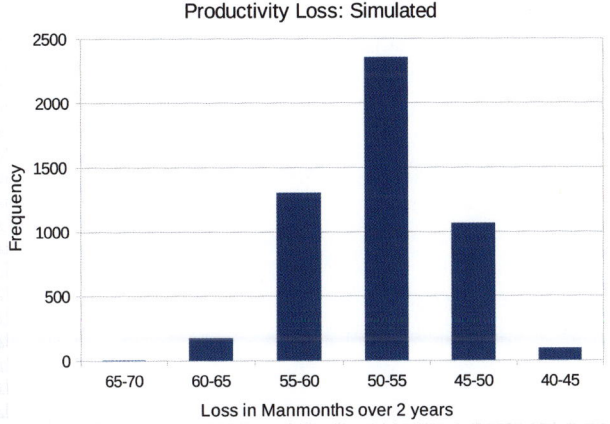

Dry-Run Simulation Results of 5000 cases

CONSULTING CONNOISSEURS          Strictly Confidential          Copyrights Reserved

**Fig. 22.12**  Expected loss of man-months in 5000 simulations

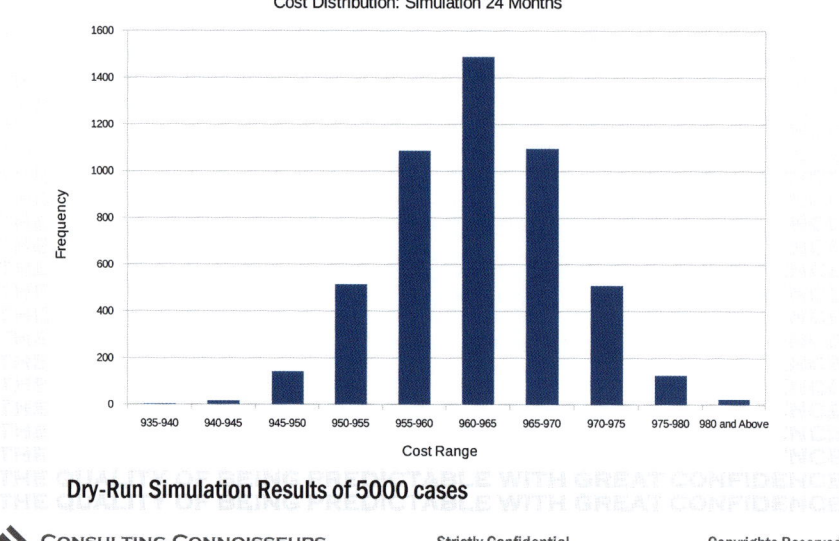

**Fig. 22.13** Expected costs in 5000 simulations

well established, different scenarios need to be built into the same to ensure a good representation of the project environment. This could be done using the existing company test data and finding out the range of the confidence level that one needs to work with. The results of the simulation would look like Figs. 22.12 and 22.13. These would immediately explain the degree of problems expected at the company with the current trends in attrition. As a project manager, therefore, it is indicative of both the cost as well as the productivity loss for his team under the conditions that would be simulated.

Once the situation is clear, several factors as described in Fig. 22.3 could be used by the project manager. Thus, the astute project manager has a bigger sense of responsibility as he needs to be *flexible* while determining potential HR strategies that need to be used in a project situation to ensure success.

## 22.3   Key Takeaways

In this chapter, we have attempted to explain the need for a more comprehensive decision-making framework for the project managers based on a firm understanding of the interplay/implications of these decisions in the overall project situation. While it is customary, in most organizations, to have an HR manager who works in tandem with the project manager to fulfill the needs of the project, it is often times necessary for the project manager to 'design' the desired project system.

In doing so, we have actually explained the situations concerning productivity differentials and the salary differentials that need to be applied to such team members. This is distinctly different from the traditional HR approach advocated by most HR professionals.

I have also taken up a crucial topic, viz., attrition management and explained how the Neo-Attrition Analysis Framework needs to be applied in project situations. Again here, the methodology is a sharp contrast to the conventional methods and enables a better handle of project phenomenon than the traditional approaches.

Productivity management models can also be developed on similar lines for the project manager to use the know-how and take pro-active measures in ensuring project success. The problem with the HR faculty is that the project manager does not normally believe that this functional expertise belongs to the 'core' of the project management faculty. However, it is evident from the discussions we have had in this chapter, that the project manager does need to have a good hold on the HR faculty as well.

# References

1. Archibald RD (2015) A global system for categorizing projects: the need for, recommended approach to, practical uses of, and description of a current project to develop the system. http://www.russarchibald.com/AGLOBALSYSTEM1104.pdf. Accessed 16 May 2015, 1100 GMT
2. Dov D, Stan L, Aaron S, Asher T (1998) In search of project classification: a non-universal approach to project success factors. Publ Res Pol 27(9):915–935
3. Slinger B (2015) Advantages and disadvantages of stereotypes. http://bethanyslinger.blogspot.in/2012/10/advantages-and-disadvantages-of.html. Accessed 16 May 2015, 1500 GMT
4. PMI Standards Committee, Duncan W (1996) A guide to the project management body of knowledge
5. Damodaran A (2015) Measuring investment returns. http://people.stern.nyu.edu/adamodar/pdfiles/ovhds/ch5.pdf. Accessed 30 May 2015, 1500 GMT
6. Roger S (2005) Pressman, software engineering: a practitioner's approach, 6th edn. McGraw Hill Higher Education, India
7. Claude H (2012) Maley, project management concepts, methods, and techniques. CRC Press, FL
8. Kenneth J, Michele Kacmar HK, Zivnuska S, Shaw JD (2007) The impact of political skill on impression management effectiveness. J Appl Psychol 92(1):278–285
9. Nalewaik A, CCE MRICS, Witt J (2014) Challenges reporting project costs and risks to owner decision makers. Published in ICEC international cost management journal. AACE International, Seattle, Washington, USA. http://www.icoste.org/ICMJ%20Papers/Nalewaik1.pdf. Accessed 05 Sept 2014, 1500 GMT
10. Hendrickson C (2008) Project management for construction: fundamental concepts for owners, engineers, architects and builders, World Wide Web Edition. http://pmbook.ce.cmu.edu/. Accessed 06 Sept 2014, 1200 GMT
11. Saaty Thomas L, Vargas Luis G (2001) Models, methods, concepts and applications of the analytic hierarchy process. Kluwer Academic Publishers, Boston
12. Olivas R (2007) Decision trees: a primer for decision-making professionals. Rev. 5.1 b. http://www.lumenaut.com/info.htm. Accessed 11 Sept 2014, 1100GMT
13. Hillier FS, Lieberman GJ (2009) Introduction to operations research: concepts and cases, 8th edn. Tata McGraw-Hill Publishing Company Limited, New Delhi
14. Roxburgh C (2014) The use and abuse of scenarios. McKinsey Insights. http://www.mckinsey.com/insights/strategy/the$_$use$_$and$_$abuse$_$of$_$scenarios. Accessed 24 Sept 2014, 1100 GMT
15. Tim B (2009) Project risk management–the commercial dimension. Viva Books Private Limited, New Delhi, Viva Thorogood Professional Insights
16. Malik P (2014) 7 Reasons to create WBS. http://www.pmbypm.com/7-reasons-to-create-wbs/. Accessed 01 Oct 2014, 0500 GMT

© Springer Science+Business Media Singapore 2017

393

N. Gurjar, *A Forward Looking Approach to Project Management*,
Lecture Notes in Management and Industrial Engineering,
DOI 10.1007/978-981-10-0782-8

17. Fugar FDK, Agyakwah-Baah AB (2010) Delays in building construction projects in Ghana. Aust J Constr Econ Build 10(1/2):103–116
18. United States Government Accountability Office (2014) GAO schedule assessment guide. http://www.gao.gov/assets/600/591240.pdf. Accessed 01 Oct 2014, 0500 GMT
19. Gurjar NS (2014) Strategizing business process outsourcing. http://www.consulting connoisseurs.com. Accessed 17 Oct 2014, 0500 GMT
20. Ryals L (2012) How to succeed at key account management. http://blogs.hbr.org/2012/07/how-to-succeed-at-key-account/. Accessed 18 Oct 2014, 1500 GMT
21. Charles VK (2009) Nothing ventured, nothing gained: successful strategies adopted by Indian states for public private partnerships. Research Paper for Masters Program, International Institute of Social Studies, The Hague, Netherlands
22. Kalidindi SN, Boeing Singh Fugar L (2009) Financing road projects in India using PPP scheme. In: Proceedings of the 2009 mid-continent transportation research symposium by Iowa State University, Ames, Iowa, August 2009
23. Rajasthan: mainstreaming public private partnerships. http://ppp.rajasthan.gov.in/. Accessed 10 Nov 2014, 500 GMT
24. De Mascia S (2012) Project psychology: using psychological models and techniques to create a successful project. Gower Publishing Limited, England
25. Leslie M, Holloway CA (2009) The sales learning curve. Published in ICEC international cost management journal. AACE International, Seattle, Washington, USA, 2009. http://www.icoste.org/ICMJ%20Papers/Nalewaik1.pdf. Accessed 05 Sept 2014, 1500 GMT
26. Gurjar NS (2014) The 1-2-3 of HR: understanding productivity. http://www.consulting connoisseurs.com/CC_123HR.pdf. Accessed 17 Oct 2014, 0500 GMT

# Index

© Springer Science+Business Media Singapore 2017
N. Gurjar, *A Forward Looking Approach to Project Management*,
Lecture Notes in Management and Industrial Engineering,
DOI 10.1007/978-981-10-0782-8

Printed by Amazon Italia Logistica S.r.l.
Torrazza Piemonte (TO), Italy

36030946R00243